ROOT ANATOMY AND MORPHOLOGY

A Guide to the Literature

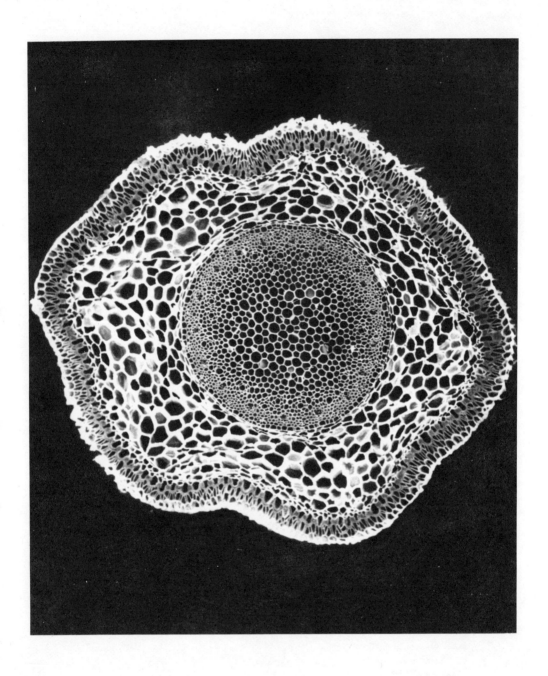

Root transection of the West African orchid <u>Solenangis</u> <u>clavata</u> (Rolfe)
Schltr. Scanning electron photomicrograph (X90). Courtesy of Dr.E.S.
Ayensu, Botany Department, Smithsonian Institution, Washington, D.C.

Root Anatomy

and Morphology

A Guide to the Literature

ROBERT H. MILLER

Staff Scientist, Agricultural Research Service
United States Department of Agriculture
Washington, D.C.

ARCHON BOOKS

1974

Library of Congress Cataloging in Publication Data

Miller, Robert Harold, 1919–
  Root anatomy and morphology.

  Includes bibliographical references.
  1. Roots (Botany)—Bibliography. I. Title.
[DNLM: 1. Plants—Anatomy and histology—Bibliography.
Z5354.R6 M649r]
Z5354.R6M48   016.5814'98   74–8606
ISBN 0–208–01452–7

©1974 by The Shoe String Press, Inc.
First published 1974 as an Archon Book,
an imprint of The Shoe String Press, Inc.
Hamden, Connecticut 06514

Printed in the United States of America

CONTENTS

❧◎❧

PREFACE

～◅⦿ꝏ⦾⮞～

The singular purpose of this bibliography is to provide a comprehensive
as well as a convenient reference that takes into account the pertinent
research over the past century on root anatomy and morphology. During the
course of its compilation several other aspects as to its possible utility
also became apparent. It could serve not only as an assemblage of the widely
scattered literature on the subject but also become a pragmatic research tool
as well as a suitable teaching aid. In addition it could provide an insight
on the research trends in this field during the past 100 years and also
present an overview of redundant research and apparent research voids.

The present work is a complete revision and much expanded version of an
earlier (1960) preliminary bibliography. While still not exhaustive it includes most of the available references from the year 1900 (2772 entries),
as well as a considerable number of earlier publications (203 entries), up
to the time that this compilation went to the publisher. No sustained effort
was made to incorporate all of the earlier work. Much of the relevant literature of the last half of the 19th century is difficult to obtain for a first-
hand examination. Furthermore, in numerous instances, article titles either
inadequately described their content or provided no indication of subject
coverage. Intuition and presumption often turned up many a reference which
upon closer examination did contain relevant material. Consequently, the
author attempted to personally verify all references cited in this bibliography. Many of the early, as well as fairly recent, foreign published and
unpublished dissertations on the subject are next to impossible to obtain. In
many instances even current foreign periodicals and books either require very
long waiting periods to be acquired or else are altogether unobtainable.
Under the circumstances it would appear to be manifestly impossible to ever
collate all of the subject references.

Originally the intention was to cite only those research papers pertinent
to the cytohistological and developmental aspects of root apical meristems.
As the work progressed it was felt that it would be intrinsically valuable to
broaden the inclusions so as to embrace other relevant studies on root anatomy
and morphology. Included are references on haustorial systems of parasites as
well as studies on mycorrhizae as they relate to root anatomy. A number of
citations are included that involve adventitious root development and root
ecology studies relative to their anatomy. Excluded for the most part are
studies on nutritional and other physiological aspects, root pathology, root
nodule formation and development, roots of fossil plants, and ultrastructural
aspects of root histology.

Authors and coauthors are arranged both alphabetically and chronologically.
The chronological arrangement provides some historical perspectives and trends.
The main body of this unannotated bibliography contains 2974 citations in some
22 languages, covers more than 240 plant families and over 875 genera. In
addition to numerous other sources, citations were also obtained from 579
different periodicals. A convenient aid to the identification of these periodicals is provided by the listing of each complete periodical title as well as

vii

its abbreviation. Their number far exceeds those that are more commonly familiar and used by researchers in the field of root anatomy and morphology. Many are probably totally unfamiliar to the average investigator, and many would not normally be considered as organs of publication that would contain material on root anatomy or morphology. During the compilation of this bibliography it became apparent that many researchers were not cognizant of pertinent literature relative to their respective investigations.

Many of the earlier foreign periodicals have individually undergone one or more title changes. No attempt was made to update an investigator's original periodical citation. For those desiring to review some of the pertinent literature prior to 1900 a selected reference listing is provided. Additional earlier references may be found among the citations listed in the main body of the bibliography.

Plant families and genera are cross-referenced to the citations in this bibliography by numerical listings. For the most part these indices were compiled from the titles of the articles. No attempt was made to correct generic names as they appeared in the original titles as to errors in spelling, validity, transference, or legitimacy. A small initial letter has been used consistently for all species names. Additional taxa are obtainable by perusing the references per se. To have provided an all-inclusive listing of genera, based upon the content of each reference, would not have been that worthwhile considering the magnitude of effort involved.

It has been said that "a bibliography is never finished and always more or less defective, even on ground long gone over" (Elliott Coues). There are undoubtedly many other relevant papers from the turn of the century to date that are either obscured by the title or because this bibliographer failed to obtain the organ of publication for one reason or another. Nevertheless, this bibliography should satisfactorily serve its purpose (multum in parvo). Hopefully users will make additions and corrections where pertinent research was inadvertently or unintentionally omitted, and by keeping it "updated" will thereby enhance its usefulness as a reference tool now that the groundwork in the main has been completed.

An insufficient, albeit sincere, word of appreciation is hereby tendered to the many who have assisted in various ways in obtaining materials and providing translation services. A special thanks for their services rendered goes to those very competent and sagacious reference librarians at the National Agricultural Library at Beltsville, Maryland, namely -- Melba M. Bruno, Shirley Gaventa, Helen P. Alexander, and Helen Boyd (retired).

Robert H. Miller

Adelphi, Maryland
April, 1974

ROOT ANATOMY AND MORPHOLOGY

A Guide to the Literature

LITERATURE BEFORE 1900

～⊗⊌⊙⊱

1. ANDREAE, E. 1894. Über abnorme Wurzelanschwellungen bei <u>Ailanthus</u> <u>glandulosa</u>. Inaugural-Dissertation. Königl. Bayer. Friederich-Alexanders-Universität zu Erlangen. Aug. Vollrath, Erlangen. 32 pp.

2. ANDREWS, W.M. 1890. Apical growth in roots of <u>Marsilea</u> <u>quadrifolia</u> and <u>Equisetum</u> <u>arvense</u>. Bot. Gaz. 15:174-177.

3. ARCULARIUS, R. 1897. Ein Fall von Wurzelkropf bei <u>Abies</u> <u>pichta</u>; zugleich als Beitrag zur Kenntnis der Maserbildung bei Coniferen. Inaugural-Dissertation. Königl. Friederich-Alexanders-Universität zu Erlangen. W. Hoppe, Leipzig. 45 pp.

4. ASCHERSON, P. 1883. Bemerkungen über das Vorkommen gefärbter Wurzeln bei den Pontederiaceen, Haemodoraceen und einiger Cyperaceen. Berich. Deutsch. Bot. Gesel. 1:498-502.

5. BACCARINI, P., and SCILLAMA, V. 1898. Contributo alla organografia ed anatomia del <u>Glinus</u> <u>lotoides</u> L. Contrib. Biol. Veg. 2:4-49.

6. BECKER, C. 1895. Beitrag zur vergliechenden Anatomi der Portulacaceen. Inaugural-Dissertation. Friederich-Alexanders-Universität Erlangen. M. Ernst, München. 38 pp.

7. BEIJERINCK, M.W. 1887. Beobachtungen und betrachtungen über Wurzelknospen und Nebenwurzeln. Verh. Konink. Akad. Weten. 25:1-146.

8. BEINLING, E. 1879. Untersuchungen über die entstehung der adventiven Wurzeln und laubknospen an blattstecklingen von <u>Peperomia</u>. Beitr. Biol. Pflanz. 3:25-50.

9. BERCKHOLTZ, W. 1891. Beiträge zur Kenntnis der Morphologie und Anatomie von <u>Gunnera</u> <u>manicata</u> Linden. Biblio. Bot. 24:1-19.

10. BERTRAND, C.E. 1885. <u>Phylloglossum</u> <u>drummondii</u>. Arch. Bot. Nord. France 2:70-223.

11. BEYSE, G. 1881. Untersuchungen über anatomischen Bau und das mechanische Princip im Aufbau einiger Arten der Gattung <u>Impatiens</u>. Inaugural-Dissertation. Georgia Augusta zu Göttingen, Halle. E. Blochmann und Sohn, Dresden. 60 pp.

12. BLOCH, O. 1880. Untersuchungen über die Verzweigung fleischiger Phanerogamen-Wurzeln. Inaugural-Dissertation. Friederich-Wilhelms-Universität zu Berlin. Mesch, Berlin. 22 pp.

13. BOODLE, L.A., and WORSDELL, W.C. 1894. On the comparative anatomy of the Casuarineae, with special reference to the Gnetaceae and Cupuliferae. Ann. Bot. 8:231-264.

14. ----------. 1899. On some points in the anatomy of the Ophioglossaceae. Ann. Bot. 13:377-393.

15. BOWER, F.O. 1881. On the germination and histology of the seedlings of Welwitschia mirabilis. Quart. Jour. Micro. Sci. 21:15-30.

16. ----------. 1881. On the further development of Welwitschia mirabilis. Quart. Jour. Micro. Sci. 21:571-594.

17. ----------. 1882. The germination and embryology of Gnetum gnemon. Quart. Jour. Micro. Sci. 22:278-298.

18. ----------. 1885. On the apex of the root of Osmunda and Todea. Quart. Jour. Micro. Sci. 25:75-103.

19. ----------. 1886. On the development and morphology of Phylloglossum drummondii. Philos. Trans. Roy. Soc. London (Ser. B) 176:665-678.

20. BRIOSI, G., and TOGNINI, F. 1897. Intorno all anatomia della canapa (Cannabis sativa L.). II. Atti Ist. Bot. Univ. Pavia (Ser. 2) 4:167-329.

21. BRUCHMANN, H. 1874. Ueber Anlage und Wachstum der Wurzeln von Lycopodium und Isöetes. Jena Zeit. Naturw. 8:522-578.

22. BUCHANAN, J. 1878. On the root-stock of Marattia fraxinea Sm. Jour. Linn. Soc. London (Bot.) 16:2-5.

23. BUCHENAU, F. 1893. Ueber den Aufbau des Palmiet-Schilfes (Prionium serratum Drège) aus dem Caplande. Biblio. Bot. 27:1-26.

24. BUCHERER, E. 1889. Beiträge zur Morphologie und Anatomie der Dioscoreaceen. Biblio. Bot. 16:1-34.

25. BUNTING, M. 1898. The structure of the cork tissues in roots of some rosaceous genera. Trans. Proc. Bot. Soc. Penn. 1:54-65. (Same in: Contr. Bot. Lab. Univ. Penn. 1904, N.S. 5, 2:54-65).

26. BURRAGE, J.H. 1897. The adhesive discs of Ercilla volubilis A. Juss. Jour. Linn. Soc. 33:95-102.

27. BÜSGEN, M. 1897. Bau und Leben unserer Waldbaume. Gustav Fischer, Jena. 230 pp.

28. CALDWELL, O.W. 1899. On the life history of Lemna minor. Bot. Gaz. 27:37-66.

29. CAMPBELL, D.H. 1886. The development of the root in Botrychium ternatum. Bot. Gaz. 11:48-53.

30. --------------. 1891. Notes on the apical growth in roots of Osmunda and Botrychium. Bot. Gaz. 16:37-43.

31. ------------. 1893. On the development of _Azolla filiculoides_ Lam. Ann. Bot. 7:155-187.

32. ------------. 1894. Observations on the development of _Marattia douglasii_ Baker. Ann. Bot. 8:1-20.

33. ------------. 1897. A morphological study of _Naias_ and _Zannichellia_. Proc. Calif. Acad. Sci. (Ser. 3, Bot.) 1:1-61.

34. ------------. 1898. The development of the flower and embryo in _Lilaea subulata_ H.B.K. Ann. Bot. 12:1-28.

35. CHATIN, A. 1856. Anatomie des plantes aériennes de l'ordre des Orchidées. I. Mémoire: Anatomie des racines. Mém. Soc. Nat. Sci. Math. Cherbourg 4:5-18.

36. CHAUVEAUD, G. 1895. Sur le développement du faisceaux liberien de la racine des Graminees. Bull. Mus. Hist. Nat. 1:209-211.

37. ------------. 1895. Sur le mode de formation des faisceaux liberiens de la racines des Cyperacées. Bull. Soc. Bot. France 42:450-451.

38. ------------. 1897. Sur la structure de la racine de l'_Hydrocharis morsus ranae_. Rev. Gén. Bot. 9:305-312.

39. COLOZZA, A. 1898. Contributo all'anatomie delle Alstroemeriee. Malpighia 12:165-198.

40. CONSTANTIN, J. 1885. Recherches sur l'influence qu'exerce le milieu sur la structure des racines. Ann. Sci. Nat.,Bot. (Sér. 7) 1:135-182.

41. CORMACK, B.G. 1896. On polystelic roots of certain palms. Trans. Linn. Soc. London (Ser. 2) 5:275-286.

42. COURCHET, M. 1884. Étude anatomique sur les Ombellifères. Ann. Sci. Nat., Bot. (Sér. 6) 17:107-129.

43. DANGEARD, P.A. 1889. Recherches sur le mode d'union de la tige et de la racine chez les Dicotylédones. Le Botaniste 1:75-125.

44. ------------. 1890. Mémoire sur la morphologie et l'anatomie des _Tmesipteris_. Le Botaniste 2:163-222.

45. DE BARY, A. 1884. _Comparative anatomy of phanerogams and ferns_. Clarendon Press, Oxford 658 pp.

46. DE CANDOLLE, A.P. 1826. Mémoires sur les lenticelles des arbres et de développement des racines qui en sortent. Ann. Sci. Nat., Bot. (Sér. 1) 7:1-26.

47. DE HAAN, J. VAN BREDA. 1891. Anatomie van het geslacht Melocactus. Doctoral dissertation. Rijks-Universitat te Leiden. Joh. Enschede en Zonen, Haarlem. 123 pp.

48. DENNERT, E. 1894. Vergleichende Pflanzenmorphologie. J.J.Weber, Leipzig. 254 pp.

49. DENNISTON, R.H. 1898. Veratrum viride Ait. and Veratrum album L. A comparative histological study. Pharm. Arch. 1:1-4.

50. DETLEFSEN, E. 1882. Versuch einer mechanischen Erklarung des excentrischen Dickenwachsthums verholzter Achsen und Wurzeln. Arb. Bot. Inst. Würzburg 2:670-688.

51. DE VRIES, H. 1880. Ueber die Kontraktion der Wurzeln. Landw. Jahrb. 9:37-80.

52. -----------. 1886. Studien overzvigwortels. Maanbl. Natuurw. 13:53-68.

53. DIXON, H.H. 1894. On the vegetative organs of Vanda teres. Proc. Roy. Irish Acad. (Ser. 3) 3:441-458.

54. ELFERT, W. 1895. Morphologie und Anatomie der Limosella aquatica. Inaugural-Dissertation. König. Friederich-Alexander-Universität zu Erlangen. Chr. Wiegler, Berlin. 44 pp.

55. ERIKSSON, J. 1878. Ueber das Urmeristem der Dikotylen-Wurzeln. Jahrb. Wiss. Bot. 2:380-436.

56. -----------. 1878. Om Meristemet I Dikotyla Vaxters Rötter. Fr. Berlings, Lund. 44 pp.

57. FALKENBERG, P. 1876. Vergleichenden Untersuchungen über den Bau der Vegetationsorgane der Monocotyledonen. Ferdinand Enke, Stuttgart. 202 pp.

58. FARMER, J.B., and FREEMAN, W.G. 1899. On the structure and affinities of Helminthostachys zeylanica. Ann. Bot. 13:421-445.

59. FIRTSCH, G. 1886. Anatomisch-physiologisch Untersuchungen über die Keimpflanze die Dattelpalme. Sitz. Kaiser. Akad. Wiss. Berlin (Math.-Nat. Kl.) 93:342:354.

60. FISCHER, A. 1884. Untersuchungen über das Siebrohren-System der Cucurbitaceen. Gebrüder Borntraeger, Berlin. 109 pp.

61. FLAHAULT, C. 1878. Recherches sur l'accroissement terminal de la racine chez les Phanérogames. Ann. Sci. Nat., Bot. (Sér. 6) 6:1-168.

62. FRIEDERICK, K. 1880. Über eine Eigenthümlichkeit der Luftwurzeln von Acanthoriza aculeata Wendl. Acta Horti Petrop. 7:533-540.

63. FRON, G. 1897. Sur la racine des <u>Suaeda</u> et des <u>Salsola</u>. Comp. Rend. Hebd. Séances Acad. Sci. 125:366-368.

64. -------. 1898. Sur la cause de la structure spiralée des racines de certaines Chénopodiacées. Comp. Rend. Hebd. Séances Acad. Sci. 127: 563-565.

65. FRON, M.G. 1899. Recherches anatomiques sur la racine et la tige des Chénopodiacées. Ann. Sci. Nat., Bot. (Sér. 8) 9:157-240.

66. FUTTERER, W. 1896. Beiträge zur Anatomie und Entwicklungsgeschichte der Zingiberaceae. Bot. Centralb. 68:241-248; 273-279; 346-356; 393-400; 417-431.

67. GEHRKE, O. 1887. Beiträge zur Kenntnis der Anatomie von Palmenkeimlingen. Inaugural-Dissertation. Friederich-Wilhelms-Universität zu Berlin. Max Bading, Berlin. 23 pp.

68. GÉNEAU DE LAMARLIÈRE, L. 1891. Structure comparée des racines renflées de certaine Ombellifères. Comp. Rend. Hebd. Séances Acad. Sci. 112: 1020-1022.

69. ---------------------. 1892. Sur le développement du <u>Conopodium denudatum</u> Koch. Assoc. Franç. Avan. Sci. (Comp. Rend. 21me Session): Part 2. 2:445-449.

70. ---------------------. 1893. Recherches sur le developpement de quelques Ombellifères. Rev. Gén. Bot. 5:159-171; 225-229; 258-264.

71. GÉRARD, R. 1881. Recherches sur le passage de la racine à la tige. Ann. Sci. Nat., Bot. (Sér. 6) 11:279-430.

72. GHEORGHIEFF, S. 1887. Beitrag zur vergleichenden Anatomie der Chenopodiaceen. Inaugural-Dissertation. Universität zu Leipzig. Friederich Scheel, Cassel. 67 pp.

73. GOFF, E.S. 1897. The roots of the strawberry plant. Trans. Wisconsin State Hort. Soc. 27:248-256.

74. ---------. 1897. A study of the roots of certain perennial plants. Rept. Wisconsin Agric. Exp. Sta. 14:286-298.

75. GRANEL, M. 1887. Note sur l'origine des sucoirs de quelques phanerogames parasites. Bull. Soc. Bot. France 34:313-321.

76. ---------. 1889. Recherches sur l'origine des sucoirs des phanerogames parasites. Jour. Bot. 3:149-153.

77. GRAVIS, A. 1898. Recherches anatomiques et physiologiques sur le <u>Tradescantia</u> <u>virginica</u> L. Mém. Cour. Mém. Étrang. Acad. Roy. Sci. Lett. Beaux-Arts Belgique 57:3-304.

78. GREGG, W.H. 1887. Anomalous thickening in the roots of <u>Cycas seemanni</u> A. Braun. Ann. Bot. 1:63-70.

79. GROOM, P. 1893. On the velamen of orchids. Ann. Bot. 7:143-151.

80. --------. 1893. On <u>Dischidia rafflesiana</u> Wall. Ann. Bot. 7:223-242.

81. GROSSE, F.E. 1895. Beiträge zur vergleichenden Anatomie der Onagraceen. Inaugural-Dissertation. König. Friederich-Alexander-Universität zu Erlangen. C. Rich, Dresden. 67 pp.

82. GRÜNWALD, R. 1897. Vergleichende Anatomie der Martyniaceae und Pedaliaceae. Inaugural-Dissertation. Friederich-Wilhelm-Universität Erlangen. Wilhelm Engel in Schotten, Hessen. 43 pp.

83. GWYNNE-VAUGHAN, D.T. 1897. VII. On some points in the morphology and anatomy of the Nymphaeaceae. Trans. Linn. Soc. London (Ser. 2, Bot.) 5:287-299.

84. HAGEN, C. 1873. Untersuchungen über die Entwicklung und Anatomie der Mesembryanthemeen. Inaugural-Dissertation. Friederich-Wilhelms-Universität zu Bonn. Carl Georgi, Bonn. 26 pp.

85. HALSTED, B.D. 1892. A century of American weeds. Their root systems tabulated. Bull. Torrey Bot. Club 19:141-147.

86. HANSTEIN, J. 1870. Die Entwicklung des Keimes der Monokotylen und Dikotylen. In: <u>Botanische Abhandlungen aus dem Gebiet der Morphologie und Physiologie</u>. (ed.) Hanstein, J. Band 1, Heft 1. Adolph Marcus, Bonn. 112 pp.

87. HARSHBERGER, J.W. 1897. Maize: A botanical and economic study. Contr. Bot. Lab. Univ. Penn. 1:75-202.

88. ----------------. 1898. Water storage and conduction in <u>Senecio praecox</u> DC. from Mexico. Contr. Bot. Lab. Univ. Penn. (N.S. 5) 2:31-40. (Also in: Trans. Proc. Bot. Soc. Penn. 1:31-40).

89. HARTIG, T. 1878. <u>Anatomie und Physiologie der Holzpflanzen</u>. J. Springer, Berlin. 412 pp.

90. HECKEL, E., and SCHLAGDENHAUFFEN, F. 1895. Sur le Bakis (<u>Tinospora baris</u> Miers) et le Sangol (<u>Cocculus leaeba</u> G.-P. et Rich.) du Sénégal du Soudan. Ann. Inst. Colon. Marseille 2:51-76.

91. HEGELMAIER, F. 1868. <u>Die Lemnaceen. Eine monographische Untersuchung</u>. Wilhelm Engelmann, Leipzig. 169 pp.

92. HEINRICHER, E. 1893. Biologische Studien an der Gattung <u>Lathraea</u>. Berich. Deutsch. Bot. Gesel. 9:1-18.

93. --------------. 1895. Anatomischer Bau und Leistung der Saugorgane der Schuppenwurz-Arten. (Lathraea clandestina Lam. und L. squamaria L.). Beitr. Biol. Pflanz. 7:315-406.

94. --------------. 1897. Die grünen Halbschmarotzer. I. Odontites, Euphrasia und Orthantha. Jahrb. Wiss. Bot. 31:77-124.

95. --------------. 1898. Die grünen Halbschmarotzer. II. Euphrasia, Alectrolophus und Odontites. Jahrb. Wiss. Bot. 46:273-376.

96. HÉRAIL, J., and BONNET, V. 1891. Manipulations de botanique médicale et pharmaceutique. J.-B. Baillière et Fils, Paris. 320 pp.

97. HERMANN, W. 1886. Morphologische und Anatomische Untersuchung einiger Arten der Gattung Impatiens mit besonderer berücksichtigung von Impatiens sultani. Inaugural-Dissertation. Universität Freiburg. J. Dilger, Freiburg. 44 pp.

98. HILDEBRAND, F. 1899. Die Keimung der Samen von Anemone apennina. Berich. Deutsch. Bot. Gesel. 17:161-166.

100. HILL. T.G. 1898. On the roots of Bignonia. Ann. Bot. 12:323-328.

101. HOFFMANN, H. 1849-1852. Ueber die Wurzeln der Doldengewächse. Flora (1849) 32:17-25; 721-728. (1850) 33:385-389; 401-405; 657-665. (1851) 34:513- 519; 529-535. (1852) 35:225-233; 241-256.

102. HOLLE, H.G. 1876. Ueber den Vegetationspunkt der Angiospermen-Wurzeln, inbesondere die Haubenbildung. Bot. Zeit. 34:241-255; 257-264.

103. HOLM, T. 1891. Notes upon Uvularia, Oakesia, Diclytra and Krigia. Bull. Torrey Bot. Club 18:1-11.

104. --------. 1895. Anatomy of Vellozieae. Bot. Gaz. 20:111-112.

105. --------. 1897. Obolaria virginica L.: A morphological study. Ann. Bot. 11:369-383.

106. --------. 1897. Studies in the Cyperaceae. III. Carex fraseri Pers., a morphological and an anatomical study. Amer. Jour. Sci. (Ser. 4) 3:121-128.

107. --------. 1897. Studies in the Cyperaceae. IV. Dulichium spathaceum Pers., a morphological and anatomical study. Amer. Jour. Sci. (Ser. 4) 3:429-437.

108. --------. 1898. Pyrola aphylla: A morphological study. Bot. Gaz. 25:246-254.

109. --------. 1899. Podophyllum peltatum. Bot. Gaz. 27:419-433.

110. -------. 1899. <u>Juncus</u> <u>repens</u> Michx.: A morphological and anatomical study. Bull. Torrey Bot. Club 26:359-364.

111. IRMISCH, T. 1857. Über die Keimung und die Erneuerungsweise von <u>Convolvulus</u> <u>sepium</u> und <u>Convolvulus</u> <u>arvensis</u>, so wie über hypokotylische Adventivknospen bei krautartigen phanerogamen Pflanzen. Bot. Zeit. 15:433-472; 481-484; 489-497.

112. ----------. 1858. <u>Beiträge zur Biologie und Morphologie der Orchideen</u>. Ambrosius Abel, Leipzig. 82 pp.

113. ----------. 1877. Beiträge zur vergleichenden Morphologie der Pflanzen. Ueber einige Aroideen. Abh. Nat. Gesel. Halle 13:161-206.

114. ITSCHERT, P. 1894. Beiträge zur anatomischen Kenntnis von <u>Strychnos</u> <u>tieute</u>. Inaugural-Dissertation. Friederich-Alexanders-Universität zu Erlangen. Aug. Vollrath, Erlangen. 24 pp.

115. JANCZEWSKI, E. DE. 1874. Recherches sur l'accroissement terminal des racines dans les Phanérogames. Ann. Sci. Nat., Bot. (Sér. 5) 20:162-201.

116. ------------------. 1874. Recherches sur développement des radicelles dans les Phanérogames. Ann. Sci. Nat., Bot. (Sér. 5) 20:208-233.

117. ------------------. 1885. Organisation dorsiventrale dans les racines des Orchidées. Ann. Sci. Nat., Bot. (Sér. 7) 2:55-81.

118. JANSE, J.M. 1897. Les endophytes radicaux de quelques plantes javanaises. Ann. Jard. Bot. Buitenz. 14:53-201.

119. JEFFREY, E.C. 1899. The devlopment, structure, and affinities of the genus <u>Equisetum</u>. Mem. Boston Soc. Nat. Hist. 5:155-190.

120. JORGENSEN, A. 1877-1879. Bidrag til rodens naturhistorie. I. Om Bromeliaceernes rødder. Bot. Tidskr. (Ser. 3) 10:144-170. (in Danish)

121. JOST, L. 1887. Ein Beitrag zur Kenntnis der Athmungsorgane der Pflanzen. Bot. Zeit. 45:600-606; 617-628; 633-642.

122. JUMELLE, H. 1897. Étude anatomique du <u>Cissus</u> <u>gongyloides</u> Burch. Rev. Gén. Bot. 9:129-149.

123. KAMIÉNSKI, F. 1878. Vergleichende Anatomie der Primulaceen. Abh. Nat. Gesel. Halle 14:143-230.

124. ------------. 1881. Die Vegetationsorgane der <u>Monotropa</u> <u>hipopitys</u> L. Bot. Zeit. 39:457-461.

125. KARSTEN, H. 1847. Die Vegetationsorgane der Palmen. Abh. Königl. Acad. Wiss. Berlin (Phys. Abh.) Jahrgang 1847, pp. 73-235.

126. ----------. 1865. Die Vegetationsorgane der Palmen. Ein Beitrag zur vergleichenden Anatomie und Physiologie. In: Gesammelte Beiträge. (ed.) Karsten, H. Ferd. Dümmler's Verlag, Berlin. 1:242-252.

127. ----------. 1865. Ueber den Bau der Cecropia peltata L. In: Gesammelte Beiträge. (ed.) Karsten, H. Ferd. Dümmler's Verlag, Berlin. 1:242-252.

128. ----------. 1890. Ueber die Bewurzelung der Palmen. In: Gesammelte Beiträge. (ed.) Karsten, H. R. Friedländer und Sohn, Berlin. pp. 17-21. (also in Linnaea 1856, 68:601-608).

129. ----------. 1891. Ueber die Mangrove-Vegetation im Malayischen Archipel. Biblio. Bot. 22:1-71.

130. KATTEIN, A. 1897. Der morphologische Werth des Centralcylinders der Wurzel. Bot. Centralb. 72:55-61; 91-97; 129-139.

131. KELLER, L. 1889. Anatomische Studien über die Luftwurzeln einiger Dikotyledonen. Inaugural-Dissertation. Ruperto-Carolinischen Universität zu Heidelberg. J. Horning, Heidelberg. 45 pp.

132. KERCKHOFF, C. 1896. Beiträge zur Kenntnis von Carlina acaulis und Atractylis gummifera. Inaugural-Dissertation. Friederich-Alexanders-Universität Erlangen. M. Ernst, München. 56 pp.

133. KHOURI, J. 1895. Contribution à l'étude botanico-chemique et thérapeutique du Goyavier (Psidium pomiferum L.). Ann. Inst. Colon. Marseille. 2:81-151.

134. KLEBAHN, H. 1891. Ueber Wurzelanlagen unter Lenticellen bei Herminiera elaphroxylon und Solanum dulcamara. Flora 74:125-139.

135. KLEIN, J., and SZABÓ, F. 1880. Zur Kenntnis der Wurzeln von Aesculus hippocastaneum L. Flora 63:163-168.

136. KOCH, L. 1874. Untersuchungen über die Entwicklung der Cucuteen. In: Botanische Abhandlungen aus dem Gebiet der Morphologie und Physiologie. (ed.) Hanstein, J. Band 2 Heft 3. Adolph Marcus, Bonn. 136 pp.

137. -------. 1883. Untersuchungen über die Entwicklung der Orobanchen. Berich. Deutsch. Bot. Gesel. 1:188-202.

138. -------. 1887. Die Entwicklungsgeschichte der Orobanchen. Carl Winter's Universitätbuchhandlung, Heidelberg. 389 pp.

139. -------. 1895. Ueber Bau und Wachstum der Wurzelspitze von Angiopteris evecta Hoffm. Jahrb. Wiss. Bot. 27:369-402.

11

140. KRAUSE, H. 1879. Beiträge zur Anatomie der Vegetations-Organe von Lathraea squamaria L. Inaugural-Dissertation. Universität Breslau. Breslauer Genossenschafts, Breslau. 36 pp.

141. KUBIN, E., and MÜLLER, J.F. 1878. Entwicklungs-Vorgänge bei Pistia stratioites und Vallisneria spiralis. In: Botanische Abhandlungen aus dem Gebiet der Morphologie und Physiologie. (ed.) Hanstein, J. Band 3 Heft 4. Adolph Marcus, Bonn. 70 pp.

142. KUHN, R. 1889. Untersuchungen über die Anatomie der Marattiaceen und anderer Gefässkryptogamen. Flora 72:457-504.

143. LEBLOIS, A. 1887. Recherches sur l'origine et le développement des canaux sécréteurs et des poches sécrétices. Ann. Sci. Nat., Bot. (Sér. 7) 6:247-330.

144. LECLERC DU SABLON. 1887. Recherches sur les organes d'absorption des plantes parasites (Rhinanthées et Santalacées). Ann. Sci. Nat., Bot. (Sér. 7) 6:90-117.

145. LEITGEB, H. 1863. Zur Kenntnis von Hartwegia commosa Nees. Sitz. Kaiser. Akad. Wiss. (Math.-Nat. Kl.) 48:138-160.

146. ----------. 1864. Ueber die Luftwurzeln der Orchideen. Ann. Mag. Nat. Hist. 14:159-160.

147. ----------. 1864. Die Luftwurzeln der Orchideen. Denk. Akad. Wiss. (Math.-Nat. Kl.) 24:179-222.

148. LEMAIRE, A. 1886. Recherches sur l'origine et le développement des racines laterales chez les Dicotylédones. Ann. Sci. Nat., Bot. (Sér. 7) 3:163-274.

149. LENFANT, G. 1897. Contribution à l'anatomie des Renonculacées. Le genre Delphinium. Arch. Inst. Bot. Univ. Liège 1:3-46.

150. LERMER, J.K., and HOLZNER, G. 1896. Die Wurzeln. Zeitschr. Gesam. Brauw. 5:57-60.

151. LESAGE, P. 1891. Sur le différenciation du liber dans la racine. Comp. Rend. Hebd. Séances Acad. Sci. 112:444-446.

152. LIERAU, M. 1887. Beiträge zur Kenntnis der Wurzeln der Araceen. Inaugural-Dissertation. Universität Breslau. Wilhelm Engelmann, Leipzig. 38 pp.

153. ---------. 1888. Über die Wurzeln der Araceen. Bot. Jahrb. 9:1-38.

154. LINDE, O. 1886. Beiträge zur Anatomie der Senegawurzel. Flora 69:1-32.

155. LINSBAUER, K. 1898. Beiträge zur vergleichenden Anatomie einiger tropischer Lycopodien. Sitz. Kaiser. Akad. Wiss. (Math.-Nat.Kl.) 107:995-1030.

156. LOPRIORE, G. 1896. Ueber die Regeneration gespaltner Wurzeln. Abh. Kaiser. Leop. Carol. Deutsch. Akad. Nat. 66:211-286.

157. LUBBOCK, J. 1892. A contribution to our knowledge of seedlings. Vol. 2. D. Appleton and Co., London. 646 pp.

158. LUND, S., and KJAERSKOU, H. 1885. Morphological-anatomical description of Brassica oleracea L., B. campestris L. and B. napus L. H. Hagerups Boghandel, Kjobenhavn. 149 pp. (in Danish)

159. MACDOUGAL, D.T. 1896. The root-tubers of Isopyrum occidentale. Bot. Gaz. 21:280-282.

160. --------------. 1897. The curvature of roots. Bot. Gaz. 23:307-366.

161. --------------. 1899. Symbiosis and saprophytism. Bull. Torrey Bot. Club 26:511-530.

162. MAGNUS, G. 1898. Beiträge zur Anatomie der Tropaeolaceen. Inaugural-Dissertation. Ruprecht-Karls-Universität zu Heidelberg. J. Horning, Heidelberg. 50 pp.

163. MANGIN, L. 1882. Origine et insertion des racines adventives et modifications corrélatives de la tige chez les Monocotylédones. Ann. Sci. Nat., Bot. (Sér. 6) 14:216-363.

164. MANSION, A. 1897. Contribution à l'anatomie des Renonculacées. Le genre Thalictrum. Arch. Inst. Bot. Univ. Liège 1:3-73.

165. MARIÉ, P. 1885. Recherches sur la structure des Renonculacées. Ann. Sci. Nat., Bot. (Sér. 7) 20:5-180.

166. MAXWELL, F.B. 1892. A comparative study of the roots of Ranunculaceae. Bot Gaz. 17:281.

167. ------------. 1893. A comparative study of the roots of Ranunculaceae. Bot. Gaz. 18:8-16; 41-47; 97-102.

168. MEINECKE, E.P. 1894. Beiträge zur Anatomie der Luftwurzeln der Orchideen. Flora 78:133-203.

169. MICHALOWSKI, J. 1881. Beitrag zur Anatomie und Entwicklungsgeschichte von Papaver somniferum L. Inaugural-Dissertation. Universität Breslau. Louis Streisand, Grätz, Pr. Posen. 52 pp.

170. MICHEELS, H. 1889. Recherches sur les jeunes Palmiers. Mém. Cour. Mém. Sav. Étrang. Acad. Roy. Sci. Lett. Beaux-Arts Belgique 51:3-126.

171. MIQUEL, F.-A.-W. 1846. Recherches sur la structure d'un tronc agé du Cycas circinalis. Ann. Sci. Nat., Bot. (Sér. 3) 5:11-24.

13

172. MIRBEL, C.F.B. DE. 1839. Nouvelles notes sur le cambium. Arch. Muséum Hist. Nat. J.-B. Baillière et Fils, Paris. pp.303-335.

173. ------------------. 1839. Nouvelles notes sur le cambium. Mém. Acad. Roy. Sci. Inst. France (Sér. 2) 18:727-805.

174. MITTEN, W. 1847. Sur le parasitisme des racines du Thesium linophyllum. Ann. Sci. Nat., Bot. (Sér. 3) 7:127-128.

175. MÖBIUS, M. 1899. Der japanische Lackbaum, Rhus vernicifera DC. Eine morphologisch-anatomisch Studie. Abh. Senck. Nat. Gesel. 20:203-247.

176. MOLISCH, H. 1883. Ueber das Langenwachsthum geköpfter und verletzter Wurzeln. Berich. Deutsch. Bot. Gesel. 1:362-366.

177. MOROT, L. 1885. Recherches sur le péricycle ou couche peripherique du cylindre centrale chez les phanérogames. Ann. Sci. Nat., Bot. (Sér. 7) 20:217-309.

178. --------. 1889. Sur les affinités anatomiques du genre Podoon. Jour. Bot. 3:388-390.

179. MÜLLER, F. 1877. Untersuchungen uber die Struktur einigen Arten von Elatine. Inaugural-Dissertation. Universität Gottingen. Neubauer'sche Buchdruckeri (F. Huber), Regensburg. 27 pp.

180. MÜLLER, F. 1895. Die Keimung einiger Bromeliaceen. Berich. Deutsch. Bot. Gesel. 13:175-182.

181. MÜLLER, H. 1875. Ueber Wachsthum und Bedeutung der Wurzeln. Landw. Jahrb. 4:999-1033.

182. NÄGELI, C. 1858. Ueber das Wachsthum des Stammes und der Wurzel bei den Gefässpflanzen und die Anordnung der Gefäss-stange im Stengel. Beitr. Wiss. Bot. 1:1-156.

183. ---------., and LEITGEB, H. 1868. Enstehung und Wachstum der Wurzeln. Beitr. Wiss. Bot. 4:73-160.

184. NICOLAI, O. 1865. Das Wachstum der Wurzel. Schr. König. Phys.-Ökon. Gesel. Könings. 7:33-76.

185. OLIVIER, L. 1881. Recherches sur l'appareil tégumentaire des racines. Ann. Sci. Nat., Bot. (Sér. 6) 11:5-133.

186. OUDEMANS, C.A. 1861. Ueber den Sitz der Oberhaut in den Luftwurzeln der Orchideen. Verh. Kononk. Akad. Wetens. 9:4-32.

187. PALLA, E. 1889. Zur Anatomie der Orchideen-Luftwurzeln. Sitz. Kaiser. Akad. Wiss. (Math.-Nat. Kl.) 98:200-207.

188. PARLATORE, F. 1884. Tavole per una "anatomia delle piante aquatiche."
Pubblicazioni del R. Istituto di Studi Superiori Pratici e de
Perfezionamento in Firenze (Sezione di Scienze, Fisische e Naturali).
24 pp.

189. PEARSON, H.W. 1898. Anatomy of the seedling of Bowenia spectabilis Hook.
f. Annals Bot. 12:475-490.

190. PEIRCE, G.J. 1893. On the structure of the haustoria of some phanero-
gamic parasites. Annals Bot. 7:291-327.

191. -----------. 1894. Das eindringen von Wurzeln in lebende gewebe. Bot.
Zeit. 52:169-176.

192. PENZIG, O. 1877. Untersuchungen über Drosophyllum lusitanicum Lk.
Inaugural-Dissertation. Universität Breslau. O. Opitz, Namslau. 46 pp.

193. ---------. 1887. Studi botanici sugli agrumi e sulle piante affini.

Ann. Agricol. 590 pp.

194. PERROT, E. 1897. Sur le tissu criblé extra-liberien et la tissu
vasculaire extra-ligneux. Comp. Rend. Hebd. Séances Acad. Sci. 125:
1115-1118.

195. ---------. 1899. Anatomie comparée des Gentianacées. Ann. Sci. Nat.,
Bot. (Sér. 8) 7:105-292.

196. PETTIT, A.S. 1895. Arachis hypogaea L. Mem. Torrey Bot. Club
4:275-296.

197. PFEFFER, W. 1871. Die Entwicklung des Keimes der Gattung Selaginella.
In: Botanische Abhandlungen aus dem Gebiet der Morphologie und Physiologie. (ed.) Hanstein, J. Band 1 Heft 3. Adolph Marcus, Bonn.
80 pp.

198. PFITZER, E. 1883. Zur Morphologie und Anatomie der Monocotylen
ähnlichen Eryngien. Berich. Deutsch. Bot. Gesel. 1:133-137.

199. PFUHL, F. 1878. Ueber die Anatomie der Gattungen Brassica, Sinapsis,
Raphanus raphanistrum. Inaugural-Dissertation. Universität Giessen.
Merzbach, Posen. 28 pp.

200. POHL, J. 1894. Botanische Mitteilung über Hydrastis canadensis.
Biblio. Bot. 29:1-12.

201. POIRAULT, G. 1891. Sur quelques points de l'anatomie des organes
végétatifs des Ophioglosées. Comp. Rend. Hebd. Séances Acad. Sci.
112:967-968.

202. POULSEN, V.A. 1888. Anatomiske studier over Eriocaulaceerne. Hos
Brödrene Salmonsen (J. Salmonsen), Københaven. 167 pp. (also in:
Viden. Medd. Nat. Foren. Kjobh. 1888. 40:221-383)

203. PRANTL, K. 1887. Beiträge zur Morphologie und Systematik der
Ranunculaceen. Bot. Jahrb. 9:225-273.

204. PRILLIEUX, E. 1856. De la structure anatomique et du mode de végétation
du Neottia nidus-avis. Ann. Sci. Nat., Bot. (Sér. 4) 5:267-282.

205. PRINGSHEIM, N. 1863. Zur Morphologie der Salvinia natans. Jahrb.
Wiss. Bot. 3:484-541.

206. PRITZEL, G.A. 1872. Thesaurus literaturae botanicae. F.A. Brockhaus,
Lipsiae. 576 pp.

207. PROLLIUS, F. 1884. Über Bau und Inhalt der Aloineenblätter, Stamme und
Wurzeln. Arch. Pharm. 22:553-578.

208. QUEVA, Ca 1899. Contributions à l'anatomie des monocotylédonées. I.
Les Uvulariées tubéreuses. Trav. Mem. Univ. Lille 7:1-162.

209. REINKE, J. 1871. Untersuchungen über Wachsthumgeschichte und Morphologie
der Phanerogamen-Wurzel. In: Botanische Abhandlungen aus dem Gebiet der
Morphologie und Physiologie. (ed.) Hanstein, J. Band 1 Heft 3. Adolph
Marcus, Bonn. 50 pp.

210. ---------. 1872. Andeutungen über den Bau der Wurzel von Pinus pinea.
Bot. Zeit. 30:49-54.

211. ---------. 1872. Zur Geschichte unserer Kenntnisse vom Bau der
Wurzelspitze. Bot. Zeit. 30:661-674.

212. RICHTER, C.G. 1899. Beiträge zur Biologie von Arachis hypogaea.
Inaugural-Dissertation. Königl. Universität Breslau. Anton Schreiber,
Breslau. 37 pp.

213. RIKLI, M. 1895. Beiträge zur vergleichenden Anatomie der Cyperaceen mit
besonderer Berücksichtigung der innern Parenchymscheide. Jahrb. Wiss.
Bot. 27:484-580.

214. RIMBACH, A. 1893. Ueber die Ursache der Zellhautwellung in der Exodermis
der Wurzeln. Berich. Deutsch. Bot. Gesel. 11:94-112; 467-472.

215. ----------. 1897. Ueber die Lebenweise des Arum maculatum. Berich.
Deutsch. Bot. Gesel. 15:178-182.

216. ----------. 1897. Lebensverhältnisse des Allium ursinum. Berich.
Deutsch. Bot. Gesel. 15:248-252.

217. ----------. 1897. Biologische Beobachtungen an Colchicum autumnale L.
Berich. Deutsch. Bot. Gesel. 15:298-302.

218. ----------. 1898. Die kontraktilen Wurzel und ihre Thätigkeit. Beitr. Wiss. Bot. 2:1-28.

219. ----------. 1898. Ueber Lillium martagon. Berich. Deutsch. Bot. Gesel. 16:104-110.

220. ----------. 1899. Beiträge zur Physiologie der Wurzeln. Berich. Deutsch. Bot. Gesel. 17:161-166.

221. ROSS, H. 1883. Beiträge zur Anatomie abnormer Monokotylenwurzeln. (Musaceen, Bambusaceen). Berich. Deutsch. Bot. Gesel. 1:331-338.

222. ROSTOWZEW, S. 1890. Beiträge zur Kenntnis der Gefässkryptogamen. I. Umbildung von Wurzeln in Sprosse. Flora 73:155-168.

223. SACHS, J. 1874. Über das Wachsthum der Haupt und Nebenwurzeln. Arb. Bot. Inst. Würzburg 1:384-474; 585-634.

224. SARAUW, G.F.L. 1891. Versuche über die Verzweigungs-Bedingungen der Stutzwurzeln von Selaginella. Berich. Deutsch. Bot. Gesel. 9:51-65.

225. SAUVAGEAU, C. 1889. Sur la racine du Najas. Jour. Bot. 3:3-11.

226. ------------. 1889. Contribution à l'étude du système mécanique dans la racine des plantes aquatiques les Potamogeton. Jour. Bot. 3:61-72.

227. ------------. 1889. Contribution à l'étude du système mécanique dans la racine des plantes aquatiques les Zostera, Cymodocea et Posidonia. Jour. Bot. 3:169-181.

228. SCHAAR, F. 1898. Über den Bau des Thallus von Rafflesia roschussenii Teysm. Binn. Sitz. Akad. Wiss. Wien (Math.-Nat. Kl.) 107:1039-1056.

229. SCHACHT, H. 1853. Beitrag zur Entwicklungs-geschichte zur Wurzel. Flora 36;257-266.

230. ----------. 1862. Ueber den Stamm und die Wurzel der Araucaria brasiliensis. Bot. Zeit. 20:409-411; 417-423.

231. SCHENCK, H. 1886. Vergleichende Anatomie der submersen Gewächse. Biblio. Bot. 1:1-67.

232. ----------. 1889. Ueber das Äerenchym, ein dem Kork homologes Gewebe bei Sumpflanzen. Jahrb. Wiss. Bot. 20:526-574.

233. ----------. 1889. Ueber die Luftwurzeln von Avicennia tomentosa und Laguncularia racemosa. Flora 72:83-88.

234. SCHENCKE, P. 1893. Über Stratioites aloides zur Familie der Hydrocharideen gehörig. Inaugural-Dissertation. Friederich-Alexanders-Universität, Erlangen. Fr. Eberhardt, Nordhausen. 28 pp.

235. SCHLICKUM, A. 1896. Morphologischer und anatomischer Vergleich der Kotyledonen und ersten Laubblätter der Keimpflanzen der Monokotylen. Biblio. Bot. 35:1-88.

236. SCHMITZ, F. 1875. Über die Anatomische Struktur der perennirenden Convolvulaceenwurzeln. Bot. Zeit. 33:677-690.

237. SCHÖNLAND, S. 1887. The apical meristem in the roots of the Pontederiaceae. Annals Bot. 1:179-182.

238. SCHULZE, W. 1899. Morphologie und Anatomie der Convallaria majalis L. Inaugural-Dissertation. Universität Basel. Jean Trapp, Bonn. 43 pp.

239. SCHWARZ, F. 1883. Die Wurzelhaare der Pflanzen. Untersuch. Bot. Inst. Tübingen Heft 2. 1:135-188.

240. SCHWENDENER, S. 1882. Ueber das Scheitwachsthum der Phanerogamen-Wurzeln. Sitz. König. Preuss. Akad. Wiss. Berlin 1:183-199.

241. SCOTT, D.H. 1882. Zur Entwicklungsgeschichte der gegliederten Milch-röhren. Arb. Bot. Inst. Würzburg 2:648-664.

242. ----------., and WAGER, H. 1888. On the floating roots of Sesbania aculeata Pers. Annals Bot. 1:307-314.

243. ----------. 1891. On some points in the anatomy of Ipomoea versicolor Meissn. Annals Bot. 5:173-179.

244. ----------., and BREBNER, G. 1891. On the internal phloem in the root and stem of Dicotyledons. Annals Bot. 5:259-300.

245. ------------ and ----------. 1893. On the secondary tissues in certain Monocotyledons. Annals Bot. 7:21-62.

246. ----------., and SARGANT, E. 1893. On the pitchers of Dischidia rafflesiana (Wall.). Annals Bot. 7:243-269.

247. ----------. 1897. On two new instances of spinous roots. Annals Bot. 11:327-332.

248. SCOTT, R., and SARGANT, E. 1898. On the development of Arum maculatum from the seed. Annals Bot. 12:399-414.

249. SELLE, H. 1894. Ueber den anatomischen Bau der Fabae Impigem und der Wurzel von Derris elliptica. Inaugural-Dissertation. Friederich-Alexander-Universität Erlangen. Paul Tschöpe, Mecklenburg. 31 pp.

250. SEWARD, A.C. 1899. The structure and affinities of Matonia pectinata. Annals Bot. 13:319-320.

251. SOLEREDER, H. 1899. Systematische Anatomie der Dicotyledonen. Ferdinand Enke, Stuttgart. 984 pp.

252. SOLMS-LAUBACH, H.G. 1867. Über den Bau und Entwicklung der Ernährungs-
organe parasitischer Phanerogamen. Jahrb. Wiss. Bot. 6:509-638.

253. ------------------. 1877. Das haustorium der Loranthaceen und der
Thallus der Rafflesiaceen und Balanophoreen. Abh. Nat. Gesel. Halle
13:239-276.

254. STERCKX, R. 1897. Contribution à l'anatomie des Renonculacées. Tribu
des Clematidées. Arch. Inst. Bot. Univ. Liège 1:3-55.

255. STRASBURGER, E. 1872. Zur Kenntnis der Archispermenwurzel. Bot.
Zeit. 30:757-763.

256. --------------. 1872. Die Coniferen und die Gnetaceen. Eine morphologische studie. Ambr.Abel, Leipzig. Text 442 pp., atlas with
26 plates.

257. TONI, J.B., and PAOLETTI, J. 1891. Beitrag zur Kenntnis des anatomischen
Baues von Nicotiana tabacum L. Berich. Deutsch Bot. Gesel. 9:42-50.

258. TRÉCUL, A. 1846. Recherches sur l'origine des racines. Ann. Sci. Nat.,
Bot. (Sér. 3) 6:303-345.

259. TREUB, M. 1875. Le mèristéme primitif de la racine dans les monocotylédones. Musée Bot. Leide 2:17-98.

260. --------. 1875. Recherches sur les organes de la végétation du
Selaginella martensii Spring. Musée Bot. Leide 2:101-126.

261. TREUB, M.M. 1883. Sur le Myrmecodia echinata Gaudich. Ann. Jard. Bot.
Buitenz. 3:129-159.

262. TSCHIRCH, A. 1889. Angewandte Pflanzenanatomie. Vol. 1. Grundriss
der Anatomie. Urban und Scwarzenberg, Wien. 548 pp.

263. TUBEUF, C. VON. 1896. Die Haarbildungen der Coniferen. II. Die Wurzelhaare der Coniferen. Nat. Zeit. Forst-Landw. 5:173-193.

264. VAN BREDA DE HAAN, J. 1891. Anatomie van het geslacht Melocactus.
Doctoral thesis. Rijks-Universiteit, Leiden. Joh. Enschedé en Zonen,
Haarlem. 123 pp.

265. VAN TIEGHEM, P. 1870. Recherches sur la symétrie de structure des
plantes vasculaires. Mémoire 1: La racine. Ann. Sci. Nat., Bot.
(Sér. 5) 13:5-314.

266. --------------. 1886. Sur la formation des racines laterales de
Monocotylédones. Bull. Soc. Bot. France 33:342-343.

267. --------------., and DOULIOT, H. 1886. Origine des radicelles et des
racines latérales chez les Legumineuses et les Cucurbitacées. Bull.
Soc. Bot. France 33:494-501.

268. ------------- and ----------. 1886. Sur la polystélie. Ann. Sci. Nat., Bot. (Sér. 7) 3:275-322.

269. --------------. 1887. Recherches sur la deposition des radicelles et des Bourgeons. Ann. Sci. Nat., Bot. (Sér. 7) 5:130-151.

270. --------------. 1887. Sur les poils radicaux gémines. Ann. Sci. Nat., Bot. (Sér. 7) 6:127-128.

271. --------------. 1887. Sur le rèseau sous-endodermique de la racine des Crucifères. Bull. Soc. Bot. France 34:125-130.

272. --------------. 1887. Sur le second bois primaire de la racine. Bull. Soc. Bot. France 34:101-105.

273. --------------. 1887. Sur les racines doubles et les Bourgeons doubles des Phanérogames. Jour. Bot. 1:19-26.

274. --------------. 1887. Structure de la racine et disposition des radicelles dans Centrolepidées, Eriocaulées, Joncées,Mayacées et Xyridées. Jour. Bot. 1:305-315.

275. --------------, and DOULIOT, H. 1888. Recherches comparative sur l'origine des membres endogenes dans les plantes vasculaires. Ann. Sci. Nat., Bot. (Sér. 7) 8:1-660.

276. --------------, and MONAL. 1888. Sur le rèseau sous-épidermiques de la racine des Geraniacées. Bull. Soc. Bot. France 35:274.

277. --------------. 1891. Sur les tubes criblés extralibériens et les vaisseaux extraligneux. Jour. Bot. 5:117-128.

278. --------------. 1891. Traite de Botanique. Part 1. Lahure, Paris. 960 pp.

279. --------------. 1894. Structure de la racine dans les Loranthacées parasites. Bull. Soc. Bot. France 41:121-127.

280. VÖCHTING, H. 1878. Über Organbildung im Pflanzenreich. Erster Theil. Max Cohen und Sohn (Fr. Cohen), Bonn. 258 pp.

281. -----------. 1884. Über Organbildung im Pflanzenreich. Zweiter Theil. Emil Strauss, Bonn. 200 pp.

282. VOGEL, A. 1863. Beiträge zur Anatomie und Histologie der unterirdischen Theile von Convolvulus arvensis L. Verh. Zoo.-Bot. Gesel. Wien 13:257-300.

283. VONHÖNE, H. 1880. Ueber das Hervorbrechen endogener Organe aus dem Mutterorgane. Flora 63:227-234; 243-257; 268-274.

284. VON MANNAGETTA, G.R.B. 1890. Monographie der Gattung _Orobanche_.
Biblio. Bot. 19:1-275.

285. VON MOHL, H. 1832. Sind die Lenticellen als Wurzelknospen zu
betrachten? Flora 15:65-74.

286. ----------. 1862. Einige anatomische und physiologische Bemerkungen
über das Holz der Baumwurzeln. Bot. Zeit. 20:225-230; 233-239;
268-278; 281-287; 289-295; 313-319; 321-327.

287. WAAGE, T. 1891. Ueber haubenlose Wurzeln der Hippocastanaceen und
Sapindaceen. Berich. Deutsch. Bot. Gesel. 9:132-162.

288. WARD, H.M. 1892. _The Oak_. Kegan Paul, Trench, Trubner and Co., Ltd.,
London. 175 pp.

289. ----------., and DALE, E. 1899. On _Craterostigma_ _pumilum_ Hochst., a
plant from Somali-land. Trans. Linn. Soc. London (Ser. 2), Bot.
5:343-355.

290. WEISS, J.E. 1880. Anatomie und Physiologie fleischig verdickter Wurzeln.
Flora 63:81-89; 97-112; 113-123.

291. WENT, F.A.F.C. 1895. Ueber Haft- und Näherwurzeln bei Kletterpflanzen
und Epiphyten. Ann. Jard. Bot. Buitenz. 12:1-72.

292. WIELER, A. 1898. Die Funktion der Pneumathoden und des Aërenchyms.
Jahrb. Wiss. Bot. 32:503-524.

293. WILSON, L.L.W. 1898. Observations on _Conopholis_ _americana_. Contr.
Bot. Lab. Univ. Penn. 2:3-19.

294. WILSON, W.P. 1889. The production of aerating organs on the roots of
swamp and other plants. Proc. Acad. Nat. Sci. Phila. pp.67-69.

295. WINKLER, C. 1872. Zur Anatomie von _Araucaria_ _brasiliensis_. Bot. Zeit.
30:597-610.

296. WORSDELL, W.C. 1895. On the comparative anatomy of certain species of
the genus _Christisonia_. Annals Bot. 9:103-136.

297. --------------. 1898. The comparative anatomy of certain genera of the
Cycadaceae. Jour. Linn. Soc. 33:437-457.

298. ZACHARIAS, E. 1891. Ueber das Wachstum der Zellhaut bei Wurzelhaaren.
Flora 74:466-491.

299. ZAWODNY, J. 1897. Beitrag zur Kenntnis der Wurzel von _Sorghum_
saccharatum Pers. Zeitschr. Naturwiss. 70:169-183.

LITERATURE AFTER 1900

୶ⲞꞪⲞꞩ

300. ABESADZE, K.I. 1927. Anatomy of Rhododendron caucasicum Pall. in the
Caucasus. Bull. Jard. (Bot.) Princ. URSS 26:209-222. (Russian
with German summary)

301. ABESSADZE, K.J., MAKAREVSKAJA, E.A., and ZCHAKAJA, K.E. 1930. Über
die verscheidenen Grade der Widerstandsfähigkeit gegen Reblaus
allgemein verbreiteter georgischer Rebensorten, bedingt durch die
Unterscheide in der anatomischen Struktur ihrer Wurzeln. Sci. Pap.
App. Sec. Tiflis Bot. Gard. 7:121-158. (Russian with German summary)

302. ABROL, B.K., and KAPOOR, L.D. 1963. Pharmacognosy of roots of Prangos
pabularia Lindl. Planta Med. 11:128-133.

303. ACQUARONE, P. 1930. The roots of Musa sapientum L. United Fruit Co.
Research Bull. No. 26.

304. ADÁMKOVÁ, A., and BENEŠ, K. 1966. Position and extent of the elongation
in the elongation zone in the root tip of the broad bean Vicia faba L.
Biol. Plant. 8:427-430.

305. ADAMSON, R.S. 1910. Notes on the roots of Terminalia arjuna Bedd.
New Phytol. 9:150-156.

306. -------------. 1926. On the anatomy of some shrubby Iridaceae. Trans.
Roy. Soc. So. Africa Vol. 13, Part 2. pp. 175-195.

307. -------------. 1934. Anomalous secondary thickening in Compositae.
Annals Bot. 48:505-514.

308. ADDOMS, R.M. 1950. Notes on the structure of elongating pine roots.
Amer. Jour. Bot. 37:208-211.

309. AHMAD, K.J. 1969. Pharmacognosy of the leaf and root of Barringtonia
acutangula Gaertn. Planta Med. 17:338-345.

310. AHN, G.B., III. 1962. Preliminary morphological study of Andropogon
gerardi Vitman, a perennial bunch grass. Proc. Penn. Acad. Sci.
35:70-76.

311. ALCONERO, R. 1968. Vanilla root anatomy. Phyton. 25:103-110.

312. ALDRICH-BLAKE, R.N. 1930. The plasticity of the root system of Corsican
pine in early life. Oxford Forestry Memoirs. No. 12. Clarendon
Press, Oxford. 64 pp.

313. ALEKSANDROV, V.G., and VISLOUKH, V.I. 1934. Principal features of
structure of different organs of opium poppy and the distribution of
latex ducts in these organs. Bot. Zhur. 19:141-162. (Russian with
German summary)

22

314. ----------------. 1966. Anatomiĩa Rasteniĭ. Sovetskaia Nauka, Moskova. 431 pp. (in Russian)

315. ALEXANDROV, W.G. 1926. Von den Eigenheiten in der lage der Kristalle und Eiweiss enthaltenden Zellen in den Wurzeln und Stengeln der Weinrebe (Vitis vinifera). Bot. Archiv 14:461-467.

316. ----------------. 1929. Beiträge zur Kenntnis der Zuckerrubenwurzel. Planta 7:124-132.

317. ALEXANDROVA, O.G. 1927. Sur l'architecture du xyleme de l'hypocotyle et des racines du tournesol. Jour. Soc. Bot. Russie 12:275-290. (Russian with French summary)

318. ----------------. 1928. Xylem of hypocotyl and roots of Helianthus annuus. Jour. Soc. Bot. Russie 12:275-290. (Russian with French summary)

319. ALFIERI, I.R. 1968. Ontogeny and structure of the tap root of Melilotus alba. Diss. Absts. 28(11):4439-B.

320. ------------., and EVERT, R.F. 1968. Analysis of meristematic activity in the root tip of Melilotus alba Desr. New Phytol. 67:641-647.

321. ALI, B. 1929. Studies in seedling anatomy of Zizyphus jujuba Lam., Acacia arabica Willd., and Vicia faba Desr. Abstract in: Proc. Indian Sci. Congr. (Madras) 16:235.

322. ALLEN, G.S. 1947. Embryogeny and development of the apical meristems of Pseudotsuga. II. Late embryogeny. Amer. Jour. Bot. 34:73-80.

323. ----------. 1947. Embryogeny and development of the apical meristems of Pseudotsuga taxifolia. III. Development of the apical meristems. Amer. Jour. Bot. 34:204-211.

324. ALQUATI, P. Studi anatomici e morfologici sull'ulivo (Olea europea). Atti Soc. Lig. Sci. Nat. Geogr. 17:128-148; 180-216; 225-239.

325. ALTEN, H. VON. 1909. Kritische Bemerkungen und neue Ansichten über die Thyllen. Bot. Zeit. 67:1-23.

326. -------------. 1909. Wurzelstudien. Bot. Zeit. 67:175-199.

327. -------------. 1910. Über den systematischen Wert der "physiologischen Scheiden" und ihrer Verstarkungen bei den Wurzeln. Bot. Zeit. 68:121-127; 137-146; 153-164.

328. AMICO, A. 1947. Osservazioni su Sternbergia lutea Ker-Gawl. Nuovo Giorn. Bot. Ital. (N.S.) 54:748-771.

329. ANCIBOR, E. 1971. Estudio anatomico y morfologico de una Crucifera andina en cojin: Lithodraba mendocinensis. Darwinia 16:519-561.

330. ANDERSON, C.E., and POSTLETHWAIT, S.N. 1960. The organization of the root apex of Glycine max. Proc. Indiana Acad. Sci. 70:61-65.

331. ANDREEVA, I.S. 1969. The anatomy of the vegetative organs of Kolopanax septemlobus (Thunb.) Koidz. (K. pictus). Uch. Zap. Omsk. Gos. Ped. Inst. 51:104-119. (in Russian)

332. ANONYMOUS. 1955. Notes on the anatomy of the oil palm. 2. The seedling. Jour. West African Inst. Oil Palm Res. 2:92-95.

333. AOBA, T., WATANABE, S., and SOMA, K. 1961. Studies on the formation of tuberous root in Dahlia: Anatomical observation of primary root and tuberous root. Jour. Jap. Soc. Hort. Sci. 30:81-88.

334. ARBER, A. 1914. On root development in Stratioides aloides L. with special reference to the occurrence of amitosis in an embryonic tissue. Proc. Cambr. Philos. Soc. 17:369-379.

335. --------. 1930. Root and shoot in angiosperms: A study of morphological categories. New Phytol. 29:297-315.

336. --------. 1934. The Gramineae. University Press, Cambridge. 480 pp.

337. --------. 1952. Monocotyledons. A morphological study. Cambridge Univ. Press, Cambridge. 258 pp.

338. --------. 1963. Water plants. In: Historiae Naturalis Classica. (eds.) Cramer, J. and Swann, K.H. Vol. 23. Gregg Associates, Brussels. 436 pp.

339. ARGO, V.N. 1964. Strangler fig, native epiphyte. Nat. Hist. 73:26-31.

340. ARMAND, L. 1912. Recherches morphologiques sur le Lobelia dortmanna. I. Rev. Gén. Bot. 24:465-478.

341. ARNOLD, B.C. 1966. Histogenesis in roots of Nothofagus solandri var. cliffortioides (Hook. f.) Poole. Pacific Sci. 20:95-99.

342. ARNOLD, C.A. 1940. A note on the origin of lateral rootlets of Eichhornea crassipes (Mart.) Solms. Amer. Jour. Bot. 27:728-730.

343. ARNOTT, H.J. 1962. The seed, germination, and seedling of Yucca. Univ. Calif. Publ. Bot. 35:1-96.

344. ARNY, A.C., and JOHNSON, I.J. 1928. The roots of flax plants. Jour. Amer. Soc. Agron. 20:373-380.

345. ARTSCHWAGER, E.F. 1918. Anatomy of the potato plant, with special reference to the ontogeny of the vascular system. Jour. Agric. Res. 14:221-252.

346. ----------------. 1924. On the anatomy of the sweet potato root, with notes on the internal breakdown. Jour. Agric. Res. 27:157-166.

347. ----------------. 1925. Anatomy of the vegetative organs of the sugar cane. Jour. Agric. Res. 30:197-221.

348. ----------------. 1926. Anatomy of the vegetative organs of the sugar beet. Jour. Agric. Res. 33:143-176.

349. ----------------. 1930. A study of the structure of sugar beets in relation to sugar content and type. Jour. Agric. Res. 40:867-915.

350. ----------------., and STARRETT, R.C. 1931. Suberization and wound-periderm formation in sweetpotato and gladiolus as affected by temperature and relative humidity. Jour. Agric. Res. 43:353-364.

351. ----------------- and -------------. 1933. Suberization and wound-cork formation in the sugar beet as affected by temperature and relative humidity. Jour. Agric. Res. 47:669-674.

352. ----------------. 1937. Observations of the effect of environmental conditions on the structure of the lateral roots of sugar beet. Jour. Agric. Res. 55:81-86.

353. ----------------. 1940. Morphology of the vegetative organs of sugarcane. Jour. Agric. Res. 60:503-549.

354. ----------------. 1943. Contribution to the morphology and anatomy of Guayule (Parthenium argentatum). U.S.D.A. Tech. Bull., No. 842. pp. 1-33.

355. ----------------. 1943. Contribution to the morphology and anatomy of the Russian dandelion (Taraxacum kok-saghyz). U.S.D.A. Tech. Bull., No. 843. pp. 1-24.

356. ----------------. 1946. Contribution to the morphology and anatomy of Cryptostegia (Cryptostegia grandiflora). U.S.D.A. Tech. Bull., No. 915. pp. 1-40.

357. ----------------. 1948. Anatomy and morphology of the vegetative organs of Sorghum vulgare. U.S.D.A. Tech. Bull., No. 957. pp. 1-55.

358. ASSAVESNA, S. 1959. Morphology and anatomy of Phormium tenax Forster. Diss. Absts. 20(3):858-859.

359. ATAL, C.K., and KHANNA, K.L. 1960. Pharmacognostic study of Valeriana pyrolaefolia Decaisne. Jour. Pharm. Pharm. 12:739-743.

360. ATWOOD, S. 1936. The anomalous root structure of Cycas revoluta. Amer. Jour. Bot. 23:336-340.

361. AVANZI, S., and D'AMATO, F. 1967. New evidence on the organization of the root apex in leptosporangiate ferns. Caryologia 20:257-264.

362. AVERS, C.J. 1957. An analysis of differences in growth rate of trichoblasts and hairless cells in the root epidermis of Phleum pratense. Amer. Jour. Bot. 44:686-690.

363. ----------. 1963. Fine structure studies of Phleum root meristem cells. II. Mitotic assymetry and cellular differentiation. Amer. Jour. Bot. 50:140-148.

364. AVERY, G.S., JR. 1930. Comparative anatomy and embryology of embryos and seedlings of maize, oats, and wheat. Bot. Gaz. 89:1-39.

365. ----------------. 1933. Structure and germination of tobacco seed and the developmental anatomy of the seedling plant. Amer. Jour. Bot. 20:309-327.

366. AYENSU, E.S. 1966. Taxonomic status of Trichopus: anatomical evidence. Jour. Linn. Soc. (Bot.). 59:425-430.

367. ----------. 1968. Comparative anatomy of the Stemonaceae (Roxburghiaceae). Bot. Gaz. 129:160-165.

368. ----------. 1968. The anatomy of Barbacenopsis, a new genus recently described in the Velloziaceae. Amer. Jour. Bot. 55:399-405.

369. ----------. 1971. Comparative anatomy of Dioscorea rotundata and Dioscorea cayenensis. In: New Research in Plant Anatomy. (eds.) Robson, N.K.B., Cutler, D.F., and Gregory, M. Academic Press, New York. pp. 127-136.

370. ----------. 1972. Dioscoreales. In: Anatomy of the Monocotyledons. (ed.) Metcalfe, C.R. Vol. 6. Clarendon Press, Oxford. 182 pp.

371. ----------. 1972. Morphology and anatomy of Synsepalum dulcificum (Sapotaceae). Bot. Jour. Linn. Soc. 65:179-187.

372. BAKKE, A.L. 1936. Leafy spurge, Euphorbia esula L. Iowa Agric. Exp. Sta. Res. Bull., No. 198. pp. 209-246.

373. BAKSHI, T.S. 1959. Ecology and morphology of Pterospora andromeda. Bot. Gaz. 120:203-217.

374. ----------., and COUPLAND, R.T. 1959. An anatomical study of the subterranean organs of Euphorbia esula in relation to its control. Canad. Jour. Bot. 37:613-620.

375. BAL, S.N., and DATTA, S.C. 1941. Senega and its substitute. Indian Jour. Pharm. 3:55-58.

26

376. ---------- and ----------. 1946. Pharmacognostic studies on Indian ipecacuanha. Indian Jour. Pharm. 7:76-80.

377. ---------., and GUPTA, B. 1956. Morphology of roots of Aristolochia indica Linn. Sci. Cult. 22:276-277.

378. BALANSARD, J., and MOREL, A. 1940. Étude morphologique et anatomique du Cassia sieberiana DC. Ann. Inst. Colon. Marseille (Sér. 5) 8:1-20.

379. BALDOVINOS, G. DE LA P. 1953. Growth of the root tip. In: Growth and Differentiation in Plants. (ed.) Loomis, W.E. Iowa State College Press, Ames. pp. 27-54.

380. BALDWIN, W.K.W. 1933. The organization of the young sporophyte of Isoëtes engelmanni A. Br. Proc. Trans. Roy. Soc. Canada (Ser. 3) Sect. 5. 27:11-30.

381. BALL, E. 1956. Growth of the embryo of Ginkgo biloba under experimental conditions. I. Origin of the first root of the seedling in vitro. Amer. Jour. Bot. 43:488-495.

382. BAMBACIONI, V. 1924. Sopra alcune anomalie delle radici di Vicia faba L. Annali Bot. 16:244-252.

383. BANERJI, I. 1943. The aerial roots of Plumeria acutifolia Por. Curr. Sci. 12:85-86.

384. BANERJI, K.G. 1929. Some observations on the anatomy and biology of the aerial roots of Vitis quadrangularis Wall. Abstract in: Proc. 16th Indian Sci. Congr. (Madras) 16:229.

385. BANERJI, M.L. 1961. On the structure of teratological seedlings. I. Cosmos bipinnatus Cav. Proc. Indian Acad. Sci. 53:10-19.

386. ------------. 1962. On the anatomy of teratological seedlings. II. Calendula officinalis Linn. Bull. Bot. Soc. Bengal 16:24-29.

387. ------------. 1964. On the anatomy of teratological seedlings. III. Tagetes erecta Linn. Proc. Indian Acad. Sci. 60:203-213.

388. BANNAN, M.W. 1941. Vascular rays and adventitious root formation in Thuja occidentalis L. Amer. Jour. Bot. 28:457-463.

389. -----------. 1941. Variability in wood structure in roots of native Ontario conifers. Bull. Torrey Bot. Club 68:173-194.

390. -----------. 1942. Notes on the origin of adventitious roots in the native Ontario conifers. Amer. Jour. Bot. 29:593-598.

391. -----------. 1962. The vascular cambium and tree-ring development. In: Tree Growth. (ed.) Kozlowski, T.T. Ronald Press, New York. pp. 3-21.

392. BARANOVA, E.A. 1951. Laws of formation of adventitious roots in plants. Trudy Glav. Bot. Sada Lenin. 2:168-193. (in Russian)

393. BARANOVA, P.A. 1957. Coleorrhiza in Myrtaceae. Phytomorph. 7:237-243.

394. BÁRÁNY, L. 1926. Beiträge zur Histologie der Vegetationsorgane von *Sibiraea croatica* Degen. Bot. Közlem. 23:44-54; (9). (Hungarian with German summary)

395. BARBER, C.A. 1905. The haustoria of Sandal roots. Indian For. 31:189-201.

396. -----------. 1906. Studies in root-parasitism. The haustorium of *Santalum album*. I. Early stages up to penetration. Mem. Dept. Agric. India (Bot. Ser.) 1:1-30.

397. -----------. 1907. Studies in root-parasitism. The haustorium of *Santalum album*. II. The structure of the mature haustorium and the inter-relations between host and parasite. Mem. Dept. Agric. India (Bot. Ser.) 1:1-58.

398. -----------. 1907. Studies in root-parasitism. III. The haustorium of *Olax scandens*. Mem. Dept. Agric. India (Bot. Ser.) 2:1-47.

399. -----------. 1907. Parasitic trees in southern India. Proc. Cambr. Philos. Soc. 14:246-256.

400. -----------. 1908. Studies in root-parasitism. IV. The haustorium of *Cansjera rheedii*. Mem. Dept. Agric. India (Bot. Ser.) 2:1-37.

401. BARDELL, E.M. 1915. Production of root hairs in water. Univ. Wash. Publ. Bot. 1:1-9.

402. BARGHOORN, E.S., JR. 1941. The ontogenetic development and phylogenetic specialization of rays in the xylem of dicotyledons. II. Modification of the multiseriate and uniseriate rays. Amer. Jour. Bot. 28:273-282.

403. BARLOW, P.W. 1969. Organization in root meristems. Doctoral dissertation. Oxford University, England.

404. BARRATT, K. 1920. A contribution to our knowledge of the vascular system of the genus *Equisetum*. Annals Bot. 34:201-235.

405. BARTELLETI, V. 1901. Studio monografico inturno alla famiglia delle Ochnaceae e specialmente delle Specie malesi. Malpighia 15:105-174.

406. BARTHOLOMEW, E.T., and REED, H.S. 1943. General morphology, histology, and physiology. In: *The Citrus Industry*. (eds.) Webber, H.J. and Batchelor, L.D. University of California Press, Berkeley. Chapt. 6, pp. 714-717.

407. BARTOO, D.R. 1929. Origin and development of tissues in root of Schizaea rupestris. Bot. Gaz. 87:642-652.

408. -----------. 1930. Origin of tissues of Schizaea pusilla. Bot. Gaz. 89:137-153.

409. BARTOŠ, V. 1930. Die Rübengefässbündel. Zeitschr. Zucker. Čechosl. Repub., Report 26. 54:289-296.

410. BARYKINA, R.P. 1967. Anatomical analysis of the vegetative organs of the steppe almond (Amygdalus nana L.). Vest. Moskov. Univ. (Ser. 6, Biologija) Počvovedenie. 22:42-56.

411. BÄSECKE, P. 1908. Beiträge zur Kenntnis der physiologischen Scheiden der Achsen und Wedel der Filicinen, sowie über den Ersatz des Korkes bei dieser Pflanzengruppe. Bot. Zeit. 66:25-87.

412. BASS-BECKLING, L.G.M. 1921. The origin of the vascular structure in the genus Botrychium, with notes on the general anatomy. Rec. Trav. Bot. Neerl. 18:333-375.

413. BATES, G.H. 1934. The relation of leaf size to root structure. Ecology 22:271-278.

414. BATESON, W. 1921. Root cuttings and chimaeras. II. Jour. Genetics 11:91-97.

415. BATTEN, L. 1918. Observations on the ecology of Epilobium hirsutum. Jour. Ecology 6:161-177.

416. BAUMGÄRTEL, O. 1917. Die Anatomie der Gattung Arthrocnemum Moqu. Sitz. Akad. Wiss. Wien 126:41-74.

417. BEAKBANE, A.B., and THOMPSON, E.C. 1939. Anatomical studies of stems and roots of hardy fruit trees. II. The anatomical structures of the roots of some vigorous and some dwarfing apple rootstocks, and the correlation of structure with vigour. Jour. Pomol. Hort. Sci. 17:141-149.

418. -------------. 1961. Structure of the plant stem in relation to adventitious rooting. Nature 192:954-955.

419. BEAL, W.J. 1900. A few observations of root hairs. Plant World 3:182-184.

420. BEARD, F.H. 1943. Root studies. X. The root system of hops on different soil types. Jour. Pomol. Hort. Sci. 20:147-154.

421. BEARD, J.B. 1973. Turfgrass: Science and Culture. Chapt. 2. Growth and development - The Root. Prentice-Hall, Inc., New Jersey. pp.28-32.

422. BEAUQUESNE, L. 1937. Recherches sur quelques Menispermacées médicinales des genres Tinospora et Cocculus. Trav. Lab. Mat. Méd. École Sup. Pharm. Paris Vol. 28. 181 pp.

423. BEAUVERIE, J. 1933. Les Gymnospermes. Bosc Frères (M. et L. Riou), Lyon. 160 pp.

424. BEAUVISAGE, L. 1920. Contribution à l'étude anatomique de la famille des Ternstroemiacées. Imprimeur E. Arrault et Cie, Tours. 470 pp.

425. BECKEL, D.K.B. 1956. Cortical disintegration in the roots of Bouteloua gracilis (H.B.K.) Lag. New Phytol. 55:183-190.

426. BECQUEREL, P. 1913. L'ontogenie vasculaire de la plantule du Lupin et ses conséquences pour certaines théories de l'anatomie classique. Bull. Soc. Bot. France 60:177-186.

427. ------------. 1922. La découverte de la phyllorhize. Rev. Gén. Sci. Pures Appl. 33:101-110.

428. BEHRE, K. 1929. Physiologische und zytologische Untersuchungen über Drosera. Planta 7:208-306.

429. BEIJERINCK, M.W. 1921. Beobachtungen und Betrachtungen über Wurzel-knospen und Nebenwurzeln. Verz. Geschr. Beijer. 2:7-121.

430. BELFORD, D.S., and PRESTON, R.D. 1961. The structure and growth of root hairs. Jour. Exp. Bot. 12:157-168.

431. BELICOVÁ, J. 1966. Anatomical-ecological studies of end-roots. Časop. Slez. Mus. Opa. (Ser. C, Dendrology) 5:51-63.

432. BELL, J.K., and MCCULLY, M.E. 1970. A histological study of lateral root initiation and development in Zea mays. Protoplasma 70:179-205.

433. BELL, P.R. 1951. Studies in the genus Elaphoglossum Schott. II. The root and bud traces. Annals Bot. (N.S.) 15:333-346.

434. BELL, W.H. 1934. Ontogeny of the primary axis of Soja max. Bot Gaz. 95:622-635.

435. BELOSTOKOV, G.P. 1964. The formation and structure of the conducting system in Quercus robur seedlings. Dokl. Akad. Nauk SSSR (Bot. Sci. Sect.) 157:1234-1238. (in Russian)

436. ---------------. 1965. Anatomical structure of seedlings of some coniferous trees. Izv. Akad. Nauk SSSR (Leningrad), (Ser. Biol.) 30:152-155. (in Russian)

437. ---------------. 1966. Structure of the conducting system of the seed-lings of leaf-bearing woody plants with epigeous germination. Bot. Zhur. 51:705-716. (in Russian)

438. ----------------. 1967. The conducting system in seedlings of coniferous plants. Bull. Princ. Bot. Gard. Moscow 64:89-94.

439. BELOW, S. 1914. Zur Kenntnis der Gattung Panicum. Angew. Bot. 7:306-324. (Russian with German summary)

440. BELZUNG, E. 1900. Anatomie et physiologie végétales. Ancienne Librarie Germer Baillière et Cie, Paris. 1320 pp.

441. BENSON, M. 1910. Root parasitism in Exocarpus. Annals Bot. 24: 667-677.

442. BERCKEMEYER, W. 1929. Ueber kontraktile Umbelliferenwurzeln. Bot. Archiv 24:273-318.

443. --------------., and ZIEGENSPECK, H. 1929. Der Mechanismus einiger kontraktiler Wurzeln von Monokotylen. Bot. Archiv 27:225-229.

444. BERGMANN, M. 1944. Vergleichende Untersuchungen über die Anatomie schweizerischer Ranunculus-Arten und deren Gehalt an Anemonol und Saponin. Berich. Schweiz. Bot. Gesel. 54:399-522.

445. BERKMAN, A.H. 1936. Seedling anatomy of Cannabis sativus L. Doctoral dissertation. University of Chicago, Chicago. 21 pp.

446. BERLYN, G.P. 1972. Seed germination and morphogenesis. In: Seed Biology. (ed.) Kozlowski, T.T. Academic Press, New York. Vol. I, pp. 223-312.

447. BERNARD, C. 1967. Germination et plantules de quelques Cactacées. Adansonia (Ser. 2) 6:593-641.

448. BERNARD, CH., and ERNST, A. 1910. Beiträge zur Kenntnis der Saprophyten Javas. II. Äussere und innere Morphologie von Thismia javanica J.J.S. Ann. Jard. Bot. Buitenz. (Ser. 2) 8:36-47.

449. -------------- and --------. 1911. Beiträge zur Kenntnis der Saprophyten Javas. V. Anatomie von Thismia clandestina Miq. und Thismia versteegii Sm. Ann. Jard. Bot. Buitenz. (Ser. 2) 9:61-69.

450. BERNATSKY, J. 1906. Systematische Anatomie der Polygonateen. Növény. Közlem. 5:111-124; (23)-(29). (Hungarian with German translation)

451. BERQUAM, D.L. 1972. Size-structure correlation in developing roots of Cissus and Syngonium. Jour. Minn. Acad. Sci. 38:42-45.

452. BERTHELOT, J. 1961. Anatomie et ontogénie de quelques plantules d'Impatiens scabrida D.C. Bull. Soc. Bot. France 108:217-237.

453. BERTHON, R. 1943. Sur l'origine des radicelles chez les Angiospermes. Comp. Rend. Hebd. Séances Acad. Sci. 216:308-309.

31

454. BERTRAND, P. 1937. Anatomie et ontogénie comparées des végétaux vasculaires. Bull. Soc. Bot. France 84:515-529.

455. -----------. 1940. Orientation des racines et radicelles diarches de fougères et des phanérogames par rapport à la tige ou à la racine mère. Bull. Soc. Bot. France 87:84-87.

456. -----------. 1947. Les végétaux vasculaires. Masson et Cie, Paris. 184 pp.

457. BESKARAVAĬNYĬ, M.M. 1955. Concrescence of roots of woody genera in the region of Kamyshin. Agrobiology 3:78-89. (in Russian)

458. BEXON, D. 1925. Observations on the anatomy of teratological seedlings. V. On the anatomy of some atypical seedlings of Sinapsis alba and Brassica oleracea. Annals Bot. 39:25-39.

459. --------. 1926. An anatomical study of the variation in the transition phenomena in the seedling of Althaea rosea. Annals Bot. 40:369-390.

460. --------., and WOOD, A.E. 1930. Observations on the anatomy of teratological seedlings. VII. The anatomy of some polycotylous seedlings of Impatiens royalei Walp. Annals Bot. 44:297-309.

461. BHAMBIE, S. 1963. Studies in Pteridophytes. IV. The development, structure and organization of root in Isoetes coromandelina L. Proc. Indian Acad. Sci. (Sect. B) 58:153-164.

462. -----------., and RAO, C.G.PRAKASA. 1969. Root-apex organization in some pteridophytes. Abstract in: Seminar on Morphology, Anatomy and Embryology of Land Plants. (eds.) Johri, B.M. et al. Centre of Advanced Study in Botany, University of Delhi.

463. BHANDARI, N.N., and NANDA, K. 1970. The endophytic system of Arceuthobium minutissimum, the Indian dwarf mistletoe. Annals Bot. (N.S.) 34:517-525.

464. BHATNAGAR, J.K., and RAINA, M.K. 1970. Pharmacognostic studies of indigenous drugs. I. Microscopic characters of the roots of Carissa spinarum Linn. Res. Bull. (Sci.) Panjab Univ. (N.S.) 21:317-321.

465. BHATNAGAR, S.P. 1965. Studies in angiospermic parasites. Bull. Nat. Bot. Gard. 112:1-90.

466. BHATTACHARYA, I.C. 1961. Pharmacognostical study of Indian Jalap. Indian Jour. Pharm. 23:186-190.

467. -----------------. 1961. A note on Aconitum chasmanthum Stapf ex Holmes. Indian Jour. Pharm. 23:276-278.

468. -----------------. 1961. Pharmacognostical study of Aconitum violaceum Jacq. Indian Jour. Pharm. 23:278-280.

469. BIEBL, R., and GERM, H. 1967. Praktikum der Pflanzenanatomie. 2nd ed. Springer-Verlag, Wien. 247 pp.

470. BIERHORST, D.W. 1969. On Stromatopteris and its ill-defined organs. Amer. Jour. Bot. 56:160-174.

471. --------------. 1971. Morphology of vascular plants. Macmillan Co., New York. 560 pp.

472. BIRGE, W.I. 1911. The anatomy and some biological aspects of the "Ball Moss," Tillandsia recurvata L. Bull. Univ. Texas, Science Ser. No. 20. No. 194. pp. 1-24.

473. BISALPUTRA, T. 1961. Anatomical and morphological studies in the Chenopodiaceae. II. Vascularization of the seedling. Austral. Jour. Bot. 9:1-19.

474. BISHT, B.S. 1963. Pharmacognosy of "Piplamul" - the root and stem of Piper longum Linn. Planta Med. 11:410-416.

475. BLAAUW, A.H. 1912. Das Wachstum der Luftwurzeln einer Cissus-Art. Ann. Jard. Bot. Buitenz. 26:266-293.

476. BLAKELY, W.F. 1922. The Loranthaceae of Australia. Proc. Linn. Soc. New So. Wales 47:1-25.

477. BLANK, F. 1939. Beitrag zur Morphologie von Caryocar nuciferum L. Berich. Schweiz. Bot. Gesel. 49:437-494.

478. BLASBERG, C.H. 1932. Phases of the anatomy of Asparagus officinalis. Bot. Gaz. 94:206-214.

479. BLOCH, E. 1910. Sur quelques anomalies de structure des plantes alpines. Rev. Gén. Bot. 22:281-290.

480. --------. 1914. Sur les modifications productés dans la structure des racines et des tiges par une compression extérieure. Comp. Rend. Hebd. Séances Acad. Sci. 158:1701-1703.

481. --------. 1919. Modifications anatomiques des racines par action mécanique. Comp. Rend. Hebd. Séances Acad. Sci. 169:195-197.

482. --------. 1924. Dissymetries de structure de rhizomes soumis a certaines actions mécaniques. Ann. Sci. Nat., Bot. (Sér. 10) 6:169-244.

483. BLOCH, R. 1926. Um differenzierungen an Wurzelgeweben nach Verwundung. Berich. Deutsch. Bot. Gesel. 4:308-316.

484. --------. 1935. Observations on the relation of adventitious root formation to the structure of air-roots of orchids. Proc. Leeds Philos. Lit. Soc. (Sci. Sect.) 3:92-101.

485. --------. 1937. Wound healing and necrosis in air roots in <u>Phoenix</u>
  <u>reclinata</u> and leaves of <u>Araucaria</u> <u>imbricata</u>. Amer. Jour. Bot.
  24:279-287.

486. --------. 1943. Differentiation in red root-tips of <u>Phalaris</u>
  <u>arundinaceae</u>. Bull. Torrey Bot. Club 70:182-183.

487. --------. 1944. Developmental potency, differentiation and pattern
  in meristems of <u>Monstera</u> <u>deliciosa</u>. Amer. Jour. Bot. 31:71-77.

488. --------. 1946. Differentiation and pattern in <u>Monstera</u> <u>deliciosa</u>.
  The idioblastic development of the trichosclereids in the air root.
  Amer. Jour. Bot. 33:544-551.

489. BLODGETT, F.H. 1910. The origin and development of bulbs in the genus
  <u>Erythronium</u>. Bot. Gaz. 50:340-373.

490. BLOKHINTSEVA, I.I. 1940. Formation of rubber in kok-saghyz. Vest.
  Sel'sk. Nauk, Tekhn Kultury, Moscow 3:50-56. (in Russian)

491. -----------------. 1940. Formation of rubber in kok-saghyz as a result
  of the functioning of the latex vessels. Bull. Acad. Sci. URSS
  (Sci. Ser., Biology) 4:608-613. (in Russian)

492. BLOMQUIST, H.L. 1922. Vascular anatomy of <u>Angiopteris</u> <u>evecta</u>. Bot.
  Gaz. 73:181-199.

493. BOBILIOFF, W. 1930. Anatomie en physiologie van <u>Hevea</u> <u>brasiliensis</u>.
  Rugrok and Co., Batavia. 288 pp. (in Dutch)

494. BÖCHER, T.W. 1964. Morphology of the vegetative body of <u>Metasequoia</u>
  <u>glypstroboides</u>. Dansk Bot. Arkiv 24:7-70.

495. -----------. 1969. Planternes morfologi. In: <u>Botanik</u>. (eds.) Böcher,
  T.W., Lange, M., and Sorenson, T. Vol. 1, No. 3. Munksgaard,
  Kobenhavn. 170 pp.

496. BOCQUILLON, H. 1901. Étude botanique et pharmacologique des Xanthoxylées.
  Doctoral thèses. Université de Paris. A. Hennuyer, Paris. 125 pp.

497. BODMER, H. 1929. Beiträge zur Anatomie und Physiologie von <u>Lythrum</u>
  <u>salicaria</u>. Beih. Bot. Centralb. 45:1-58.

498. BOEHM, R., and KUBLER, K. 1908. Ueber Kawarwurzel. Arch.Pharm.
  246:663-666.

499. BOEKE, J.E. 1940. On the origin of the intercellular channels and
  cavities in the rice root. Ann. Jard. Bot. Buitenz. 50:199-208.

500. BOESHORE, I. 1920. The morphological continuity of Scrophulariaceae
  and Orobanchaceae. Contr. Bot. Lab. Univ. Penn. 5:139-177.

501. BOEWIG, H. 1904. The histology and development of <u>Cassytha</u> <u>filiformis</u> L. Contr. Bot. Lab. Univ. Penn. 2:399-416). (Same in: Trans. Proc. Bot. Soc. Penn. 1904. 1:399-416)

502. BOGAR, G.D., and SMITH, F.H. 1965. Anatomy of seedling roots of <u>Pseudotsuga</u> <u>menziesii</u>. Amer. Jour. Bot. 52:720-729.

503. BONATI, G.-H. 1918. Le genre <u>Pedicularis</u> L. Doctoral thèses. Université de Nancy. Imprimeur Berger-Levrault, Nancy. 168 pp.

504. BOND, G. 1930. The occurrence of cell division in the endodermis. Proc. Roy. Soc. Edinb. 50:38-50.

505. BONDOIS, G. 1913. Contribution à l'étude de l'influence du milieu aquatique sur les racines des arbres. Ann. Sci. Nat., Bot. (Sér. 9) 18:1-24.

506. BONGA, J.M. 1969. The morphology and anatomy of holdfasts and branching radicles of <u>Arceuthobium</u> <u>pusillum</u> cultured in vitro. Canad. Jour. Bot. 47:1935-1938.

507. BONNET, A.L.M. 1955. Contribution à l'étude des Hydropteridées: Recherches sur <u>Salvinia</u> <u>auriculata</u> Aubl. Ann. Sci. Nat., Bot. (Sér. 11) 16:529-600.

508. --------------. 1955. Contribution à l'étude des Hydropteridées. I. Recherches sur <u>Pilularia</u> <u>globulifera</u> L. et <u>P</u>. <u>minuta</u> Dur. La Cellule 57:129-239.

509. BONNETT, H.T., Jr., and TORREY, J.G. 1966. Comparative anatomy of endogenous buds and lateral root formation in <u>Convolvulus</u> <u>arvensis</u> roots cultured in vitro. Amer. Jour. Bot. 53:496-507.

510. -----------------. 1968. The root endodermis; fine structure and function. Jour. Cell. Biol. 37:199-205.

511. -----------------. 1969. Cortical cell death during lateral root formation. Jour. Cell. Biol. 40:144-159.

512. BONNETT, O.T. 1961. The oat plant: its histology and development. Bull. Ill. Agric. Exp. Sta. No. 672. 112 pp.

513. BONNIER, G. 1900. Sur l'ordre le formation des eléments du cylindre centrale dans la racine et la tige. Comp. Rend. Hebd. Séances Acad. Sci. 131:781-789.

514. ----------. 1903. Influence de l'eau sur la structure des racines aériennes d'Orchidées. Comp. Rend. Hebd. Séances Acad. Sci. 137: 505-510.

515. ----------. 1903. Sur des formations secondaires anormales du
     cylindre central dans les racines aériennes d'Orchidées. Bull. Soc.
     Bot. France 50:291-295.

516. ----------., and FRIEDEL, J. 1912. Les vaisseaux spiralés et la
     croissance en longeur. Rev. Gén. Bot. 24:385-391.

517. BOODLE, L.A. 1900. Comparative anatomy of the Hymenophyllaceae,
     Schizaeaceae and Gleicheniaceae. I. On the anatomy of the Hymenophyll-
     aceae. Annals Bot. 14:455-496.

518. ----------. 1901. Comparative anatomy of the Hymenophyllaceae,
     Scizaeaceae and Gleicheniaceae. II. On the anatomy of the Schizaeaceae.
     Annals Bot. 15:359-421.

519. ----------. 1901. Comparative anatomy of the Hymenophyllaceae,
     Schizaeaceae and Gleicheniaceae. III. On the anatomy of the Gleichen-
     iaceae. Annals Bot. 15:703-747.

520. ----------. 1905. The anatomy of palm roots. (A review of: Drabble,
     E. 1904. On the anatomy of the roots of palms. Trans. Linn. Soc.
     London Ser. 2. 6:427-510). New Phytol. 4:19-23.

521. ----------. 1906. Lignification of phloem in Helianthus. Annals Bot.
     20:319-321.

522. ----------., and FRITSCH, F.E. 1908. Solereder's Systematic Anatomy
     of the Dicotyledons. (English translation) Revision by Scott, D.H.
     Clarendon Press, Oxford. Vol. 1, 644 pp; Vol. 2, 1183 pp.

523. ----------. 1913. The root and haustorium of Buttonia natalensis.
     Kew Bull. No. 6. pp. 240-242.

524. BOOTH, W.E. 1933. Comparative anatomy of Mentzelia oligoperma and M.
     decapetala. Univ. Kans. Sci. Bull. 21:439-461.

525. BORCHERT, R. 1966. Innere Wurzeln als Festingungselement der Epiphyti-
     schen Bromeliaceae Tillandsia incarnata H.B.K. Berich. Deutsch. Bot.
     Gesel. 79:253-258.

526. BORESCH, K. 1908. Über Gummifluss bei Bromeliaceen bebst Beiträgen zu
     ihrer Anatomie. Sitz. Kaiser. Akad. Wiss. Wien 117:1033-1080.

527. BORISSOW, G. 1924. Über die eigenatrigen Kieselkörper in der Wurzel-
     endodermis bei Andropogon-Arten. Berich. Deutsch. Bot. Gesel.
     42:366-380.

528. ----------. 1928. Weiteres über die Rasdorskychen Körperchen. Weitere
     Untersuchungen über die Kieselkörperchen in der Wurzelendodermis den
     Andropogoneen. Berich. Deutsch. Bot. Gesel. 46:463-480.

529. BORTHWICK, A.W. 1905. The production of adventitious roots and their relation to bird's-eye formation (Maser-holz) in the wood of various trees. Notes Roy. Bot. Gard. Edinb. No. 16. pp. 15-36.

530. BOSE, S.R. 1957. Lenticels in earth roots of Crinum asiaticum L. Sci. Cult. 22:695-696.

531. BOSELLI, E. 1904. Contributo allo studio dell'influenza dell'ambiente aqueo sull forma e sulla struttura delle piante. Annali Bot. 1:255-274.

532. BOTTOMLEY, W.B. 1912. The root nodules of Myrica gale. Annals Bot. 26:111-117.

533. BOTTUM, F.R. 1941. Histological studies on the root of Melilotus alba. Bot. Gaz. 103:132-145.

534. BOUET, M. 1954. Études cytologiques sur le développement des poils absorbants. Rev. Cytol. Biol. Vég. 15:261-305.

535. BOUILLENE, R. 1925. Les racines-échasses de Iriartea exorrhiza Mart. (Palmiers) et de Pandanus div. sp. (Pandanacées). Mém. Acad. Roy. Belgique (Sér. 2), Cl. Sci. 8:1-45. (Same in: Arch. Inst. Bot. Univ. Liège, 1927. 6:3-45)

536. ------------. 1928. Anatomical material for the study of growth differentiation in higher plants. Plant Phys. 3:459-471.

537. BOUNAGA, D. 1964. Anatomie du genre Eryngium en Afrique du Nord. Bull. Soc. Hist. Nat. Afrique Nord 54:7-80.

538. BOUREAU, E. 1937. De la presence de tracheides dans le bois alterne de la radicule des Conifères. Comp. Rend. Hebd. Séances Acad. Sci. 204:1579-1581.

539. ----------. 1939. Recherches anatomiques et expérimentales sur l'ontogenie des plantules des Pinacées et ses rapports avec la phylo-génie. Ann. Sci. Nat., Bot. (Sér. 11) 1:1-219.

540. ----------. 1952. L'evolution des végétaux et l'anatomie des plantules. Ann. Biol. (Sér. 3) 28:163-191.

541. ----------. 1954. Anatomie végétale, l'appareil végétatifs des Phanerogames. Tome premier. Presses Universitaires de France, Paris. 330 pp.

542. BOUVIER, W. 1915. Beiträge zur vergleichenden Anatomie der Asphodel-doideae. (Tribus: Asphodeleae und Hemerocallideae). Denk. Kaiserl. Akad. Wiss. Wien (Math.-Nat. Kl.) 91:540-577.

543. BOUVRAIN, G. 1939. Sur la présence de faisceau libéro-ligneux dans la partie supérieure de la racine de l'Helianthus annuus (var. uniflorus). Comp. Rend. Hebd. Séances Acad. Sci. 208:920-923.

544. -----------. 1942. Au sujet de quelques acquisitions nouvelles sur l'ontogenie des dicotylédones. Ann. Sci. Nat., Bot. (Sér. 11) 3:1-73.

545. BOWDEN, B.N. 1964. Studies on Andropogon gayanus Kunth. II. An outline of the morphology and anatomy of Andropogon gayanus var. bisqualmulatus (Hochst.) Hack. Jour. Linn. Soc., (Bot.) 58:509-519.

546. BOWER, F.O. 1904. Ophioglossum simplex Ridley. Annals Bot. 18: 205-216.

547. ----------. 1919. Botany of the living plant. Macmillan and Co., Ltd., London. 580 pp.

548. BOWMAN, H.H.M. 1917. Ecology and physiology of the red mangrove. Proc. Amer. Philos. Soc. 56:589-672.

549. --------------. 1918. Ecology and physiology of the red mangrove. Doctoral thesis. University of Pennsylvania, Philadelphia. 83 pp.

550. --------------. 1921. Histological variations in Rhizophora mangle. Rept. Mich. Acad. Sci. 22:129-134.

551. BOYD, L. 1930. Developmental anatomy of monocotylous seedlings. 1. Paris polyphylla; 2. Costus speciosus. Trans. Proc. Bot. Soc. Edinb. 30:218-229.

552. -------. 1932. Monocotylous seedlings: Morphological studies in the post-seminal development of the embryo. Trans. Proc. Bot. Soc. Edinb. 31:1-224.

553. -------., and AVERY, G.S., Jr. 1936. Grass seedling anatomy: the first internode of Avena and Triticum. Bot. Gaz. 97:765-779.

554. BOYLE, L.W. 1934. Histological characters of flax roots in relation to resistance to wilt and root rot. U.S.D.A. Tech. Bull. No. 458. 18 pp.

555. BRABEC, F. 1953. Über polysomatie in der Wurzel von Bryonia. Chromosoma 6:135-141.

556. BRACEGIRDLE, B., and MILES, P.H. 1971. An atlas of plant structure. Vol. 1. Heinemann Educational Books Ltd., London. 123 pp.

557. BRANDZA, G. 1908. Recherches anatomiques sur le germination des Hypericacées et des Guttifères. Ann. Sci. Nat., Bot. (Sér. 9) 8:221-300.

558. BRAUNE, W., LEMAN, A., and TAUBERT, H. 1971. <u>Pflanzenanatomisches</u> <u>Praktikum</u>. 2nd ed. Gustav Fischer, Stuttgart. 331 pp.

559. BRAY, W.L. 1910. The mistletoe pest in the Southwest. U.S.D.A. Bur. Plant Ind. Bull. No. 166. 39 pp.

560. BREBNER, G. 1902. On the anatomy of <u>Danaea</u> and other Marattiaceae. Annals Bot. 16:517-552.

561. BREMEKAMP, C.E. 1914. De anatomische bouw van de wortelschors bij het suikerreit. Arch. Suiker. Nederl.-Indië 22:508-514.

562. BRENNER, W. 1902. Ueber die Luftwurzeln von <u>Avicennia tomentosa</u>. Berich. Deutsch. Bot. Gesel. 20:175-188.

563. BREYER-BRANDWIJK, M.G. 1934. Contributions to our knowledge of the anatomy of South African poisonous plants. So. African Jour. Sci. 31:240-253.

564. BRITTAN, N.H. 1970. A preliminary survey of the stem and leaf anatomy of <u>Thysanotus</u> R.Br. (Liliaceae). In: New research in plant anatomy. (eds.) Robson, N.K.B., et al. (Suppl. 1) Jour. Linn. Soc. (Bot.) 63:57-70.

565. BRITTLEBANK, C.C. 1908. The life-history of <u>Loranthus exocarpi</u> Behr. Proc. Linn. Soc. New So. Wales 33:650-656.

566. BRITTON, E.G., and TAYLOR, A. 1901. Life history of <u>Schizaea pusilla</u> Bull. Torrey Bot. Club 28:1-19.

567. BROOK, P.J. 1951. Vegetative anatomy of <u>Carpodetus serratus</u> Forst. Trans. Roy. Soc. New Zeal. 79:276-285.

568. BROOKS, E.R., and GUARD, A.T. 1952. Vegetative anatomy of <u>Theobroma</u> <u>cacao</u>. Bot. Gaz. 113:444-454.

569. BROUGH, P., and TAYLOR, M.H. 1940. An investigation of the life-cycle of <u>Macrozamia spiralis</u> Miq. Proc. Linn. Soc. New So. Wales 65:494-524.

570. BROUWER, W. 1970. <u>Handbuch des Speziellen Pflanzenbaues in 2 Bänden</u>. Band 1. Wurzeln und Wurzelwachstum. 1. Lieferung, pp. 22-26; Lieferung, pp. 492-496. Paul Parey, Berlin und Hamburg.

571. BROWN, A.B. 1935. Cambial activity, root habit, and sucker development in two species of poplar. New Phytol. 34:163-179.

572. BROWN, R., and BROADBENT, D. 1952. The development of cells in the growing zones of the root. Jour. Exp. Bot. 1:249-263.

573. --------. 1963. Cellular differentiation in the root. Symposia of the Society for Experimental Biology. No. 17. Cell differentiation. pp. 1-17.

574. BROWN, W.V. 1965. The grass embryo - a rebuttal. Phytomorph. 15:274-284.

575. BRUCH, H. 1954. Anatomische und entwicklungsgeschichtliche Untersuchungen an der Wurzel von Foeniculum vulgare Miller. Doctoral dissertation. Universität Mainz. Duncker und Humblot, Berlin (1955). 26 pp.

576. --------. 1955. Beiträge zur Morphologie und Entwicklungsgeschichte der Fenchelwurzel (Foeniculum vulgare Mill.). Beitr. Biol. Pflanz. 32:1-26.

577. BRUCHMANN, H. 1905. Von der Wurzeltragern der Selaginella kraussiana A.Br. Flora 95:150-166.

578. ------------. 1909. Von den Vegetationsorganen der Selaginella lyalli Spring. Flora 99:436-467.

579. ------------. 1910. Über Selaginella preissiana Spring. Flora 100:288-295.

580. BRÜCKNER, G. 1927. Beiträge zur Anatomie, Morphologie und Systematik der Commelinaceen. Bot. Jahrb. 61:1-70.

581. BRUHN, W. 1910. Beiträge zur experimentallen Morphologie, zur Biologie und Anatomie der Luftwurzeln. Flora 101:98-166.

582. BRUMFIELD, R.T. 1942. Cell growth and division in living root meristems. Amer. Jour. Bot. 29:533-543.

583. ---------------. 1943. Cell lineage studies in root meristems by means of chromosome rearrangements induced by x-rays. Amer. Jour. Bot. 30:101-110.

584. BRUYNE, C. 1922. Idioblasts et diaphragmes des Nymphéacées. Comp. Rend. Hebd. Séances Acad. Sci. 175:452-455.

585. BRYAN, G.S. 1936. Fascicled roots of Cycas revoluta. Amer. Jour. Bot. 23:334-335.

586. BRYANT, A.E. 1934. A demonstration of the connection of the protoplasts of the endodermal cells with the Casparian strips in the roots of barley. New Phytol. 33:231.

587. -----------. 1934. Comparison of anatomical and histological differences between roots of barley grown in aerated and non-aerated culture solutions. Plant Phys. 9:389-391.

588. BÜCHER, H., and FICKENDEY, E. 1919. Die Ölpalme. Ausland. Einzeldar., Auswart. Amt. 2:1-124.

589. BUCHOLZ, J.T. 1925. The embryogeny of Cephalotaxus fortunei. Bull. Torrey Bot. Club 52:311-423.

590. ------------., and OLD, E.M. 1933. The anatomy of Cedrus in the dormant stage. Amer. Jour. Bot. 20:35-44.

591. BUGNON, P. 1926. La dichotomie cotylédonaire, caractère ancestral. Bull. Soc. Bot. France 72:1088-1094.

592. BUGNON, S.K. 1935. Racine et différenciation vasculaire. Comp. Rend. Assoc. Franç. Avan. Sci. 51:231-233.

593. BULGAKOV, S.W. 1944. Modified anatomical structure in kok-saghyz roots and its biological import. Comp. Rend. (Doklady) Acad. Sci. URSS (N.S.) 45:35-38.

594. BUNNING, E. 1951. Über die Differenzierungsvorgänge in der Cruciferenwurzel. Planta 39:126-153.

595. ----------. 1952. Weitere Untersuchungen über die Differenierungsvorgänge in Wurzeln. Zeitschr. Bot. 40:385-406.

596. ----------. 1952. Morphogenesis in plants. In: Survey of Biological Progress. (ed.) Avery, G.S., Jr. Academic Press, New York. 2:105-140.

597. BUNTING, M. 1904. The structure of the cork tissues in roots of some rosaceous genera. Contr. Bot. Lab. Univ. Penn. (N.S. 5) 2:54-65. (Same in: Trans. Proc. Bot. Soc. Penn. 1898. 1:54-65)

598. BUNTON, L. 1910. Histology of Townsendia exscapa and Lesquerella spathulata. Kans. Univ. Sci. Bull. 5:183-205.

599. BURKILL, I.H. 1960. Organography and evolution of Dioscoreaceae, the family of the Yams. Jour. Linn. Soc. (Bot.) 56:319-412.

600. BURSTRÖM, H.G. 1959. Growth and formation of intercellularies in root meristems. Physiol. Plant. 12:371-385.

601. -------------., and ODHNOFF, C. 1963. Vegetative anatomy of plants. Svenska Bokförlaget, Stockholm. 149 pp.

602. BUSCALIONI, L. 1901. Sull'anatomia del cilidro central nelle radici delle Monocotiledoni. Malpighia 15:277-296.

603. -------------., and MUSCATELLO, G. 1909. Sulle radici avventizie nell' interno del fusto del Rhus viminalis Ait. e su alcune alterazioni del sistema radicale di questa specie. Malpighia 23:447-469.

41

604. --------------., and LOPRIORE, G. 1909. Il pleroma tubuloso, l'endo-
     dermide midollare, la frammentazione stelare e la schizorrizia nelle
     radici delle Phoenix dactylifera L. Atti Accad. Gioenia Sci. Nat.
     Catania (Ser. 5) Mem. 10. 2:1-14.

605. -------------- and ------------. 1910. Il pleroma tubuloso, l'endo-
     dermide midollare, la frammentazione desmica e la schizorrizia nelle
     radici della Phoenix dactylifera L. Atti. Accad. Gioenia Sci. Nat.
     Catania (Ser. 5) 3:1-102.

606. --------------. 1921. Sulle radici aeree fasciate de Carallia
     integerrima D.C. Malpighia 29:81-96.

607. BÜSGEN, M. 1901. Einiges über Gestalt und Wachsthumsweise der Baum-
     wurzeln. Allgem. Forst-Jagdt Zeit. 77:273-278.

608. ----------. 1905. Studien über die Wurzelsysteme einiger dicotyler
     Holzpflanzen. Flora 95:58-94.

609. ----------., MÜNCH, E., and THOMSON, T. 1931. Structure and life of
     forest trees. 3rd ed. John Wiley and Sons, Inc., New York. 436 pp.

610. BUTTERS, F.K. 1909. The seeds and seedlings of Caulophyllum
     thalictroides. Minn. Bot. Studies Part 1. 4:11-32.

611. BUVAT, R. 1944. Recherches sur le dédifférenciation des cellules
     végétales. I. Plantes entières et boutures. Ann. Sci. Nat., Bot.
     (Sér. 11) 5:1-130.

612. --------., and GENEVES, L. 1951. Sur l'inexistence des initiales
     axiales dans la racine d'Allium cepa L. (Liliacées). Comp. Rend.
     Hebd. Séances Acad. Sci. 232:1579-1581.

613. --------. 1952. Structure, évolution et fonctionnement du méristème
     apicale de quelques Dicotylédones. Ann. Sci. Nat., Bot. (Sér. 2)
     13:198-300.

614. --------., and LIARD, O. 1953. Nouvelle constation de l'inerte des
     sois-disant initiales dans le méristème radiculaire de Triticum
     vulgare L. Comp. Rend. Hebd. Séances Acad. Sci. 236:1193-1195.

615. ---------- and --------. 1954. La proliferation cellulaire dans le
     méristème radiculaire d'Equisetum arvense L. Comp. Rend. Hebd.
     Séances Acad. Sci. 238:1257-1258.

616. BUXBAUM, F. 1925. Vergleichende Anatomie der Melanthoideae. Rep.
     Spec. Nov. Reg. Veg. (Beihefte) 29:1-80.

617. ----------. 1927. Nachträge zur vergleichenden Anatomie der
     Melanthoideae. Beih. Bot. Centralb. (Abt. 1) 44:255-263.

618. ----------. 1937. Allgemeine Morphologie der Kakteen. Die Wurzel. Cactaceae (Teil I) pp. 3-9.

619. ----------. 1950. Morphology of Cacti. Section I. Roots and stems. (ed.) Kurtz, E.B., Jr. Abbey Garden Press, Pasadena, Calif. 87 pp.

620. BYRNE, J.M., and HEIMSCH, C. 1968. The root apex of Linum. Amer. Jour. Bot. 55:1011-1019.

621. ----------. 1969. The root apex of Malva: The structural development of the quiescent center in selected species. Doctoral Dissertation. Miami University, Oxford, Ohio. 86 pp.

622. ----------., and HEIMSCH, C. 1970. The root apex of Malva sylvestris. I. Structural development. Amer. Jour. Bot. 57:1179-1184.

623. CALVIN, C.L. 1966. Anatomy of mistletoe (Phoradendron flavescens) seedlings grown in culture. Bot. Gaz. 127:171-183.

624. ----------. 1967. Anatomy of the endophytic system of the mistletoe, Phoradendron flavescens. Bot. Gaz. 128:117-137.

625. CAMEFORT, H., and PANIEL, J. 1962. Morphologie et anatomie des végétaux vasculaires. In: "Biologie." (ed.) Obré, M.A. G. Doin et Cie, Paris. 371 pp.

626. CAMP, R.R. 1966. The root apex of the Malvaceae. M.A. thesis. Miami University, Oxford, Ohio. 50 pp.

627. CAMPBELL, D.H. 1910. The embryo and young sporophyte of Angiopteris and Kaulfussia. Ann. Jard. Bot. Buitenz. (Suppl. 3) Part 1. pp. 69-82.

628. -------------. 1911. The eusporangiatae. The comparative morphology of the Ophioglossaceae and Marattiaceae. Publ. Carn. Inst. Wash. No. 140. 229 pp.

629. -------------. 1918. The structure and development of mosses and ferns. Macmillan Co., New York. 708 pp.

630. -------------. 1921. The eusporangiate ferns and the stelar theory. Amer. Jour. Bot. 8:303-314.

631. CAMPBELL, G.K.G. 1936. The anatomy of Potamogeton pectinatus. Trans. Proc. Bot. Soc. Edinb. 32:179-186.

632. CANNON, W.A. 1901. The anatomy of Phoradendron villosum Nutt. Bull. Torrey Bot. Club 28:374-390.

633. ----------. 1904. Observations on the germination of Phoradendron villosum and P. californicum. Bull. Torrey Bot. Club 31:435-443.

634. -----------. 1909. The parasitism of Orthocarpus purpurascens Benth. Plant World 12:259-261.

635. -----------. 1910. The root habits and parasitism of Krameria canescens Gray. In: The conditions of parasitism in plants. (eds.) Macdougal, D.T. and Cannon, W.A. Publ. Carn. Inst. Wash. No. 131. 96 pp.

636. -----------. 1912. Deciduous rootlets of desert plants. Science 35:632-633.

637. -----------. 1912. Structural relations in xenoparasitism. Amer. Nat. 46:675-681.

638. -----------. 1913. Notes on root variations in some desert plants. Plant World 16:323-341.

639. -----------. 1949. A tentative classification of root systems. Ecology 30:542-548.

640. -----------. 1954. A note on the grouping of lateral roots. Ecology 35:293-295.

641. CANTLON, J.E., CURTIS, E.J.C., and MALCOM, W.M. 1963. Studies of Melampyrum lineare. Ecology 44:466-474.

642. CARANO, E. 1903. Sulla particolare struttura delle radici. Annali Bot. 1:199-205.

643. ---------. 1906. Ricerche sulla morfologia delle Pandanacee. Annali Bot. 5:1-46.

644. ---------. 1906. Ricerche sulle Pandanacee. Atti Reale Accad. Lincei (Ser. 5) 15:243-246.

645. CAREY, G., and FRASER, L. 1932. The embryology and seedling development of Aegiceras majus Gaertn. Proc. Linn. Soc. New So. Wales 57:341-360.

646. CARLQUIST, S. 1957. The genus Fitchia (Compositae). Univ. Calif. Publ. Bot. (1959) 29:1-72.

647. -------------. 1960. Anatomy of Guayana Xyridaceae: Abolboda, Orectanthe, and Achlyphila. Mem. New York Bot. Gard. 10:65-117.

648. -------------. 1961. Comparative plant anatomy. Holt, Rinehart and Winston, New York. pp. 94-101.

649. -------------. 1966. Anatomy of Rapataceae - roots and stems. Phytomorph. 16:17-38.

650. CARLSON, M.C. 1929. Origin of adventitious roots in Coleus cuttings. Bot. Gaz. 87:119-126.

651. ------------. 1933. Comparative anatomical studies of Dorothy Perkins and American Pillar roses. I. Anatomy of canes. II. Origin and development of adventitious roots in cuttings. Contr. Boyce Thomp. Inst. Plant Res. 5:313-330.

652. ------------. 1938. The formation of nodal adventitious roots in Salix cordata. Amer. Jour. Bot. 25:721-725.

653. ------------. 1938. Origin and development of shoots from the tips of roots of Pogonia ophioglossoides. Bot. Gaz. 100:215-225.

654. ------------. 1943. The morphology and anatomy of Calopogon pulchellus. Bull. Torrey Bot. Club 70:349-368.

655. ------------. 1950. Nodal adventitious roots in willow stems of different ages. Amer. Jour. Bot. 37:555-561.

656. CARMIELLO, R. 1966. Struttura dell'apice radicale di Quercus pedunculata Ehrh. in natura ed in vitro. Allionia 12:157-168.

657. CARO, H. 1903. Beiträge zur Anatomie der Commelinaceen. Inaugural-Dissertation. Heidelberg Universität. 85 pp.

658. CARPENTER, A.M. 1941. Seedling anatomy of certain Pinaceae. Bull. Pittsburgh Univ. 16:62-69.

659. CARPENTER, D.C. 1937. Anatomy of Urginea maritima (L.) Baker. Doctoral dissertation. University of Michigan, Ann Arbor. 30 pp. (Also in: Papers Mich. Acad. Sci. Arts Let. Vol. 25, 1939; published 1940.)

660. CARPENTER, I.W., and GUARD, A.T. 1954. Anatomy and morphology of the seedling roots of four species of the genus Quercus. Jour. Forestry 52:269-274.

661. CARTELLIERI, E. 1926. Das Absorbtionssystem der Rafflesiaceae: Brugmansia. Bot. Archiv 14:284-311.

662. ---------------. 1928. Das Haustorium von Cassytha pubescens R.Br. Planta 6:162-182.

663. CATALANO, G. 1911. Morfologia interna delle radici di alcune Palme e Pandanacee. Atti Reale Accad. Lincei (Ser 5) 20:725-729.

664. -----------. 1912. Morfologia interna delle radici di alcune Palme e Pandanacee. Annali Bot. 10:65-99.

665. -----------. 1913. Interno alla struttura delle radici di Chamaedorea elatior. Annali Bot. 12:151.

666. -----------. 1914. Sulle funzione delle radici contrattili. Atti Reale Accad. Lincei (Ser. 5) 23:970-976.

667. -----------. 1915. Intorno alla struttura e alla funzione di alcune radici contrattili. Nuovo Giorn. Bot. Ital. (N.S.) 22:148-174.

668. ČERNOHORSKÝ, Z. 1967. Practical plant morphology. Státní Pedagogické Nakladatelstvi, Praha. 217 pp. (in Czechoslovakian)

669. CHADEFAUD, M. 1944. Travaux récents et remarques sur l'anatomie des racines. Rev. Sci. 82 (Fasc. I, No. 3228):47-50.

670. CHAINAYE, R. 1927. Monographie du Soliva anthemidifolia R.Br. Arch. Inst. Bot. Univ. Liège 6:3-33.

671. CHAKRAVARTI, S.C. 1953. Anatomical study of the radicle of mustard with special reference to vascular differentiation. Proc. 40th Indian Sci Congr. Assn. Part 3: Absts., Sect. 6: Botany. 40:98.

672. CHAMBERLAIN, C.J. 1935. Gymnosperms: Structure and evolution. University of Chicago Press, Chicago. 484 pp.

673. CHAN, T.T. 1951-1952. The development of the Narcissus plant. Daff. Tulip Yearb., Roy. Hort. Soc. London 17:72-100.

674. CHANDLER, B. 1909. Aerial roots of Tibouchina moricandiana Baill. Notes Roy. Bot. Gard. Edinb. No. 20. pp. 245-250.

675. -----------. 1912. Aerial roots of Acanthoriza aculeata. Trans. Proc. Bot. Soc. Edinb. 24:20-24.

676. CHANDLER, R. 1911. Notes on Donatia Novae-zelandiae Hook f. Notes Roy. Bot. Gard. Edinb. 22:43-47.

677. CHANDLER, S.E. 1905. On the arrangement of the vascular strands in the 'seedlings' of certain leptosporangiate ferns. Annals Bot. 19:365-410.

678. CHANDLER, W.H. 1926. Polarity in the formation of scion roots. Proc. Amer. Soc. Hort. Sci. 22:218-222.

679. CHANDHOKE, K.R., and CHAWALA, S. 1967. Studies in the family Convolvulaceae. I. Apical organization, seedling anatomy and floral anatomy of Ipomoea fistulosa Mart. Jour. Birla Inst. Tech. Sci. 1:134-142.

680. CHANDRA, V. 1956. Studies on Rauvolfia. Part III. Morphological studies of Rauvolfia canescens L. Indian Jour. Pharm. 18:140-146.

681. CHANG, C.Y. 1927. Origin and development of tissues in rhizome of Pteris aquilina. Bot. Gaz. 83:288-306.

682. CHAPMAN, V.J. 1944. The morphology of Avicennia nitida Jacq. and the function of its pneumatophores. Jour. Linn. Soc. (Bot.) 52:487-533.

683. CHARLES, G.M. 1911. The anatomy of the sporeling of <u>Marattia alata</u>. Bot. Gaz. 51:81-101.

684. CHARLIER, A. 1905. Contribution à l'étude anatomique des plantes à gutta-percha et d'autres Sapotacées. Jour. Bot. 19:127-152.

685. ------------. 1906. À l'étude anatomique des plantes a gutta-percha et d'autres Sapotacees (fin.). Jour. Bot. 20:22-77.

686. CHARLTON, W.A. 1966. The root system of <u>Linaria vulgaris</u> Mill. I. Morphology and anatomy. Canad. Jour. Bot. 44:1111-1116.

687. ------------. 1967. The root system of <u>Linaria vulgaris</u> Mill. II. Differentiation of root types. Canad. Jour. Bot. 45:81-91.

688. CHATELIER, G.G. DU. 1940. Recherches sur les Sterculiacées. Rev. Gén. Bot. 52:211-233.

689. CHAUDHURI, H., and AKHTAR, A.R. 1931. The coral-like roots of <u>Cycas revoluta</u>, <u>Cycas circinalis</u> and <u>Zamia floridana</u> and the alga inhabiting them. Jour. Indian Bot. Soc. 10:43-59.

690. -------------- and -----------. 1931. A study of the root-tubercles of <u>Podocarpus chinensis</u>. Jour. Indian Bot. Soc. 10:92-99.

691. CHAUVEAUD, G. 1900. Recherches sur le mode de formation des tubes criblés dans la racine des Dicotylédones. Ann. Sci. Nat., Bot. (Sér. 8) 12:333-394.

692. ------------. 1901. Sur la structure des plantes vasculaires. Comp. Rend. Hebd. Séances Acad. Sci. 132:93-95.

693. ------------. 1901. Sur le passage de la structure primaire à la structure secondaire dans le Haricot. Bull. Mus. Hist. Nat. 7:23-26.

694. ------------. 1901. Sur la passage de la disposition alterne des éléments liberiéns et ligneux à leur disposition superposée dans le Trocart (<u>Triglochin</u>). Bull. Mus. Hist. Nat. 7:124-130.

695. ------------. 1901. De la formation du péricycle de la racine dans les Fougères. Bull. Mus. Hist. Nat. 7:277-280.

696. ------------. 1901. Sur la structure de la racine de l'<u>Azolla</u>. Bull. Mus. Hist. Nat. 7:366-372.

697. ------------. 1901-1902. Recherches sur le développement de l'appareil conducteur dans la racines des Equisetacées. Bull. Soc. Philom. Paris 4:26-45.

698. ------------. 1902. Passage de la position alterne à la position super-posée de l'appareil conducteur, avec destruction des vaisseaux centri-pètes primitifs, dans le cotylédon de l'Oignon (<u>Allium cepa</u>). Bull. Mus. Hist. Nat. 8:52-59.

699. ------------. 1902. De la variation de structure existant à l'état normal entre les racines et les radicelles de la Marsilie (Marsilia). Bull. Mus. Hist. Nat. 8:114-127.

700. ------------. 1902. De la répartition des épaississements extra-cellulaires dans les lacunes corticales de la racine des prêles (Equisetum). Bull. Mus. Hist. Nat. 8:127-129.

701. ------------. 1902. Développement des éléments précurseurs des tubes criblés dans le Thuia orientalis. Bull. Mus. Hist. Nat. 8:447-454.

702. ------------. 1902. Passage de la disposition primitive à la disposition secondaire dans le cotylédons du pin maritime (Pinus maritima). Bull. Mus. Hist. Nat. 8:549-559.

703. ------------. 1903. Développement des tubes précurseurs et des premiers tubes criblés dans l'Ephedra altissima. Bull. Mus. Hist. Nat. 9:94-96.

704. ------------. 1903. Recherches sur le mode de formation des tubes criblés dans la racine des cryptogames vasculaires et de Gymnospermes. Ann. Sci. Nat., Bot. (Sér. 8) 18:165-277.

705. ------------. 1904. Le liber précurseur dans le Sapin Pinsapo. Ann. Sci. Nat., Bot. (Sér. 8) 19:321-333.

706. ------------. 1910. Recherches sur les tissus transitores du corps végétatif des plantes vasculaires. Ann. Sci. Nat., Bot. (Sér. 9) 12:1-70.

707. ------------. 1911. L'appareil conducteur des plantes vasculaires et les phases principales de son évolution. Ann. Sci. Nat., Bot. (Sér.9) 13:113-438.

708. ------------. 1911. Sur une interpretation recente de la structure attribuée à la racine de l'Azolla filiculoides. Bull. Soc. Bot. France 58:79-82.

709. ------------. 1911. Sur l'évolution des faisceaux vasculaires dans les différentes parties de la plantule des Phanérogames. Bull. Soc. Bot. France 58:705-712.

710. ------------. 1921. La constitution des plantes vasculaires révélee par leur ontogenie. Payot et Cie, Paris. pp. 15-155.

711. ------------. 1931. Il faut chercher, dans la racine, le point de départ de l'évolution vasculaire, mais non pas le point de départ de la plante vasculaire. Bull. Soc. Bot. France 78:195-201.

712. CHAUVEL, F. 1903. Recherches sur la familie des Oxalidacées. Doctoral thèses. Université de Paris, École Superieure de Pharmacie. A. Joanin et Cie, Paris. 205 pp.

713. CHEADLE, V.I. 1937. Secondary growth by means of a thickening ring in certain monocotyledons. Bot. Gaz. 98:535-555.

714. ------------. 1940. Investigations of the vascular system in the Monocotyledoneae. Amer. Philos. Soc. Yearb. pp. 140-141.

715. ------------. 1941. Investigations of the vascular system in the Monocotyledoneae. Amer. Philos. Soc. Yearb. pp. 149-152.

716. ------------. 1942. The occurrence and types of vessels in the various organs of the plant in the Monocotyledoneae. Amer. Jour. Bot. 29:441-450.

717. ------------. 1943. The origin and certain trends of specialization of the vessel in the Monocotyledoneae. Amer. Jour. Bot. 30:11-17.

718. ------------. 1943. Vessel specialization in the late metaxylem of the various organs in the Monocotyledoneae. Amer. Jour. Bot. 30:484-490.

719. ------------. 1944. Specialization of vessels within the xylem of each organ in the Monocotyledoneae. Amer. Jour. Bot. 31:81-92.

720. ------------. 1953. Independent origin of vessels in the Monocotyledons and Dicotyledons. Phytomorph. 3:23-44.

721. ------------. 1968. Vessels in the Haemodorales. Phytomorph. 18:412-420.

722. ------------. 1968. Vessels in the Amaryllidaceae and Tecophilaeaceae. Phytomorph. 19:8-16.

723. ------------. 1970. Vessels in Pontederiaceae, Ruscaceae, Smilacinaceae and Trilliaceae. In: New Research in Plant Anatomy. (eds.) Robson, N.K.B., Cutler, D.F., and Gregory, M. Academic Press, New York. pp. 45-50.

724. ------------. 1971. Vessels in the Liliaceae. Phytomorph. 21:320-333.

725. CHEMIN, E. 1920. Observations anatomiques et biologiques sur le genre Lathraea. Ann. Sci. Nat., Bot. (Sér. 10) 2:125-272.

726. CHEN, S. 1969. The contractile roots of Narcissus. Annals Bot. (N.S.) 33:421-426.

727. CHEVALIER, A. 1901. Monographie des Myricacées. Mem. Soc. Nat. Sci. Nat. Math. Cherbourg 32:85-340.

728. CHI, H.H. 1942. Histogenesis in the roots of Holcus sorghum L. Iowa State Coll. Jour. Sci. 16:189-205.

729. CHIANG, S.- H.T. 1968. Microscopic and submicroscopic structure of developing root cap of Ceratopteris thalictroides. Taiwania 14:29-41.

49

730. ---------------. 1970. Development of the root of Dendrobium
     kwashotense Hay. with special reference to the cellular structure
     of its exodermis and velamen. Taiwania 15:1-16.

731. ---------------. 1971. Sequence of tissue differentiation in the root
     of Ceratopteris thalictroides. Taiwania 16:31-47.

732. ---------------., and CHOU, T. 1971. Histological studies on the
     roots of orchids from Taiwan. Taiwania 16:1-29.

733. CHIANG, Y.- L., and CHIANG, S.- H. 1962. The sporeling of Ceratopteris.
     Taiwania 8:35-50.

734. --------------. 1970. Macro- and microstructure of the root of
     Ceratopteris pteridoides (Hook.) Hieron. Taiwania 15:31-49.

735. CHIBBER, H.M. 1912. The morphology and histology of Piper betle Linn.
     (the betelvine). Jour. Linn. Soc. (Bot.) 41:357-383.

736. CHICK, E. 1903. The seedling of Torreya myristica. New Phytol. 2:83-91.

737. CHIFFLOT, J. 1903. Sur le symetrie bilaterale des radicelles de
     Pontederia crassipes Mart. Comp. Rend. Hebd. Séances Acad. Sci.
     136:1701-1703.

738. CHIFFLOT, M.F. 1920. Sur les canaux gommifères des racines de
     Cycadacées, plus particulierèment ceux du Stangeria paradoxa T. Moore.
     Comp. Rend. Hebd. Séances Acad. Sci. 171:257-258.

739. CHILVERS, G.A., and PRYOR, L.D. 1965. The structure of eucalypt
     mycorrhizas. Austral. Jour. Bot. 13:245-259.

740. CHODAT, R. 1926. La végétation du Paraguay. XIV. Amarantacées. Bull.
     Soc. Bot. Genève 18:247-294.

741. CHOLODNY, N. 1928. Über eine vermeintliche Anomalie im Wachstumodus der
     Wurzeln von Lupinus albus. Berich Deutsch. Bot. Gesel. 46:247-254.

742. CHOPRA, G.L. 1963. Gymnosperms. 9th ed. S. Nagin and Co., Jullundur
     City, India. 140 pp.

743. CHOPRA, R.L. 1929. Germination and seedling anatomy of Ephedra foliata
     Boiss. Kotschy (E. peduncularis Boiss.). Abstract, Proc. 16th Indian
     Sci. Congr. Assn. (Madras). 16:225.

744. ------------. 1929. A morphological and anatomical study of Ephedra
     foliata Boiss. and Kotschy (E. peduncularis Boiss.) and E. distachya
     (E. vulgaris Rich.). Abstract, Proc. 16th Indian Sci. Congr. Assn.
     (Madras). 16:226-227.

745. CHOUARD, P. 1926. Germination et formation des jeunes bulbes de quelques Liliiflores (Endymion, Scilla, Narcissus). Ann. Sci. Nat., Bot. (Sér. 10) 8:299-353.

746. ----------. 1931. Types de développement de l'appareil végétatif chez les Scillées. Ann. Sci. Nat., Bot. (Sér. 10) 13:131-323.

747. CHOUINARD, L. 1959. Sur l'existence d'un centre quiescent au niveau de l'apex radiculaire juvénile de Pinus banksiana Lamb. Contr. Univ. Laval, Fonda Rech. For. 4:27-31.

748. CHOUX, P. 1913. De l'influence de l'humidité et de la sécheresse sur la structure anatomique de deux plantes tropicales. Rev. Gén. Bot. 25:153-172.

749. CHRTEK, J., and JIRÁSEK, V. 1965. Über den Bau der Wurzelendodermiszellen bei Gräsern (Poaceae) in der Tschechoslowakei. Preslia 37:396-406.

750. CHRYSLER, M.A. 1926. Vascular tissues of Microcycas calocoma. Bot. Gaz. 82:233-252.

751. --------------. 1937. Persistent juveniles among the cycads. Bot. Gaz. 98:696-710.

752. -------------. 1941. The structure and development of Ophioglossum palmatum. Bull. Torrey Bot. Club 68:1-19.

753. CHUANG, T.- I. 1970. A systematic anatomical study of the genus Perideridia (Umbelliferae - Apioideae). Amer. Jour. Bot. 57:495-503.

754. -------------. 1971. Observations on root-parasitism in Cordylanthus (Scrophulariaceae). Amer. Jour. Bot. 58:218-228.

755. CHURCH, M.B. 1919. Root contraction. Plant World 22:337-340.

756. ----------. 1919. The development and structure of the bulb in Cooperia drummondii. Bull. Torrey Bot. Club 46:337-362.

757. CIAMPI, C., and GELLINI, R. 1958. Studio anatomico sui rapporti struttura e capacità di radicazione in tallee di olivo. Nuovo Giorn. Bot. Ital. (N.S.) 65:417-424.

758. ---------- and ----------. 1963. Formation and development of adventitious roots in Olea europaea L.: Significance of the anatomical structure for the development of radicles. Nuovo Giorn. Bot. Ital. 70:62-74.

759. CLARKE, H.M. 1936. The morphology and anatomy of Lygodium japonicum. Amer. Jour. Bot. 23:405-413.

51

760. CLOWES, F.A.L. 1950. Root apical meristems of <u>Fagus sylvatica</u>. New Phytol. 49:248-269.

761. ------------. 1951. The structure of mycorrhizal roots of <u>Fagus</u> <u>sylvatica</u>. New Phytol. 50:1-16.

762. ------------. 1953. The cytogenerative centre in roots with broad columellas. New Phytol. 52:48-57.

763. ------------. 1954. The promeristem and the minimal constructional centre in grass root apices. New Phytol. 53:108-116.

764. ------------. 1954. The root cap of ectotrophic mycorrhizas. New Phytol. 53:525-529.

765. ------------. 1957. Chimeras and meristems. Heredity 11:141-148.

766. ------------. 1958. Development of quiescent centres in root meristems. New Phytol. 57:85-88.

767. ------------. 1959. Apical meristems in roots. Biol. Rev. 34:501-529.

768. ------------. 1959. Cellular organization in root apical meristems. In: <u>Recent Advances in Botany</u>. University of Toronto Press I:791-794.

769. ------------. 1959. Cell development and differentiation in root apices. In: <u>Recent Advances in Botany</u>. University of Toronto Press II:1272-1274.

770. ------------. 1961. Apical meristems. In: <u>Botanical Monographs</u>. (ed.) James, W.O. Vol. II. Blackwell Scientific Publications, Oxford. 217 pp.

771. ------------. 1964. The quiescent center in meristems and its behavior after irradiation. In: Brookhaven Symposia in Biology. Brookhaven National Laboratory, Upton, New York. pp. 46-58.

772. ------------., and JUNIPER, B.E. 1964. The fine structure of the quiescent centre and neighboring tissues in root meristems. Jour. Exp. Bot. 15:622-630.

773. ------------. 1967. The quiescent centre. Phytomorph. 17:132-140.

774. ------------. 1969. Anatomical aspects of structure and development. In: <u>Root Growth</u>. (ed.) Whittington, W.J. Plenum Press, New York. pp. 3-19.

775. COCHRAN, H.L., and COWART, F.F. 1937. Anatomy and histology of the transition region in <u>Capsicum frutescens</u>. Jour. Agric. Res. 54:695-700.

776. COERTZE, A.F., and VAN DER SCHIJFF, H.P. 1969. The transition from root to stem and the anatomy of the stem of Podocarpus henkelii Stapf ex Dallim. et Jacks. Tydsk. Natuurwetensk. 9:72-88.

777. COHEN, L.I. 1954. The anatomy of the endophytic system of the dwarf mistletoe, Arceuthobium campylopodum. Amer. Jour. Bot. 41:840-847.

778. ----------. 1962. A developmental study of Arceuthobium. Diss. Absts. 23(30):807-808.

779. ----------. 1963. Studies on the ontogeny of the dwarf mistletoes, Arceuthobium. I. Embrogeny and histogenesis. Amer. Jour. Bot. 50:400-407.

780. COL, M. 1901. Sur l'existence de laticifères à contenu special dans les Fusains. Comp. Rend. Hebd. Séances Acad. Sci. 132:1354-1356.

781. COLANI, M. 1912. Sur les premiers stades du développement du Terminalia catappa. Rev. Gén. Bot. 24:267-270.

782. COLAS, R. 1937. Les plantes amazoniennes designées sous le nom de 'Chuchuhuasha.' Trav. Lab. Mat. Méd. École Sup. Pharm. Paris 28(5):1-134.

783. COLIN, H., and PICAULT, M. 1935. Détails de structure chez diverses sortes de betteraves cultivées. In: Compte Rendu Definitif de la V. Assemblée de l'Institut International de Recherches Betteravières Réunie à Bruxelles. pp. 172-174.

784. ---------- and ----------. 1937. La structure du cordon de la betterave. In: Compte Rendu Definitif de la VII. Assemblée de l'Institut Internationale de Recherches Betteravières Réunie à Bruxelles. pp. 13-17.

785. --------. 1937. La structure du cordon de la betterave. Publ. Inst. Belge Amelior. Better. 5:27-34.

786. COLLINS, J.L. 1960. The pineapple. In: World Crops Series. (ed.) Polunin, N. Leonard Hill (Books) Ltd., London. 294 pp.

787. COLOZZA, A. 1901. Nuova contribuzoni all'anatomia delle Alstroemeriee. Nuovo Giorn. Bot. Ital. 8:477-491.

788. ----------. 1909. Note anatomiche sulle Calyceraceae. Bull. Soc. Bot. Ital. No. 1. pp. 7-14.

789. ----------. 1910. Contributo allo studio anatomico delle Burmanniaceae. Bull. Soc. Bot. Ital. No. 7. pp. 106-115.

790. COMPTON, R.H. 1909. The anatomy of Matonia sarmentosa Baker. New Phytol. 8:299-310.

791. ------------. 1912. Theories on the anatomical transition from root to stem. New Phytol. 11:13-25.

792. ------------. 1912. An investigation of seedling structures in the Leguminosae. Jour. Linn. Soc. (Bot.) 41:1-122.

793. ------------. 1913. An anatomical study of syncotyly and schizocotyly. Annals. Bot. 27:793-821.

794. CONANT, G.H. 1927. Histological studies of resistance in tobacco to Thielavia basicola. Amer. Jour. Bot. 14:457-480.

795. CONARD, H.S. 1901. Fasciation in the sweet potato. Contr. Bot. Lab. Univ. Penn. 2:205-215.

796. ------------. 1905. The waterlilies: A monograph of the genus Nymphaea. Publ. Carn. Inst. Wash. No. 4. 279 pp.

797. ------------. 1908. The structure and life-history of the hay-scented fern. Publ. Carn Inst. Wash. No. 94. pp. 5-56.

798. CONDIT, I.J. 1969. Ficus: The exotic species. University of California, Division of Agricultural Science. 363 pp.

799. CONNARD, M.H., and ZIMMERMAN, P.W. 1931. The origin of adventitious roots in cuttings of Portulaca oleracea L. Contr. Boyce Thomp. Inst. Plant Res. 3:337-346.

800. CONWAY, V.M. 1936. Studies in the autecology of Cladium mariscus R.Br. I. Structure and development. New Phytol. 35:177-204.

801. COOK, O.F. 1917. Seedling morphology in palms and grasses. Jour. Wash. Acad. Sci. 7:420-425.

802. COOKE, E., and SCHIVELY, A.F. 1904. Observations on the structure and development of Epiphegus virginiana. Contr. Bot. Lab. Univ. Penn. 2:352-398.

803. COOKE, F.W. 1911. Observations on Salicornia australis. Trans. Proc. New Zeal. Inst. 44:349-362.

804. COPELAND, H.F. 1938. The structure of Allotropa. Madroño 4:137-153.

805. --------------. 1939. The structure of Monotropis and the classification of the Monotropideae. Madroño 5:105-119.

806. CORMACK, R.G.H. 1935. Investigations of the development of root hairs. New Phytol. 34:30-54.

807. --------------. 1937. The development of root hairs by Elodea canadensis. New Phytol. 36:19-25.

808. --------------. 1944. The effect of environmental factors on the development of root hairs in Phleum pratense and Sporobolus cryptandrus. Amer. Jour. Bot. 31:443-449.

809. --------------. 1945. Cell elongation and the development of root hairs in tomato roots. Amer. Jour. Bot. 32:490-496.

810. --------------. 1947. A comparative study of developing epidermal cells in white mustard and tomato roots. Amer. Jour. Bot. 34: 310-315.

811. --------------. 1949. The development of root hairs in Angiosperms. Bot. Rev. 15:583-612.

812. --------------. 1959. The development of root hairs. In: Recent Advances in Botany. University of Toronto Press. I:812-815.

813. --------------. 1962. The development of root hairs in Angiosperms. II. Bot. Rev. 28:446-464.

814. CORNER, E.J.H. 1966. The Natural History of Palms. University of California Press, Berkeley. 393 pp.

815. CORRINGTON, J.D. 1945. Under the microscope; roots. Nat. Mag. 38:221-223.

816. CORTÉSI, R. 1939. Recherches biologiques sur le Laurier-Rose. Doctoral thèses. Université de Lausanne. Société Genevoise d'Editions et Impressions, Genève. 91 pp.

817. COSSMANN, K.F. 1940. Citrus roots: Their anatomy, osmotic pressure and periodicity of growth. Palestine Jour. Bot. (Rehovot Series). 3:65-103.

818. COSTER, C. 1932. Wortelstudiën in de tropen. I, II, III. Landb. Tijds. Neder.- Indië 8:147-194; 369-464.

819. ---------. 1932. Wortelstudiën in de tropen. I. Tectona 25:828-872.

820. ---------. 1933. Wortelstudiën in de tropen. IV. Wortelconcurrentie. Tectona 26:450-497. (Also as: Wurzelstudien in den tropen. IV. Wurzelkonkurrenz. In: Landb. Tijds. Neder.- Indië 9:1-43. Dutch with German summary)

821. ---------. 1935. Wortelstudiën in de tropen. V. Gebergtehoutsoorten. Tectona 28:861-876. (Dutch with German summary)

822. COULON, J. 1923. Nardus stricta. Étude physiologique, anatomique et embryologique. Mém. Soc. Vaud. Sci. Nat. 1:245-331.

823. COULTER, J.M., and CHAMBERLAIN, C.J. 1901. Morphology of Spermato-
     phytes. D. Appleton and Co., New York. 188 pp.

824. -------------- and ----------------. 1903. Morphology of Angiosperms.
     D. Appleton and Co., New York. 348 pp.

825. -------------- and ----------------. 1917. Morphology of Gymnosperms.
     University of Chicago Press, Chicago. 466 pp.

826. COUPIN, H., JODIN, H., and DAUPHINÉ, A. 1942. Atlas de Botanique
     Microscopique. 4th ed. Vigot Frères, Editeurs, Paris. 116 pp.

827. CROCKART, I.B. 1938. The root system of Ranunculus monspeliacus.
     Trans. Proc. Bot. Soc. Edinb. 32:415-423.

828. CROOKS, D.M. 1933. Histological and regenerative studies on the flax
     seedling. Bot. Gaz. 95:209-239.

829. CROSS, G.L. 1931. Meristem in Osmunda cinnamomea. Bot. Gaz.
     91:65-76.

830. CUMMING, N.M. 1925. Notes on strand plants. 1. Atriplex babingtonii
     Woods. Trans. Proc. Bot. Soc. Edinb. 29:171-175.

831. CUNNINGTON, H.M. 1912. Anatomy of Enhalus acoroides (L.f.) Zoll.
     Trans. Linn. Soc. London 7:355-371.

832. CURTIS, K.M. 1917. The anatomy of the six epiphytic species of the
     New Zealand Orchidaceae. Annals. Bot. 31:133-149.

833. CUTLER, D.F. 1965. Vegetative anatomy of Thurniaceae. Kew Bull.
     19:431-441.

834. -----------. 1969. Juncales. In: Anatomy of the Monocotyledons. (ed.)
     Metcalfe, C.R. Vol. 4. Clarendon Press, Oxford. 357 pp.

835. CUTTER, E.G. 1967. Differentiation of trichoblasts in roots of
     Hydrocharis. Abstract in: Amer. Jour. Bot. 54:632.

836. -----------. 1969. Plant Anatomy: Experiment and Interpretation.
     Part I. Cells and tissues. Edward Arnold Publishers Ltd., London.
     168 pp.

837. -----------. 1971. Plant Anatomy: Experiment and Interpretation.
     Part II. Organs. Edward Arnold Publishers Ltd., London. 343 pp.

838. CZAJA, A.T. 1958. Untersuchungen über die submikroscopische Struktur
     der Zellwande von Parenchymzellen in Stengelorganen und Wurzeln.
     Planta 51:329-377.

839. CZAPEK, F. 1909. Beiträge zur Morphologie und Physiologie der epiphytischen Orchideen Indiens. Sitz. Akad. Wiss. Wien 118: 1555-1580.

840. DALBEY, N.E. 1914. A comparative anatomical study of some species of Xanthium. Kansas Univ. Sci. Bull. 9:57-65.

841. DALE, E. 1901. On the origin, development, and morphological nature of the aerial tubers in Dioscorea sativa Linn. Annals Bot. 15:491-501.

842. DALLEMAGNE-BERTHELOT, J. 1964. Observation d'une (structure tige) inhabituelle, vers la base d'un hypocotyle d'Impatiens scabrida D.C. Comp. Rend. Hebd. Séances Acad. Sci. 259:199-1201.

843. D'ALMEIDA, J.F.R. 1926. Some notes on the structure and life history of Nymphaea pubescens Willd. Jour. Indian Bot. Soc. 5:62-71.

844. ----------------. 1941. A contribution to the study of the biology and physiological anatomy of Indian marsh and aquatic plants. Jour. Bombay Nat. Hist. Soc. 42:298-304.

845. ----------------. 1943. A contribution to the study of the biology and physiological anatomy of Indian marsh and aquatic plants. Part II. Jour. Bombay Nat. Hist. Soc. 43:92-96.

846. ----------------. 1946. On the aerenchyma of Sesbania aculeata Poir. Indian Ecol. 1:47-54.

847. ----------------., and CORREA, J.P. 1948. A contribution to the study of the root habits and anatomy of Indian plants. I. Jour. Bombay Univ. (Sect. B, Biol. Sci.) N.S. 17:39-60.

848. ---------------- and -----------. 1949. A contribution to the study of the root habits and anatomy of Indian plants. Part II. Jour. Bombay Univ. (Sect. B, Biol. Sci.) N.S. 17:15-49.

849. DANGEARD, P.A. 1913. Observations sur la structure des plantules chez les Phanérogames dans ses rapports avec l'évolution vasculaire. Bull. Soc. Bot. France 60:73-80; 113-120.

850. DANIEL, J. 1916. Influence du mode de vie sur la structure secondaire des dicotylédones. Libraire Générale de l'Ensiegnement, Paris. 350 pp.

851. DANILOVA, M.F., and DUERTEVA, E.U. 1964. Anatomical and physiological data on the translocation of water and dissolved substances through the root tissues. Bot. Zhur. 49:1347-1365.

852. DANOS, B. 1966. Beiträge zur Kenntnis von Hyoscyamus niger L. Ann. Univ. Sci. Budap. Rol. Eöt. Nom. (Sect. Biol.) 8:47-61.

853. DARBISHIRE, O.V. 1904. Observations on Mamillaria elongata. Annals Bot. 18:375-416.

854. DARIEV, A.S. 1963. Structure of root apical meristems in some cotton varieties. Bot. Zhur. 48:430-436.

855. DARNELL-SMITH, G.P. 1911. The structure of root-hairs and the absorption of water. Agric. Gaz. New So. Wales 22:657-661.

856. DASGUPTA, J., and KUNDU, B.C. 1963. Pharmacognostic studies of Xanthium strumarium Linn. I. Root. Planta Med. 11:471-478.

857. DA SILVA, J.B. 1969. Contribuição ao estudo farmacognóstice da raiz Byrsonima intermedia Ad. Jussieu forma latifolia Grisebach. Rev. Fac. Farm. Bioq. Univ. São Paulo 7:313-323. (in Portuguese)

858. DASTUR, R.H., and KAPADIA, G.A. 1931. The anatomy of climbing plants. Jour. Indian Bot. Soc. 10:110-121.

859. DATTA, A., and MAJUMDAR, G.P. 1943. Root initiation in the adult axes of a few dicotyledonous species. Proc. Indian Acad. Sci. (Sect. B) 18:109-118.

860. DATTA, S.C. 1949. Pharmacognostic studies on commercial varieties of Rauwolfia. Indian Jour. Pharm. 11:105-117.

861. DAUPHINÉ, A. 1911. Contribution à l'étude anatomique du genre Kalanchoe. Ann. Sci. Nat., Bot. (Sér. 9) 13:195-219.

862. -----------. 1912. De l'évolution de l'appareil conducteur dans le genre Kalanchoe. Ann. Sci. Nat., Bot. (Sér. 9) 15:153-163.

863. -----------. 1913. Sur le développement de l'appareil conducteur chez quelques Centrospermées. Bull. Soc. Bot. France 60:312-322.

864. -----------. 1923. Sur le présence de vaisseaux primaires, superposés et centrifuges, dans la racine. Bull. Soc. Bot. France 70:73-77.

865. -----------. 1923. Polytomie, polystélie et accélération vasculaire dans les racines traumatisées. Bull. Soc. Bot. France 70:502-508.

866. -----------. 1925. Demonstration expérimentale du rapport vasculaire entre la feuille et la racine. Comp. Rend. Hebd. Séances Acad. Sci. 181:1159-1160.

867. -----------. 1926. Nouvelles expériences sur le rapport vasculaire entre la feuille et la racine. Comp. Rend. Hebd. Séances Acad. Sci. 182:1484-1485.

868. DAVEY, A.J. 1914. Seedling anatomy of certain Amentiferae. Annals Bot. 30:575-599.

869. ----------. 1946. On the seedling of _Oxalis_ _hirta_ L. Annals Bot. (N.S.) 10:237-256.

870. DAVIES, J., BRIARTY, L.G., and RIELEY, J.O. 1973. Observations on the swollen lateral roots of the Cyperaceae. New Phytol. 72:167-174.

871. DAVIS, T.A., and ANANDAN, A.P. 1957. The first root of the coconut. Indian Coco. Jour. 10:9-14.

872. DEAN, B.E. 1933. Effect of soil type and aeration upon root systems of certain aquatic plants. Plant Phys. 8:203-223.

873. DE BALSAC, R.H. 1925. Contribution à l'étude biologique des riz. Riz. Rizic. 1:39-(47).

874. DEBRAY, M.M. 1966. Contribution à l'étude du genre _Epinetrum_ (Menispermacées) de Côte-d'Ivoire. Office de la Recherche Scientifique et Technique Outre-Mer, Paris. 74 pp.

875. DEBRUYNE, C. 1925. Anatamo-Physiologische beschouwingen over de endodermis van den Wortel. Bot. Jahrb. 19:147-155.

876. DE CAMARGO, P.N. 1960. Contribution to the knowledge of the anatomy of _Hymenaea_ _stilbocarpa_ Hayne. Bol. Fac. Filos. Cienc. Letras Univ. Sao Paulo (Bot.) 17:11-105.

877. DE CAPITE, L., and MILLETTI, G. 1967. Anatomia della radici aeree e terrestri in _Anthurium_ _andreanum_ Lindl. Ann. Fac. Agraria R. Univ. Perugia 22:307-315.

878. DECOUX, L., and ERNOULD, L. 1944. L'anatomie du pivot de la betterave sucriere. Publ. Inst. Belge Amelior. Better. 12:257-319.

879. DECROCK, E. 1901. Anatomie des Primulacées. Ann. Sci. Nat., Bot. (Ser. 8) 13:1-199.

880. DE FRAINE, E. 1910. The seedling structure of certain Cactaceae. Annals Bot. 24:125-175.

881. ------------. 1912. The anatomy of the genus _Salicornia_. Jour. Linn. Soc. (Bot.) 41:317-348.

882. ------------. 1916. The morphology and anatomy of the genus _Statice_ as represented at Blakeney Point. Annals Bot. 30:239-282.

883. DE HAAN, I. 1941. De anatomische bouw van de theeplant. II. Bast enhout van tak en wortel. Arch. Theecult. Nederl.- Indië 15:213-234.

884. DEHAY, C. 1937. Racines intracaulinaires accidentales chez le Peuplier. Bull. Soc. Bot. France 84:529-533.

885. DE LAVISON, J.R.   1910.   Du mode de pénétration de quelques sels dans la plante vivante; role de l'endoderme.   Rev. Gén. Bot.   22:225-241.

886. DEMAGGIO, A.E.   1961.   Morphogenetic studies on the fern Todea barbara (L.) Moore.   II. Development of the embryo.   Phytomorph.   11:64-79.

887. DEMALSY, P.   1953.   Études sur les Hydroptéridales.   III. Le sporophyte d'Azolla nilotica Decaisne.   La Cellule   56:7-60.

888. DENIS, M.   1919.   Recherches anatomiques sur quelques plantes littorales de Madagascar.   Rev. Gén. Bot.   31:33-52.

889. --------.   1919.   Les suçoirs du Cassytha filiformis L.   Bull. Soc. Bot. France   66:398-403.

890. DE OLIVEIRA, F., and DE SOUZA GROTTA, A.   1965.   Contribuição ae estudo morfólogico e anatômico de Schinus terebinthifolius Raddi.   Anacardiaceae.   Rev. Fac. Farm. Bioq. Univ. São Paulo   3:271-293.   (in Portuguese)

891. DESCHAMPS, R.   1967.   Étude histologique, cytologique et ultrastructurale du méristèm radiculaire du Lin, Linum usitatissimum L.   Rev. Gén.Bot.   74:159-175.

892. DESHPANDE, B.D.   1953.   Presence of velamen in the earth roots of Crinum latifolium Linn.   Sci. Cult.   19:198-199.

893. --------------.   1955.   Velamen in roots of some species of Amaryllidaceae.   Proc. Rajas. Acad. Sci.   5:38-46.

894. --------------.   1955.   Velamen in the Liliaceae and Amaryllidaceae.   Sci. Cult.   21:41-42.

895. --------------.   1955.   The genera Chlorophytum Ker-Gawl. and Anthericum L. of Liliaceae.   Sci. Cult.   21:214-215.

896. --------------.   1956.   Root apices, ontogeny and morphology of velamen in some members of Liliaceae and Amaryllidaceae.   Doctoral thesis.   University of Rajasthan (India).

897. --------------.   1956.   Velamen in terrestrial Araceae.   Sci. Cult.   21:686-687.

898. --------------.   1960.   Root apical meristems in monocots.   I. Root apex organization in some members of the Amaryllidaceae.   Jour. Indian Bot. Soc.   39:126-132.

899. --------------.   1960.   Velamen in terrestrial monocots.   II. Ontogeny and morphology of velamen in the Amaryllidaceae with a discussion on the exodermis in Amaryllidaceae and the Liliaceae.   Jour. Indian Bot. Soc.   39:593-600.

60

900. ------------., and BHATNAGAR, P. 1961. Apical meristems of Ephedra foliata. Bot. Gaz. 122:279-284.

901. ------------. 1961. Root apical meristems in monocots. II. Root apex organization in some members of the Liliaceae. Jour. Indian Bot. Soc. 40:535-541.

902. ------------., and JONEJA, P. 1962. Seedling anatomy of certain members of the Umbelliferae. Proc. Indian Acad. Sci. (Sect. B) 56:332-338.

903. ------------. 1966. Seedling anatomy of certain members of the Cucurbitaceae. Proc. Indian Acad. Sci. (Sect. B) 64:62-67.

904. ------------., and SINGH, A. 1967. Seedling and floral anatomy of Withania somnifera Dunal. Jour. Indian Bot. Soc. 46:76-81.

905. ------------., and KUMAR, A. 1968. Seedling anatomy and primary vascular system in some members of Compositae. Jour. Indian Bot. Soc. 47:68-78.

906. DE SOUZA GROTTA, A. 1963. Contribuição ao estudo morfológico e anatômico de Boerhaavia coccinea Mill. Nyctaginaceae. Rev. Fac. Far,. Bioq. Univ. São Paulo 1:9-35. (in Portuguese)

907. DETZNER, H. 1910. Beiträge zur vergleichenden Anatomie der Amentaceen-Wurzeln mit Rücksicht auf die Systematik. Doctoral dissertation. George-August-Universität zu Göttingen. Göttingen Univ. Publication.

908. DEVAUX, H. 1900. Recherches sur les lenticelles. Ann. Sci. Nat., Bot. (Sér. 8) 12:1-124.

909. DE VRIES, H. 1918. Ueber die kontraktion der Wurzeln. In: Opera e periodici collata. A. Oosthoek, Utrecht. 2:26-84.

910. DEYSSON, G. 1954. Elements d'anatomie des plantes vasculaires. SEDES, Paris. 266 pp.

911. DICKISON, W.C. 1971. Comparative morphological studies in Dilleniaceae. VII. Additional notes on Acrotrema. Jour. Arnold Arb. 52:319-333.

912. DIELS, L. 1906. Blattrhizoiden bei Drosera. Berich. Deutsch. Bot. Gesel. 24:189-191.

913. DIEPENBROCK, F. 1928. Beiträge zur botanischen und pharmacognostischen Kenntnis von Corolla Illipis latifolae, Radix Tribuli cistoides, Rhizoma Araliae racemosae und Calyx Hibisci sabdariffa. Angew. Bot. 10:1-66.

914. DIETZ, J. 1930. Morphologisch-Anatomisch Untersuchungen der unterirdischen Organe tropischer Erdorchideen. Ann. Jard. Bot. Buitenz. 41:1-25.

915. DITTMER, H.J., and SPENSLEY, R.D. 1947. The developmental anatomy of _Descurainia pinnata ochroleuca_ (Woot.) Detling. Univ. New Mexico Publ. Biol. No. 3. 47 pp.

916. ------------. 1948. A comparative study on the number and length of roots produced in 19 angiosperm species. Bot. Gaz. 109:345-358.

917. ------------. 1949. Root hair variation in plant species. Amer. Jour. Bot. 36:152-155.

918. ------------., and ROSER, M.L. 1963. The periderm of certain members of the Cucurbitaceae. Southw. Nat. 8:1-19.

919. ------------., and TALLEY, B.P. 1964. Gross morphology of tap roots of desert cucurbits. Bot. Gaz. 125:121-126.

920. DOBRYNIN, G.M. 1958. Corn roots and some of their biological character-istics. Doklady Acad. Nauk SSSR (Bot. Sci. Sect.) 119:94-97.

921. DOCTERS VAN LEEUWEN, W. 1911. Über die Ursache der wiederholten Ver-zweigung der Stützwurzeln von _Rhizophora_. Berich. Deutsch. Bot. Gesel. 29:476-478.

922. ----------------------. 1920. On the vegetative propagation of two species of _Taeniophyllum_ from Java. Ann. Jard. Bot. Buitenz. 31:46-56.

923. DOGGETT, H. 1970. _Sorghum_. Longmans, Green and Co., Ltd., London. 403 pp.

924. DOMIN, K. 1908. Morphologische und phylogenetische Studien über die Familie der Umbelliferen. I. Teil. Acad. Sci. Bohême, Bull. Int'l. (Sci. Math., Nat. Med. Cl.) 13:108-153.

925. DORETY, H.A. 1908. The seedling of _Ceratozamia_. Bot. Gaz. 46:203-220.

926. -----------. 1909. Vascular anatomy of the seedling of _Microcycas calocoma_. Bot. Gaz. 47:139-147.

927. -----------. 1919. Embryo and seedling of _Dioon spinulosum_. Bot. Gaz. 67:251-257.

928. DORMER, K.J. 1946. Anatomy of the primary vascular system in dicoty-ledonous plants. Nature 158:737-739.

929. DOROFEEV, V.F. 1960. The structural characteristics of the upper adventitious roots of wheat. Bot. Zhur. 45:276-279.

930. DOUTT, M.T. 1932. Anatomy of _Phaseolus vulgaris_ L. var. Black Valentine. Tech. Bull. Mich. (State Coll.) Agric. Exp. Sta. No. 128. pp. 3-31.

931. DRABBLE, E. 1903. On the anatomy of the roots of palms. Trans. Linn. Soc. London (Ser. 2) 6:427-510.

932. ----------. 1906. The transition from stem to root in some palm seedlings. New Phytol. 5:56-66.

933. DRAHEIM, W. 1929. Beiträge zur Kenntnis des Wurzelwerkes von Iridaceen, Amaryllidaceen und Liliaceen. Bot. Archiv 23:385-440.

934. DUARTE, L.S. DO NASCIMENTO, and GROSZMANN, A. 1958. Híbridos de Zea x Euchlaena. I. Análises comparativas em raiz e folha. Rev. Agric. 33:99-114. (in Portuguese)

935. DUBARD, M. 1903. Recherches sur les plantes à bourgeons radicaux. Ann. Sci. Nat., Bot. (Sér. 8) 17:109-204.

936. ---------., and VIGUIER, R. 1905. Le systeme radiculaire de l'Euphorbia intisy. Rev. Gén. Bot. 17:260-271.

937. DUBIANSKAIA, E.A. 1956. A Guide to Practical Applications in Botany. The anatomy and morphology of plants. Government Publication of Medical Literature, Moscow. 327 pp. (in Russian)

938. DUCELLIER, L. 1914. Note sur la végétation de l'Oxalis cernua Thunb. en Algérie. Rev. Gén. Bot. 25:217-227.

939. DUCHAIGNE, A. 1948. L'ontogénie de l'appareil conducteur du Calendula officinalis L. Comp. Rend. Hebd. Séances Acad. Sci. 226:946-948.

940. ------------. 1950. Une nouvelle étude ontogénique de l'appareil conducteur des Dicotyledones. Rev. Gén. Bot. 57:129-156.

941. ------------. 1951. Le passage de la racine à la tige. Doctoral thèses. Université de Poitiers, France.

942. DUDLEY, M.G. 1937. Morphological and cytological studies of Calla palustris. Bot. Gaz. 98:556-571.

943. DUFFY, R.M. 1951. Comparative cellular configurations in the meristematic and mature cortical cells of the primary root of tomato. Amer. Jour. Bot. 38:393-408.

944. DUNCAN, J.F. 1925. "Pull roots" of Oxalis esculenta. Trans. Proc. Bot. Soc. Edinb. 29:192-196.

945. DUNLOP, D.W., and SCHMIDT, B.L. 1964. Biomagnetic - I. Anomalous development of the root of Narcissus tazetta L. Phytomorph. 14:333-342.

946. ------------ and ------------. 1965. Biomagnetics - II. Anomalies found in the root of Allium cepa L. Phytomorph. 15:400-419.

947. DUNN, G.A. 1921. Note on the histology of grain roots. Amer. Jour. Bot. 8:207-211.

948. DURIN, E. 1913. Contributions à l'étude des Moringées. Rev. Gén. Bot. 25:449-471.

949. DUSÉN, P., and NEGER, F.W. 1921. Über Xylopodien. Beih. Bot. Centralb. 38:258-317.

950. DUTT, B.S.M. 1954. Some anatomical peculiarities of the roots of Crinum asiaticum L. and C. defixum Ker. Sci. Cult. 20:183-184.

951. -----------. 1964. Anomalous secondary thickening in the roots of Gnetum ula Bromg. Curr. Sci. 33:655-656.

952. DUTTA, B. 1961. Some interesting features in the root anatomy of the seedling of Borassus flabellifer Linn. Sci. Cult. 27:578-579.

953. DYCUS, A.M., and KNUDSON, L. 1957. The role of the velamen of the aerial roots of orchids. Bot. Gaz. 119:78-87.

954. DYE, C.A. 1901. Entwicklungsgeschictliche Untersuchungen über die unterirdischen Organe von Valeriana, Rheum und Inula. Inaugural-Dissertation. Universität Bern. Büchler and Co., Bern. 96 pp.

955. EAMES, A.J. 1936. Morphology of Vascular Plants. Lower groups. McGraw-Hill Book Co., New York. 433 pp.

956. -----------., and MCDANIELS, L.H. 1947. Introduction to Plant Anatomy. 2nd ed. McGraw-Hill Book Co., New York. 427 pp.

957. -----------. 1961. Morphology of Angiosperms. McGraw-Hill Book Co., New York. 518 pp.

958. EASTERLY, N.W. 1956. Some aspects of the seedling anatomy of Ptilimnium capillaceum (Michx.) Raf. Proc. West Virg. Acad. Sci. 28(7-2): 16-19.

959. EBERSTALLER, R. 1916. Beiträge zur vergleichende Anatomie der Narcisseae. Denk. Kaiserl. Akad. Wiss. Wien (Math.- Nat. Kl.) 92:87-105.

960. EKAMBARAM, T., and PANJE, R.R. 1935. Contributions to our knowledge of Balanophora. I. The morphology and parasitism of the tuber. Proc. Indian Acad. Sci. (Sect. B) 1:452-470.

961. EKDAHL, I. 1953. Studies on the growth and the osmotic conditions of root hairs. Symb. Bot. Upsal. 11:5-83.

962. ELISEI, F.G. 1934. Un caso di identita tra fasci conduttori alterni radicali e fasci conduttori sovrapposti cotiledonari. Nuovo Giorn. Bot. Ital. 41:674-692.

963. -----------. 1936. Nuovo studio sul passagio dalla struttura vascolare alterna alla struttura vascolare sovrapposta. Atti Ist. Bot.
Univ. Pavia 7:103-116.

964. -----------. 1940. Convergente e legge della differenziazione vascolare in Allium odorum L. Atti Ist. Bot. "Giov. Briosi" Lab. Critt.
Univ. Pavia (Ser. 4) 12:221-227.

965. -----------. 1940. Teoria della circolazione mediana, laterale e
intercotiledonare nella plantula dicotile diarca. Atti Ist. Bot.
"Giov. Briosi" Lab. Critt. Univ. Pavia (Ser. 4) 12:257-265.

966. -----------. 1941. Ricerche microfluorospiiche sui punti di Caspary.
Atti Ist. Bot. "Giov. Briosi" Lab. Critt. Univ. Pavia (Ser. 4)
13:3-66.

967. EMBERGER, L. 1952. Tige, racine, feuille. Ann. Biol. 28:109-128.

968. -----------. 1960. Les végétaux vasculaires. In: Traite de Botanique
(Systematique). (eds.) Chadefaud, M. and Emberger, L. Vol. 2.
Fascicle 1. Masson, Paris. 753 pp.

969. EMERSON, F.W. 1921. Subterranean organs of bog plants. Bot. Gaz.
72:359-374.

970. EMERY, A.E.H. 1955. The formation of buds on roots of Chamaenerion
angustifolium (L.) Scop. Phytomorph. 5:139-145.

971. EMOULD, M. 1921. Recherches anatomiques et physiologiques sur les
racines respiratoires. Mém. Acad. Roy. Belgique (Cl. Sci. 2) Sér.
6. 52 pp.

972. ENGARD, C.J. 1944. Morphological identity of the velamen and exodermis
in orchids. Bot. Gaz. 105:457-462.

973. -----------. 1944. Organogenesis in Rubus. Univ. Hawaii Res. Publ.
No. 21. pp. 1-234.

974. ENGLER, A. 1903. Untersuchungen über das Wurzelwachstum der Holzarten.
Mitt. Schweiz. Cent. Forst. Versuchs. 7:247-317.

975. -----------., and KRAUSE, K. 1908. Über die lebensweise von Viscum
minimum Harvey. Berich. Deutsch. Bot. Gesel. 26a:524-530.

976. -----------. 1911. Araceae-Lasioideae. In: Das Pflanzenreich. (ed.)
Engler, A. W. Engelmann, Leipzig. 130 pp.

977. -----------., and KRAUSE, K. 1911. Über den anatomischen Bau der baum-
artigen Cyperacee Schoenodendron bucheri Engl. aus Kamerun. Abh.
Königl. Preuss. Akad. Wiss. (Math.- Phys. Kl.) Jahrgang 1911. Abh.
1. pp. 1-14.

978. ENGLERT, O. 1925. Beiträge zur vergleichenden Anatomie südamerikani-scher Lycopodium-Arten und deren Stellung im System. Bot. Archiv 11:314-360.

979. ENRICO, C. 1904. VI. Sulla particolare struttura delle radici tuberi-zzati di Thrincia tuberosa D.C. Annali Bot. 1:199-205.

980. ERICKSON, R.O., and GODDARD, D.R. 1951. An analysis of root growth in cellular and biochemical terms. Growth. Supplement to Vol. 15, Tenth Symposium on Development and Growth. 15:89-116.

981. --------------., and SAX, K.B. 1956. Elemental growth rate of the primary root of Zea mays. Proc. Amer. Philos. Soc. 100:487-498.

982. ---------------- and --------. 1956. Rates of cell division and cell elongation in the growth of the primary root of Zea mays. Proc. Amer. Philos. Soc. 100:499-514.

983. -------------. 1961. Probability of division of cells in the epidermis of the Phleum root. Amer. Jour. Bot. 48:268-274.

984. ERIKSSON-HELSINGFORS, E. 1910. Über der Alkannawurzel und die Entstehung des Farbstoffes in derselben. Berich. Deutsch. Pharm. Gesel. 20:202-208.

985. ERITH, A.G. 1924. White clover - a monograph. Duckworth, London. 150 pp.

986. ERNOULD, M. 1921. Recherches anatomiques et physiologiques sur les racines respiratoires. Mém. Acad. Roy. Belgique (Sér. 2) Class D, Sci. 6:1-52.

987. ERNST, A., and BERNARD, C. 1911. Beiträge zur Kenntnis der Saprophyten Javas. VIII. Äussere und innere Morphologie von Burmannia candida Engl. und Burmannia championii Thw. Ann. Jard. Bot. Buitenz. 9:84-97.

988. ---------- and ----------. 1912. Beiträge zur Kenntnis der Saprophyten Javas. XI. Äussere und innere Morphologie von Burmannia coelestis Don. Ann. Jard. Bot. Buitenz. (Ser. 2) 11:223-233.

989. ESAU, K. 1939. Studies of the development of the carrot root. Amer. Jour. Bot. 26:671. (Abstract)

990. --------. 1940. Developmental anatomy of the fleshy storage organ of Daucus carota. Hilgardia 13:175-226.

991. -------. 1940. Primary development of tobacco root. Amer. Jour. Bot. 27:705. (Abstract)

992. -------. 1943. Vascular differentiation in the pear root. Hilgardia 15:299-324.

993. -------. 1943. Origin and development of primary vascular tissues in seed plants. Bot. Rev. 9:125-206.

994. -------. 1953. Anatomical differentiation in root and shoot axes. In: Growth and Differentiation in Plants. (ed.) Loomis, W.E. Iowa State College Press, Ames. pp. 69-100.

995. -------. 1954. Primary vascularization in plants. Biol. Rev. 29:46-86.

996. -------. 1960. Anatomy of Seed Plants. John Wiley and Sons, New York. 376 pp.

997. -------. 1961. Plants, Viruses, and Insects. Harvard University Press, Cambridge. 110 pp.

998. -------. 1965. Plant Anatomy. 2nd ed. John Wiley and Sons, New York. 767 pp.

999. -------. 1969. The phloem. In: Handbuch der Pflanzenanatomie. (eds.) Zimmermann, W., et al. Histologie. Band 5 Teil 2. Gebrüder Borntraeger, Berlin. 505 pp.

1000. ESCHRICH, W. 1963. Invers gelagerte Leitbündelsysteme (Masern) im Wurzelholze von Angelica archangelica L. Öster. Bot. Zeitschr. 110:428-443.

1001. EUKER, R. 1906. Zum leitbündelverlaufe von Convallaria majalis L. Berich. Deutsch. Bot. Gesel. 24:330-339.

1002. EVANS, H. 1938. Studies on the absorbing surface of sugar-cane systems. I. Method of study with some preliminary results. Annals Bot. (N.S.) 2:159-182.

1003. FAGERLIND, F. 1940. Das Vorkommen von Pneumatophoren bei Pandanus. Svensk. Bot. Tidsk. 34:1-6.

1004. ------------. 1948. Bau und Entwicklung der vegetativen Organe von Balanophora. Kung. Svensk. Vet. Hand. 25:1-72.

1005. FAHN, A. 1954. Metaxylem elements in some families of the Monocotyledoneae. New Phytol. 53:530-540.

1006. -------. 1954. The anatomical structure of the Xanthorrhoeaceae Dumort. Jour. Linn. Soc. (Bot.) 55:158-184.

1007. -------. 1961. The anatomical structure of Xanthorrhoeaceae Dumort. and its taxonomic position. In: Recent Advances in Botany. University of Toronto Press. I:155-160.

1008. -------. 1964. Some anatomical adaptations of desert plants. Phytomorph. 14:93-102.

1009. -------. 1967. Plant Anatomy. Translation by S. Broido-Altman. Pergamon Press, New York. 534 pp.

1010. FARMER, J.B., and HILL, T.G. 1902. On the arrangement and structure of the vascular strands in Angiopteris evecta, and some other Marattiaceae. Annals Bot. 16:371-402.

1011. FARR, C.H. 1928. Root hairs and growth. Quart. Rev. Biol. 3:343-376.

1012. FARWELL, O.A. 1911. The sleepy grass of New Mexico: A histological study. Merck's Rept. 20:271-273.

1013. ------------. 1914. Gossypium: A histological study. Merck's Rept. 23:133-138.

1014. FAULL, J.H. 1901. The anatomy of the Osmundaceae. Bot. Gaz. 32: 381-420. (Also in: Univ. Toronto Stud., Biol. Ser. No. 2. pp. 3-39. 1902)

1015. FAURE, A. 1924. Étude organographique, anatomique et pharmacologique la famille des Cornacées (groupe des Cornales). Doctoral thèses. Université de Lille. Imprimerie Central du Nord, Lille. 221 pp.

1016. FAUST, E.C.F. 1917. Resin secretion in Balsamorhiza sagittata. Bot. Gaz. 64:441-478.

1017. FAYLE, D.C.F. 1968. Patterns of radial growth in tree roots. Diss. Absts. 28(10):4017-B.

1018. ------------. 1968. Radial growth in tree roots. Distribution - timing - anatomy. Tech. Rept. Fac. For. Univ. Toronto No. 9. 183 pp.

1019. FEDEROV, A.A. 1962. An Illustrated Atlas of Vascular Plants: (Roots and Stems). Academia Scientarum URSS, Moscow. 348 pp. (in Russian)

1020. ------------., KIRPICHNIKOV, M.E., and ARTIUSHENKO, Z.T. 1962. A Pictorial Work on the Descriptive Morphology of Vascular Plants; Stems and Roots. Izd-vo Akademik Nauk SSSR, Moskova. 352 pp. (in Russian)

1021. FEGEL, A.C. 1941. Comparative anatomy and varying physical properties of trunk, branch, and root wood in certain northeastern trees. Bull. New York State Coll. For., Tech. Publ. No. 55. Vol. XIV(2b).

1022. FEHÉR, D. 1924. Anatomie der vegetativen Organe der Robinia pseudo-acacia L. Anatomie der Wurzel. Teil 3. Erdész. Lapok 63:83-100.

1023. FELL, K.R., RAMSDEN, D., and TREASE, G.E. 1965. The morphology
and anatomy of the flowering plant of Colchicum steveni. Planta
Med. 13:158-177.

1024. FERRARINI, E. 1950. Il parassitismo di Osyris alba L. Nuovo
Giorn. Bot. Ital. 57:351-381.

1025. FILIPESCU, G. 1969. Comparative anatomical research in species of
the genus Adonis L. Anal. Stiint. Univ. "Al. I. Cuza," Iasi
Sect. II(A). Biol. 15(1):63-67. (in Romanian with French summary)

1026. ------------. 1969. Comparative anatomical research on some species
of the genus Thalictrum L. Anal. Stiint. Univ. "Al. I. Cuza," Iasi
Sect. II(A). Biol. 15(1):69-74. (in Romanian with French summary)

1027. FINERAN, B.A. 1962. Studies in the root parasitism of Exocarpus
bidwillii Hook. f. I. Ecology and root structure of the parasite.
Phytomorph. 12:339-355.

1028. ------------. 1963. Studies on the root parasitism of Exocarpus
bidwillii Hook. f. II. External morphology, distribution, and
arrangement of haustoria. Phytomorph. 13:30-41; III. Primary
structure of the haustorium. 13:42-54; IV. Structure of the mature
haustorium. 13:249-267.

1029. ------------. 1965. Studies on the root parasitism of Exocarpus
bidwillii Hook. f. V. Early development of the haustorium.
Phytomorph. 15:10-25; VI. Haustorial attachment to non-living
objects and the phenomenon of self-parasitism. 15:387-399.

1030. ------------. 1966. Fine structure of meristem and differentiating
root cap cells in Ranunculus hirtus Hook. f. Phytomorph. 16:1-16.

1031. FISCHER, W. 1937. Beiträge zur Morphologie von Datura stramonium L.
Inaugural-Dissertation. Universität Basel. A. Schudel-Bleiker,
Basel. 68 pp.

1032. FISK, E.L., and MILLINGTON, W.F. 1959. Atlas of Plant Morphology.
Portfolio I. Burgess Publishing Co., Minneapolis.

1033. ----------- and ----------------. 1962. Atlas of Plant Morphology.
Portfolio II. Burgess Publishing Co., Minneapolis.

1034. FLASKAMPER, P. 1910. Untersuchungen über die Abhängigkeit der
Gefass- und Sklerenchymbildung von äussern Faktoren nebst einigen
Bemerkungen über die angebliche Heterorhizie bei Dikotylen. Flora
101:181-219.

1035. FLOUS, F. 1934. Un cas anormal d'évolution vasculaire. Comp. Rend.
Hebd. Séances Acad. Sci. 198:2111-2114.

1036. --------. 1934. La notion de phyllorhize chez le chéne-liège. Comp. Rend. Hebd. Séances Acad. Sci. 198:2193-2195.

1037. FOARD, D.E., HABER, A.H., and FISHMAN, T.N. 1965. Initiation of lateral root primordia without completion of mitosis and without cytokinesis in uniseriate pericycle. Amer. Jour. Bot. 52:580-590.

1038. FONTELL, C.W. 1909. Anatomischer Bau der Potamogeton-Arten. Öfver. Finska Vetens.- Soc. Förhand. Afd. 14. 51:1-91.

1039. FORD, S.O. 1902. The anatomy of Ceratopteris thalictroides (L.) Annals Bot. 16:95-123.

1040. ---------. 1904. The anatomy of Psilotum triquetrum. Annals Bot. 18:589-605.

1041. FOSTER, A.S., and GIFFORD, E.M., Jr. 1959. Comparative Morphology of Vascular Plants. W.H. Freeman and Co., San Francisco. 555 pp.

1042. FOURCROY, M. 1935. Modifications de l'insertion des radicelles dans les racines traumatisées. Comp. Rend. Hebd. Séances Acad. Sci. 200:2213-2214.

1043. -----------. 1936. Accélération evolutive des radicelles dans des racines traumatisées. Comp. Rend. Hebd. Séances Acad. Sci. 202:1081-1083.

1044. -----------. 1936. Atténuation progressive de l'accélération trans- mise à une radicelle pa une racine traumatisée. Comp. Rend. Hebd. Séances Acad. Sci. 202:1527-1529.

1045. -----------. 1938. Influence de divers traumatismes sur la structure des organes végétaux a l'évolution vasculaire complète. Ann. Sci. Nat., Bot. (Sér. 10) 20:1-239.

1046. -----------. 1939. Sur la non-continuité de la zone génératrice interne des racines jeunes. Comp. Rend. Hebd. Séances Acad. Sci. 208:1336-1338.

1047. -----------. 1939. Sur l'existence des arcs extra-ligneux dans les racines des Monocotylédones. Comp. Rend. Hebd. Séances Acad. Sci. 209:841-842.

1048. -----------. 1939. Complements à l'étude du phenomene de la réduction du nombre des convergents chez Pinus pinea. Rev. Gén. Bot. 51:257- 281.

1049. -----------. 1940. La zone génératrice libéro-ligneuse de la racine jeune n'est pas sineuse et continue. Rev. Gén. Bot. 52:375-412.

1050. -----------. 1941. Près d'une insertion radicellaire, des vaisseaux échappent aux lois fondamentales de l'évolution vasculaire. Comp. Rend. Hebd. Séances Acad. Sci. 212:1166-1168.

1051. -----------. 1942. Perturbations anatomiques interessant le faisceau vasculaire de la racine au voisinage des radicelles. Ann. Sci. Nat., Bot. (Sér. 11) 3:177-200.

1052. -----------. 1943. Sur le présence de radicelles internes dans le tubercle du céleri. Bull. Soc. Bot. France 90:51-54.

1053. -----------. 1944. L'arc extra-ligneaux dans la racine du Trades-cantia virginica. Bull. Soc. Bot. France 91:86-88.

1054. -----------. 1946. Extinction des convergents dans la stele surnuméraire d'une racine tératologique. Bull. Soc. Bot. France 93:345-348.

1055. -----------. 1947. Second example de bistélie dans la racine de Ficaire. Bull. Soc. Bot. France 94:365-373.

1056. -----------. 1955. Différentes possibilités de bistélie dans la racine des Ficaires. Retour à la monostélie. Bull. Soc. Bot. France 102:505-519.

1057. -----------. 1956. Nouvel example de racine bistélique: Tradescantia virginica. Bull. Soc. Bot. France 103:590-595.

1058. -----------. 1957. Remarque au sujet de la systematique de trois Tradescantia. Bull. Soc. Bot. France 104:605-608.

1059. FRANCINI, E. 1931. Lo sviluppo del sistema conduttore in plantule di Hydrastis canadensis L. Nuovo Giorn. Bot. Ital. 38:336-357.

1060. FRANCIS, W.D. (1924) 1925. The development of buttresses in Queensland trees. Proc. Roy. Soc. Queensl. 36:21-37.

1061. FRANCKE, A. 1927. Zur Kenntnis der exodermis der Asclepiadaceen. Planta 3:1-26.

1062. FRANCOIS, L. 1908. Recherches sur les plantes aquatiques. Ann. Sci. Nat., Bot. (Sér. 9) 7:59-110.

1063. FRASER, L. 1931. An investigation of Lobelia gibbosa and Lobelia dentata. I. Mycorrhiza, latex system and general biology. Proc. Linn. Soc. New So. Wales 56:497-525.

1064. FRAYSSE, A. 1905. Sur la biologie et l'anatomie des sucoirs de l'Osyris alba. Comp. Rend. Hebd. Séances Acad. Sci. 140:270-271.

1065. FREDON, J.J. 1966. Sur le presence chez le <u>Lepidium</u> <u>virginicum</u> L. de phloeme interxylemien dans la racine et de phloeme intraxylemien dans la tige. Bull. Soc. Bot. France 113:283-287.

1066. ----------. 1966. Les anomalies de l'appareil conducteur chez <u>Raphanus</u> <u>raphanistrum</u> L. et <u>Sinapsis</u> <u>arvensis</u> L.; leur signification en rapport avec les phenomenes de tuberisation chez <u>Raphanus</u> <u>sativus</u> L. Bull. Soc. Bot. France 113:288-290.

1067. FREEMAN, T.P. 1969. The developmental anatomy of <u>Opuntia</u> <u>basilaris</u>. I. Embryo, root, and transition zone. Amer. Jour. Bot. 56:1067-1074.

1068. FREIDENFELT, T. 1902. Studien über die Wurzeln krautiger Pflanzen. I. Über die Formbildung vom biologischen Gesichtspunkte. Flora 91:115-208.

1069. --------------. 1904. (Studien über die Wurzeln krautiger Pflanzen. II.). Der anatomische Bau der Wurzel in seinem Zusammenhange mit dem Wassergehalt des Bodens. Biblio. Bot. 12(61):1-118.

1070. FREY-WYSSLING, A., and MÜHLETHALER, K. 1949. Über den Feinbau der Zellwand von Wurzelhaaren. Mikroscopie 4:257-266.

1071. ------------------ and --------------. 1950. Bau und funktion der Wurzelhaare. Landw. Monats. 28:212-219.

1072. FRIDALVSKY, L. 1958. Uber die mikroscopische Untersuchung der ätherischen Öle in der Wurzel von <u>Valeriana</u> <u>officinalis</u> L. Acta Biol. Acad. Sci. Hung. 8:81-89.

1073. FRIES, R.E. 1906. Morphologisch-anatomische Notizen über zwei südamerikanische Lianen. Bot. Studier pp. 89-101.

1074. FRIESNER, R.C. 1920. Daily rhythms of elongation and cell division in certain roots. Amer. Jour. Bot. 7:380-407.

1075. FRITSCH, F.E., and SALISBURY, E.J. 1944. <u>Plant Form and Function</u>. G. Bell and Sons, Ltd., London. 668 pp.

1076. FRITSCH, K. 1904. <u>Die Keimpflanzen der Gesneriaceen</u>. Gustav Fischer, Jena. 188 pp.

1077. FRITSCHÉ, E. 1914. Recherches anatomiques sur la <u>Corydalis</u> <u>solida</u> Sm. Bull. Soc. Bot. Belgique 47:17-34.

1078. ----------. 1914. Recherches anatomiques sur le <u>Taraxacum</u> <u>vulgare</u> Schrk. Arch. Inst. Bot. Univ. Liège 5:3-25.

1079. ----------. 1945-1952. Contribution à l'étude morphologique de <u>Streptocarpus</u> <u>wendlandii</u> Hort. In: <u>Travaux de Botanique et de Pharmacognosie</u> (dédiés Fernand Sternon), Liège Exemplaire No. 65. pp. 133-160.

1080. ------------. 1959. Quelques observations sur la biologie de
    Cuscuta europea L. Arch. Inst. Bot. Univ. Liège 26:163-187.

1081. FUCHS, A., and ZIEGENSPECK, H. 1922. Aus der Monographie des Orchis
    traunsteineri Saut. Bot. Archiv 2:238-248.

1082. ---------- and --------------. 1924. Aus der Monographie des Orchis
    traunsteineri Saut. III. Entwicklungsgeschichte einiger deutscher
    Orchideen. Bot. Archiv 5:120-132.

1083. ---------- and --------------. 1925. Bau und form der Wurzeln der
    einheimischen Orchideen in Hinblick auf ihre Aufgaben. Bot. Archiv
    12:290-379.

1084. ---------- and --------------. 1926. Entwicklungsgeschichte der Axen
    der einheimischen Orchideen und ihre Physiologie und Biologie. Teil
    I. Cypripedium, Helleborine, Limodorum, Cephalanthera. Bot. Archiv
    14:165-260.

1085. ---------- and --------------. 1926. Entwicklungsgeschichte der Axen
    der einheimischen Orchideen und ihre Physiologie und Biologie. Teil
    II. Listera, Neottia, Goodyera. Bot. Archiv 16:360-413.

1086. ---------- and --------------. 1927. Entwicklungsgeschichte der Axen
    der einheimsichen Orchideen und der Bau ihrer Axen. III. Bot.
    Archiv 18:378-475.

1087. FUCHSIG, H. 1911. Vergleichende Anatomie der Vegetationsorgane der
    Lilioideen. Sitz. Kaiser. Akad. Wiss. (Math.- Nat. Kl.) 120:
    957-999.

1088. FUNKE, G.L. 1929. Einige Bemerkungen über das Wachstum und die
    Wurzelbildung bei Syngonium podophyllum. Ann. Jard. Bot. Buitenz.
    40:75-86.

1089. FUNKE, H. 1939. Beiträge zur Kenntnis von Keimung und Bau der Mistel.
    Beih. Bot. Centralb. 59:235-274.

1090. FYSON, P.F., and BALASUBRAHMANYAM, M. 1919. Note on the oecology of
    Spinifex squarrosus L. Jour. Indian Bot. 1:19-24.

1091. GAGE, A.T. 1901. On the anatomy of the roots of Phoenix paludosa Roxb.
    Sci. Mem. Off. Med. Sanit. Dept. Govt. India 12:103-112.

1092. GALGANO, M. 1930. Lo sviluppo del sistema conduttore nelle plantule
    di Opuntia vulgaris Mill. Nuovo Giorn. Bot. Ital. 37:527-591.

1093. GALLAGHER, W.J. 1906 (1907). Contributions to the root anatomy of the
    Cupuliferae and of the Meliaceae. Rept. Brit. Assn. Adv. Sci. No.
    76. pp. 749-750.

1094. GANDARA, G. 1910. Morfologia de las raices de las plantas. Mem. Soc. Cient. "Antonio Alzate" 30:7-10.

1095. GARD, M. 1903. Études anatomiques sur les vignes et leurs hybrides artificels. Actes Soc. Linn. Bord. (Sér. 6) 58:185-319.

1096. GARJEANNE, A.J.M. 1948. Veronica. Levende Natuur 51:101-108.

1097. GARVER, S. 1922. Alfalfa root studies. U.S.D.A. Bull. No. 1087. 28 pp.

1098. GATIN, C.L. 1904. Observations sur la germination et la formation de la première racine de quelques Palmiers. Rev. Gén. Bot. 16: 177-187.

1099. ----------. 1906. Recherches anatomiques et chimiques sur la germination des Palmiers. Ann. Sci. Nat., Bot. (Sér.9) 8:113-146.

1100. ----------. 1907. Sur le développement des pneumathodes des Palmiers et sur la véritable nature de ces organes. Comp. Rend. Hebd. Séances Acad. Sci. 144:649-651.

1101. ----------. 1907. Observations sur l'appareil respiratoire des organes souterraines des Palmiers. Rev. Gén. Bot. 19:193-207.

1102. ----------. 1908. Recherches anatomiques sur l'embryon et la germination des Cannacées et des Musacées. Ann. Sci. Nat., Bot. (Sér 9) 8:113-146.

1103. ----------. 1912. Les Palmiers. O. Doin et Fils, Paris. 338 pp.

1104. GAUME, R. 1912. Germination, développement et structure anatomique de quelques Cistinées. Rev. Gén. Bot. 24:273-295.

1105. GAUTIER, L. 1908. Sur la parasitisme du Melampyrum pratense. Rev. Gén. Bot. 20:67-84.

1106. GAVAUDAN, P., and LAMARDELLE, P. 1953. Le rôle organogène des cellules initiales du méristème radiculaire de Triticum vulgare. Comp. Rend. Trav. Lab. Biol. Vég. Fac. Sci. Poitiers 5:1-3.

1107. GAYRAL, P., and VINOT, J. 1961. Anatomie des Vegetaux Vasculaires. Fascicles 1 and 2. G. Doin and Cie, Paris. 289 pp.

1108. GEHLEN, R. 1929. Stelar anatomy of Cicer arietinum and Glottidium floridanum. Amer. Jour. Bot. 16:781-788.

1109. GEIGER, M.J.T. 1929. A study of the origin and development of the cormlet of Gladiolus. Doctoral dissertation. Catholic University of America, Washington, D.C. 48 pp.

1110. GELLINI, R. 1964. Studio sulla radicazione di talee di _Ficus carica_
L. Osservazioni sulla struttura e sullo sviluppo di radici pre-
formate. Atti delle giornate di studio su la propagzaione della
specie legnose. Università di Pisa, Instituto di Coltivazioni
Arboree. pp. 198-219.

1111. GELLOS, G.J. 1966. The developmental morphology of _Phalaris
arundinacea_ L. Diss. Absts. (Sect. B) 27(3):703B.

1112. GEMMELL, A.R. 1969. _Developmental Plant Anatomy_. Edward Arnold,
Ltd., London. 60 pp.

1113. GENEVES, L. 1966. Recherches sur les infrastructures des leucoplasts
dans les jeunes racines de _Raphanus sativus_ L. (Crucifères). Rev.
Gén. Bot. 73:105-141.

1114. GENEVES, M.L. 1964. Origine des espaces intercellulaires dans les
méristèmes radiculaires de _Raphanus sativus_ L. Comp. Rend. Hebd.
Séances Acad. Sci. 258:302-304.

1115. GENTILE, A.C. 1971. _Plant Growth_. Natural History Press, Garden City,
New York. 144 pp.

1116. GEORGESCU, C.C. 1958. Quelques considérations sur l'applatissement
mécaniques des racines. Anal. Inst. Cerc. Exp. Forest. (Sér. 1)
19:25-30.

1117. GERSTUNG, R.A. 1960. Some aspects of the vascular development in
maize roots. Master's thesis. Miami University, Oxford, Ohio.
35 pp.

1118. GERTRUDE, M.- Th. 1937. Action du milieu extérieur sur le métabolisme
végétal. Métabolisme et morphogènése en milieu aquatique. Ph. D.
thèses. Université de Paris. Librairie Générale de l'Enseignement,
Paris. 122 pp.

1119. GERTZ, O. 1920. Untersuchungen über Haustorienbildung bei _Cuscuta_.
Centralb. Bakt. Parasit. Infek. 51:287-313.

1120. GESCHWIND, L. 1901. Sur les relations existant chez la betterave
entre la gènese du saccharose et la structure de la racine. Bull.
Assn. Chem. Sucr. Dist. Ind. Agric. France Colon. 18:785-796.

1121. GESSNER, F. 1956. Der Wasserhaushalt der Epiphyten und Lianen. In:
Handbuch der Pflanzenphysiologie. (ed.) Ruhland, W. Springer-
Verlag, Berlin. 3:915-950.

1122. GEWIRTZ, M., and FAHN, A. 1960. The anatomy of the sporophyte and
gametophyte of _Ophioglossum lusitanicum_ ssp. _lusitanicum_. Phytomorph.
10:342-351.

1123. GHESQUIÈRE, L. 1925. Note sur les racines tabulaires ou accotements
ailés de quelques arbres Congolais. Rev. Zool. Afric. 13:131-132.

1124. GIERSCH, C. 1934. The anatomy of Ranunculus asiaticus L. var.
superbissmus (Hort.). Amer. Midl. Nat. 15:343-357.

1125. GILG, E. 1902. Über einige Strophanthus-Drogen. Berich. Deutsch.
Pharm. Gesel. 12:182-194.

1126. -------. 1930. Entwicklungsgeschichtliche Untersuchungen über Radix
Saponariae. Arch. Pharm. Berich. Deutsch. Pharm. Gesel. 268:
476-485.

1127. -------., and SCHÜRHOFF, P.N. 1930. Die Entwicklung der Sekretbehälter
bei den Umbelliferen und Rutaceen. Arch. Pharm. Berich. Deutsch.
Pharm. Gesel. 268:7-13.

1128. GILL, A.M. 1969. The ecology of an elfin forest in Puerto Rico.
Aerial roots. Jour. Arnold Arb. 50:197-209.

1129. ---------., and TOMLINSON, P.B. 1969. Studies on the growth of red
mangrove (Rhizophora mangle L.). 1. Habit and general morphology.
Biotropica 1:1-19.

1130. ----------- and --------------. 1971. Studies on the growth of red
mangrove (Rhizophora mangle L.). 2. Growth and differentiation of
aerial roots. Biotropica 3:63-77.

1131. GILLAIN, G. 1900. Beiträge zur Anatomie der Palmen- und Pandanaceen-
Wurzeln. Bot. Centralb. 83:337-345; 369-380; 401-412.

1132. GILLES, M. 1905. Étude morphologique et anatomique du Sablier (Hura
crepitans L.). Ann. Inst. Colon. Marseille (Sér. 2) 3:41-120.

1133. GINIEIS, C. 1950. Contribution à l'étude anatomique des plantules
de Palmiers. I. La plantule de Chamaerops humilis L. Bull. Mus.
Hist. Nat. (Sér. 2) 22:510-517.

1134. ----------. 1951. Contribution à l'étude anatomique de Palmiers.
II. La plantule de Phoenix canariensis. Bull. Mus. Hist. Nat.
(Sér. 2) 23:410-415.

1135. ----------. 1952. Contribution à l'étude anatomique des plantules
de Palmiers. III. Les variations des structures dans les plantules
de Chamaerops humilis L. Bull. Mus. Hist. Nat. (Sér. 2) 24:
100-107.

1136. ----------. 1952. Contribution à l'étude des plantules de Palmiers.
IV. La plantules de Washingtonia gracilis Parish. Bull. Mus. Hist.
Nat. (Sér. 2) 24:392-399.

1137. ----------. 1965. Structure anatomique de la plantule de <u>Calamus</u> <u>deerratus</u> (Palmaceae). Bull. Inst. Franç. Afrique Noire (Sér. A), Sci. Nat. 27:1221-1236.

1138. ----------. 1969. Étude anatomique de la plantule de <u>Pandanus</u> sp. Bull. Inst. Fond. Afrique Noire (Sér. A), Sci. Nat. 31:325-339.

1139. GINZBURG, C. 1964. Ecological anatomy of roots. Doctoral thesis. Hebrew University, Jerusalem.

1140. ----------. 1966. Xerophytic structures in the roots of desert shrubs. Annals Bot. (N.S.) 30:403-418.

1141. ----------. 1967. Organization of the adventitious root apex in <u>Tamarix</u> <u>aphylla</u>. Amer. Jour. Bot. 54:4-8.

1142. GIROUARD, R.M. 1967. Initiation and development of adventitious roots in stem cuttings of <u>Hedera</u> <u>helix</u>. Anatomical studies of the juvenile growth phase. Canad. Jour. Bot. 45:1877-1881; Anatomical studies of the mature growth phase. 45:1883-1886.

1143. GIROUX, J., and SUSPLUGAS, J. 1935. Étude anatomique du <u>Grindelia</u> <u>robusta</u> Nutt. Bull. Sci. Pharm. 42:89-102.

1144. GLEICHGEWICHTOWNA, E. 1922. La structure anatomique des organes végétatifs chez le <u>Chaenomeles</u> <u>japonica</u> Ldl. (<u>Cydonia</u> <u>japonica</u> Pers.) <u>rubra</u>, <u>rosiflora</u> et le <u>Cydonia</u> <u>vulgaris</u> Pers. Kosmos 47:361-370.

1145. GOEBEL, K. 1905. <u>Organography of Plants</u>. Part II. Special organography. Clarendon Press, Oxford. 707 pp.

1146. ----------. 1905. Die Knollen der Dioscoreen und die Wurzelträger der Selaginellen, organe, welche zwischen Wurzeln und Sprossen stehen. Flora 95:167-212.

1147. ----------. 1915. Induzierte oder autonomie Dorsiventrailtät bei Orchideenluftwurzel? Biol. Centralb. 35:209-225.

1148. ----------. 1915-1918. Organographie der Pflanzen, inbesondere der Archegoniaten und Samenpflanzen. Zweite Auflage. Zweiter Teil. Spezielle Organographie. Gustav Fischer, Jena. pp. 1008-1017.

1149. ----------. 1922. Erdwurzeln mit velamen. Flora 15:1-26.

1150. ----------. 1923. Organographie de Pflanzen, inbesondere der Archegoniaten und Samenpflanzen. Zweite Auflage. Zweiter Teil. Spezielle Organographie der Samenpflanzen. Gustav Fischer, Jena. pp. 1256-1318.

1151. ----------., and SÜSSENGUTH. 1924. Beiträge zur Kenntnis der südamerikanischen Burmanniaceen. Flora 117:55-90.

1152. ────────. 1928. Organographie der Pflanzen, inbesondere der Archegoniaten und Samnepflanzen. Dritter Auflage. Erster Teil. Allgemeine Organographie. Gustav Fischer, Jena. pp. 118-126.

1153. ────────., and SANDT, W. 1930. Untersuchungen an Luftwurzeln. Bot. Abhandl. 2:3-124.

1154. GOETTE, W. 1931. Untersuchungen über die Beeinflusung des anatomischen Baues einiger Liliaceen durch Standortsfaktoren und experimentelle Eingriffe (Balusäure - Begasung). Beitr. Biol. Pflanz. 19:35-66.

1155. GOLINSKA, H. 1929. Einige Beobachtungen über die Morphologie und Anatomie der Radieschenknolle. Gartenbauwiss. 1:488-499.

1156. GOLOUBEVA, M. 1923. Agrostis prorepens (Koch) Golub. Jour. Soc. Bot. Russie 8:111-122.

1157. GOLUB, S.J., and WETMORE, R.H. 1948. Studies of development in the vegetative shoot of Equisetum arvense L. II. The mature shoot. Amer. Jour. Bot. 35:767-781.

1158. GOLUBINSKI, I.N. 1949. Root formation peculiarities in cuttings of hops. Dokl. Akad. Nauk SSSR 60:1065-1067.

1159. GOODWIN, R.H., and STEPKA, W. 1945. Growth and differentiation in the root tip of Phleum pratense. Amer. Jour. Bot. 32:36-46.

1160. ────────────., and AVERS, C.J. 1956. Studies on roots. III. An analysis of root growth in Phleum pratense using photomicrographic records. Amer. Jour. Bot. 43:479-487.

1161. GOOSSENS, A.P., and THERON, J.J. 1934. An anatomical study of Themeda triandra Forsk. (Rooigras). So. African Jour. Sci. 31:254-278.

1162. ─────────────. 1935. Notes on the anatomy of grass roots. Trans. Roy. Soc. So. Africa 23:1-21.

1163. GORE, U.R., and TAUBENHAUS, J.J. 1931. Anatomy of normal and acid injured cotton roots. Bot. Gaz. 92:436-441.

1164. GÖRGÉNYI-MÉSZÁROS, J. 1955. Die Gewebsentwicklung und Wurzelbildung des plagiotropen Sprosses von Cotinus coccygria Scop. Acta Bot. Acad. Sci. Hung. 1:27-56.

1165. ────────────────────. 1960. Gewebsentwicklung der sprossbürtigen Wurzeln der obsttragenden Ribes-Arten. I. Acta Bot. Acad. Sci. Hung. 6:221-256.

1166. --------------------. 1961. Gewebsentwicklung der sprossbürtigen Wurzeln der obsttragenden Ribes-Arten. II. Acta Bot. Acad. Sci. Hung. 7:7-35.

1167. GORIS, A. 1901. The structure of various Aconite roots. Pharm. Jour. Pharm. (Ser 4) 67:576.

1168. --------. 1939. Ranunculus thora Linn. Contribution à l'étude botanique, chimique et physiologique. Trav. Lab. Mat. Méd. École Sup. Pharm. Paris No. 30. 149 pp.

1169. GOURLEY, J.H. 1931. Anatomy of the transition region of Pisum sativum. Bot. Gaz. 92:367-383.

1170. GOVIL, C.M. 1969. Morphological nature of the tuber of Ipomoea batatas Lamk. Abstract in: Seminar on Morphology, Anatomy and Embryology of Land Plants. (eds.) Johri, B.M., et al. Centre of Advanced Study in Botany, University of Delhi.

1171. GOVINDARAJALU, E. 1966. The systematic anatomy of South Indian Cyperaceae: Bulbostylis Kunth. Jour. Linn. Soc. (Bot.) 59: 289-304.

1172. ----------------. 1967. Further contributions to the anatomy of the Alismataceae: Sagittaria guayanensis H.B.K., ssp. lappula (D. Don) Bogin. Proc. Indian Acad. Sci. (Sect. B) 65:142-152.

1173. ----------------. 1968. Further contributions to the anatomy of Marantaceae: Schumannianthus virgatus (Roxb.) Rolfe. Proc. Indian Acad. Sci. (Sect. B) 68:250-260.

1174. GRAEBNER, P., and FLAHAULT, M. 1908. Potamogetonaceae. In: Lebensgeschichte der Blutenpflanzen Mitteleuropas. (eds.) von Kirchner, O., Loew, E., and Schröter, C. Band 1. Eugen Ulmer, Stuttgart. pp. 394-714.

1175. GRAHAM, B.J., Jr., and BORMANN, F.H. 1966. Natural root grafts. Bot. Rev. 32:255-292.

1176. GRASSLEY, F.E. 1932. The anatomy of the primary body of Raphanus sativus L. Master's thesis. University of Chicago. 18 pp.

1177. GRAVES, A.H. 1908. The morphology of Ruppia maritima. Trans. Conn. Acad. Arts Sci. 14:59-170.

1178. GRAVIS, A., and DONCEEL, P. 1900. Anatomie comparée du Chlorophytum elatum Ait. et du Tradescantia virginica L. Mém. Soc. Roy. Sci. Liège (Sér. 3) 2:3-58.

1179. ----------., and CONSTANTINESCO, A. 1907. Contribution à l'anatomie des Amarantacées. Arch. Inst. Bot. Univ. Liège 4:3-65.

79

1180. ----------. 1909. Contribution à l'anatomie des Amarantacées. Mém. Soc. Roy. Sci. Liège (Sér. 3) 8:1-67.

1181. ----------. 1919. Connexions anatomiques de la tige et de la racine. Bull. Acad. Roy. Belgique (Classe D, Sci.) 4:227-236. (Also in: Arch. Inst. Bot. Univ. Liège, 1927. 6:1-10.)

1182. ----------. 1926. Contribution à l'étude anatomique de raccourcisse-ment des racines. Bull. Acad. Roy. Belgique (Cl. Sci.) Sér. 5. 12:48-69. (Also in: Arch. Inst. Bot. Univ. Liège, 1927. 6:2-23.)

1183. ----------. 1943. Observations anatomiques sur les embryons et les plantules. Lejeunia No. 3. pp. 7-180.

1184. GRAY, A. 1907. Gray's Botanical Textbook. 6th ed. Vol. 1. Structural botany. American Book Co., New York 442 pp.

1185. GRIEBEL, C. 1922. Die Zellelemente des Maniokmehles. Zeitschr. Unter. Nahr.- Genuss. 43:168-171.

1186. GRIGOR'EVA, V.G. 1949. On the anatomical structure of the primary roots of barley and oats grown at low soil temperature. Dokl. Akad. Nauk SSSR 67:1135-1138.

1187. GRINTESCU, I. 1915. Contribuţiuni anatomice aspura speciilor şi varietăţilor de tutun cultivate in România. Nicotiana rustica L. var. cult. "kapa." Institutel de Arte Grafice c. Sfetea, Bucureşti 58 pp. (in Hungarian)

1188. GRIST, D.H. 1959. Rice. Longmans, Green and Co., Ltd., London. 472 pp.

1189. GROOM, P., and WILSON, S.E. 1925. On the pneumatophores of paludal species of Amoora, Carapa, and Heritiera. Annals Bot. 39:9-24.

1190. GROTH, B.H.A. 1911. The sweet potato. Contrib. Bot. Lab. Univ. Penn. 4:1-104.

1191. GRUBB, P.J. 1970. Observations on the structure and biology of

Haplomitrium and Takakia, hepatics with roots. New Phytol. 69: 303-326.

1192. GRÜSS, J. 1927. Die Haustoren der Nymphaeaceen. Berich. Deutsch. Bot. Gesel. 45:459-466.

1193. GUÉRIN, P. 1910. Cellules à mucilage chez les Urticées. Bull. Soc. Bot. France 57:399-406.

1194. GUIGNARD, J.- L. 1961. Recherches sur l'embryogenie des Graminées: rapports des Graminées avec les autres monocotylédones. Ann. Sci. Nat., Bot. (Sér. 12) 2:491-610.

1195. GUILLAUMIN, A. 1909. Recherches sur la structure et le développement des Burseracées. Ann. Sci. Nat., Bot. (Sér. 9) 10:201-302.

1196. GUIR, L.J., and BURRESS, R.M. 1942. Anatomy of Taraxacum officinale
'Weber.' Trans. Kansas Acad. Sci. 45:94-97.

1197. GUPTA, K.M. 1962. Marsilea. Botanical Monograph No. 2. Council of
Scientific and Industrial Research, New Delhi. 113 pp.

1198. GÜSSOW, F. 1900. Beiträge zur vergleichenden Anatomie der Aralia-
ceen. Inaugural-Dissertation. Universität Breslau. Gustav Fock,
Leipzig. 67 pp.

1199. GUTTENBERG, H. VON. 1941. Der primäre Bau der Gymnospermenwurzel.
In: Handbuch der Pflanzenanatomie. (ed.) Linsbauer, K. Band 8.
Lief 41. Gebrüder Borntraeger, Berlin.

1200. ------------------. 1943. Die physiologischen Scheiden. In:
Handbuch der Pflanzenanatomie. (ed.) Linsbauer, K. Gebrüder
Borntraeger, Berlin. 5(4):viii-217.

1201. ------------------. 1947. Studien über die Entwicklung des Wurzelvegetationspunktes der Dikotyledonen. Planta 35:360-396.

1202. ------------------., HEYDEL, H.R., and PANKOW, H. 1954. Embryologische Studien an Monokotyledonen. I. Die Entstehung der Primärwurzel bei Poa annua L. Flora 141:298-311.

1203. ------------------. 1955. Die Entwicklung des Wurzelvegetationspunktes. Naturw. Runds. 10:385-388.

1204. ------------------., BURMEISTER, J., and BROSELL, H.- J. 1955.
Studien über die Entwicklung des Wurzelvegetationspunktes der
Dikotyledonen. II. Planta 46:179-222.

1205. ------------------., and SEMLOW, A. 1957. Die Entwicklung des
Embryos und der Keimpflanze von Cyperaceen. Bot. Studien Heft 7.
pp. 127-141.

1206. ------------------., and RIEBE, I. 1957. Die Entwicklung des Embryos
und der Keimpflanze von Bromeliaceen. Bot. Studien Heft 7. pp.
142-157.

1207. ------------------., and JAKUSZEIT, C. 1957. Die Entwicklung des
Embryos und der Primärwurzel von Galtonia candicans Decne. nebst
Untersuchungen über die Differenzierung des Wurzelvegetationspunktes
von Alisma plantago L. Bot. Studien Heft 7. pp. 91-126.

1208. ------------------., and MÜLLER-SCHRÖDER, R. 1958. Untersuchungen
über die Entwicklung des Embryos und der Keimpflanze von Nuphar
luteum Smith. Planta 51:481-510.

1209. ------------------. 1960. Grundzüge der Histogenese Höhrer Pflanzen. I. Die Angiospermen. In: Handbuch der Pflanzenanatomie. (eds.) Zimmerman, W., and Ozenda, P.G. Band 3. Teil 3. Gebrüder Borntraeger, Berlin. 315 pp.

1210. ------------------. 1961. Grundzüge der Histogenese Höhrer Pflanzen. II. Gymnospermen. In: Handbuch der Pflanzenanatomie. (eds.) Zimmerman, W., and Ozenda, P.G. Band 8. Teil 4. Gebrüder Borntraeger, Berlin. 172 pp.

1211. ------------------. 1964. Die Entwicklung der Wurzel. Phytomorph. 14:265-287.

1212. ------------------. 1965. Lehrbuch der Allgemeinen Botanik. Akademia-Verlag, Berlin. 735 pp.

1213. ------------------. 1966. Histogenese der Pteridophyten. In: Handbuch der Pflanzenanatomie. (eds.) Zimmerman, W., Ozenda, P.G., and Wulff, H.D. Band 7. Teil 2. Die Entwicklung der Wurzel. Gebrüder Borntraeger, Berlin. pp. 233-295.

1214. ------------------. 1968. Der primäre Bau der Angiospermenwurzel. In: Handbuch der Pflanzenanatomie. (eds.) Zimmerman, W., Ozenda, P.G., and Wulff, H.D. Zweite Auflage. Spezieller Teil. Gebrüder Borntraeger, Berlin. Band 8. Teil 5. 472 pp.

1215. GWYNNE-VAUGHAN, D.T. 1901. Observations on the anatomy of solenostolic ferns. I. Loxsoma. Annals Bot. 15:71-98.

1216. HABERLANDT, G. 1914. Physiological Plant Anatomy. Macmillan and Co., Ltd., London. 777 pp.

1217. --------------. 1915. Über Drusenhaaren Wurzeln. Sitz. König. Preuss. Akad. Wiss. Berlin (Phys.- Math. Kl.) 1:222-226.

1218. --------------. 1915. Das Vorkommen von Drüsenhaaren an Wurzeln. Berich. Deutsch. Bot. Gesel. 33:63-64.

1219. HACCIUS, B. 1953. Histogenetische Untersuchungen an Wurzelhaube und Kotyledonarscheide geophiler Keimpflanzen (Podophyllum und Eranthis). Planta 41:439-458.

1220. ----------., and REH, K. 1956. Morphologische und anatomische Untersuchungen an Umbelliferen-Keimpflanzen. Beitr. Biol. Pfalnz. 32:185-218.

1221. ----------., and TROLL, W. 1961. Über die sogenannten Wurzelhaare an den Keimpflanzen von Drosera- und Cuscuta-Arten. Beitr. Biol. Pflanz. 36:139-157.

1222. HAGEMANN, R. 1956. Untersuchungen über die Mitosenhäufigkeit in Gerstenwurzeln. Die Kulturpf. 4:46-82.

1223. ----------. 1957. Anatomische Untersuchungen an Gerstenwurzeln. Die Kulturpf. 5:75-107.

1224. ----------. 1959. Vergleichende morphologische, anatomische und Entwicklungsgeschichtliche Studien an Cyclamen persicum Mill. sowie einigen weitern Cyclamen-Arten. Bot. Studien Heft 9. pp. 1-88.

1225. HAGERUP, O. 1921. The structure and biology of arctic flowering plants. 10. Caprifoliaceae. Linnea borealis L. Medd. Grönl. 37:153-164.

1226. ----------. 1922. Om Empetrum nigrum L. En naturhistorik studie. Bot. Tidskr. 37:253-304.

1227. HAHMANN, C. 1920. Beiträge zur anatomischen Kenntnis der Brunfelsia hopeana Benth. im besonderen deren Wurzel, Radix Manaca. Angew. Bot. 2:113-133; 179-191.

1228. ----------. 1921. Die verwendung der Copernicia cerifera Mart., mit einem Beitrag zur anatomischen Kenntnis von deren Wurzel, Frucht und Samen. Arch. Pharm. 259:176-192.

1229. HAHN, G.G., HARTLEY, C., and RHOADS, A.S. 1920. Hypertrophied lenticels on the roots of conifers and their relation to moisture and aeration. Jour. Agric. Res. 20:253-265.

1230. HALKET, A.C. 1927. Observations on the tubercles of Ranunculus ficaria L. Annals Bot. 41:731-753.

1231. HALL, T.F. 1940. The biology of Saururus cernuus L. Amer. Midl. Nat. 24:253-260.

1232. ----------., and PENFOUND, W.T. 1944. The biology of the American lotus, Nelumbo lutea (Willd.) Pers. Amer. Midl. Nat. 31:744-758.

1233. HALLQUIST, S. 1914. Ein Beitrag zur Kenntnis der Pneumatophoren. Svensk Bot. Tidskr. 8:295-307.

1234. HAMANN, U. 1966. Embryologische, morphologische-anatomische und systematische Untersuchungen an Philydraceen. Willdenowia Beiheft 4. 178 pp.

1235. HAMDOUN, A.M. 1970. The anatomy of subterranean structures of Cirsium arvense (L.) Scop. Weed Res. 10:284-287.

1236. HAMIDI, A., and HUMMEL, K. 1962. Über den anatomischen Bau von Gynandropsis gynandra L. Planta Med. 10:9-13.

1237. HAMILTON, M.W. 1970. Seedling development of Grusonia bradtiana (Cactaceae). Amer. Jour. Bot. 57:599-603.

1238. HAMILTON, S.G., and BARLOW, B.A. 1963. Studies in Australian Loranthaceae. II. Attachment structures and their interrelationships. Proc. Linn. Soc. New So. Wales 88:74-90.

1239. HAMMOND, B.L. 1937. Development of Podostemon ceratophyllum. Bull. Torrey Bot. Club 64:17-36.

1240. HANDA, T. 1937. Anomalous secondary growth in Bauhinia japonica Maxim. Jap. Jour. Bot. 9:37-53.

1241. HANES, C.S. 1927. Resin canals in seedling conifers. Jour. Linn. Soc. (Bot.) 47:613-636.

1242. HAQUE, M.A. 1968. Vascular differentiation in seed and seedling of Phaseolus mungo. Diss. Absts. (1969) 29(5):3220B.

1243. HARE, C.L. 1950. The structure and development of Eriocaulon septangulare With. Jour. Linn. Soc. (Bot.) 53:422-448.

1244. HARLEY, J.L. 1959. The biology of mycorrhiza. In: Plant Science Monographs. (ed.) Polunin, N. Leonard Hill Ltd., London. 233 pp.

1245. HARLING, G. 1958. Monograph of the Cyclanthaceae. Hakan Ohlssons Boktryckeri, Lund. 428 pp. (Also in: Acta Horti Bergiani Vol. 18(1). 1958.)

1246. HARMS, H. 1936. Nepenthaceae. In: Engler and Prantl, Die Naturlichen Pflanzenfamilien. Auflage 2. Band 17b. Wilhelm Engelmann, Leipzig. pp. 728-765.

1247. HARRIS, J.A., SINNOTT, E.W., PENNYPACKER, J.Y., and DURHAM, G.B. 1921. The vascular anatomy of dimerous and trimerous seedlings of Phaseolus vulgaris. Amer. Jour. Bot. 8:63-102.

1248. HARTLEY, C.W.S. 1967. The Oil Palm. Longmans, Greene and Co., London. 706 pp.

1249. HARTSEMA, A.M. 1927. Untersuchungen über die Luftwurzeln von einigen Jussieua-Arten. Flora (N.F.) 22:242-263.

1250. HARTWICH, C. 1902. Beiträge zur Kenntnis der Sarsaparillwurzeln. Arch. Pharm. 240:325-335.

1251. HARVEY-GIBSON, R.J. 1902. Contributions towards a knowledge of the anatomy of the genus Selaginella. Part IV. The root. Annals Bot. 16:449-466.

1252. ------------------. 1913. Observations on the morphology and anatomy of the genus Mystropetalon Harv. Trans. Linn. Soc. London (Ser. 2, Bot.) 8:143-154.

1253. HASLINGER, H. 1914. Vergleichende Anatomie der Vegetationsorgane der Juncaceen. Sitz. Akad. Wiss. Wien (Math.- Nat. Kl.) 123 (Halb. 2):1147-1194.

1254. HASMAN, M., and INANÇ, N. 1957. Investigations on the anatomical structure of certain submerged, floating and amphibious hydrophytes. Rev. Fac. Sci. Univ. Istanbul (Ser. B) 22:137-153.

1255. HATCH, A.B., and DOAK, K.D. 1933. Mycorrhiza and other features of the root system of Pinus. Jour. Arnold Arb. 14:85-99.

1256. HATFIELD, E.J. 1921. Anatomy of the seedling and young plant of Macrozamia fraseri. Annals Bot. 35:565-583.

1257. HAUPT, A.W. 1953. Plant Morphology. McGraw-Hill Book Co., Inc., New York. 464 pp.

1258. HAURI, H. 1912. Anabasis aretioïdes Moq. et Coss., eine Polsterpflanze der algerischen Sahara. Beih. Bot. Centralb. 28:323-421.

1259. HAUSEN, E. 1901. Ueber Morphologie und Anatomie der Aloïneen. Verh. Bot. Ver. Prov. Brand. 42:1-50.

1260. HAUSMANN, E. 1908. Anatomische Untersuchungen an Nolina recurvata Hemsley. Beih. Bot. Centralb. 23:43-80.

1261. HAVIS, L. 1935. The anatomy and histology of the transition region of Tragopogon porrifolius. Jour. Agric. Res. 51:643-654.

1262. --------. 1937. The morphology and anatomy of the Dahlia seedling. Proc. Amer. Soc. Hort. Sci. 34:592-594.

1263. --------. 1939. Anatomy of the hypocotyl and roots of Daucus carota L. Jour. Agric. Res. 58:557-564.

1264. HAYAT, M.A. 1962. Developmental and comparative anatomy of ranalian seedlings (with special emphasis on the Annonaceae). Diss Absts. 24(1):44-45.

1265. ----------. 1963. Apical organization in roots of the genus Cassia. Bull. Torrey Bot. Club 90:123-136.

1266. ----------., and HEIMSCH, C. 1963. Some aspects of vascular differenciation in roots of Cassia. Amer. Jour. Bot. 50:965-971.

1267. ----------. 1965. Xylem differentiation in roots of Magnolia grandiflora L. Proc. North Dakota Acad. Sci. 18:81-82.

1268. ----------. 1966. Certain aspects of primary tissue development in roots of the Annonaceae. Phytomorph. 16:443-453.

1269. ----------., and CANRIGHT, J.E. 1968. The developmental anatomy of
the Annonaceae. II. Well-developed seedling structure. Bot. Gaz.
129:193-205.

1270. HAYDEN, A. 1919. The ecologic subterranean anatomy of some plants
of a prairie province in central Iowa. Amer. Jour. Bot. 6:87-105.

1271. HAYWARD, H.E. 1932. The seedling anatomy of Ipomoea batatas. Bot.
Gaz. 93:400-420.

1272. ------------. 1942. The anatomy of the seedling and roots of the
Valencia orange. U.S.D.A. Tech. Bull. No. 786. 82 pp.

1273. ------------. 1948. The Structure of Economic Plants. Macmillan Co.,
New York. 674 pp.

1274. HECKARD, L.R. 1962. Root parasitism in Castilleja. Bot. Gaz.
124:21-29.

1275. HECKEL, E. 1902. Nouvelles observations sur le Tanghin du Ménabé
(Menabea venenata Baill.) et sur sa racine toxique et médicamenteuse.
Comp. Rend. Hebd. Séances Acad. Sci. 134:441-443.

1276. HECTOR, J.M. 1936. Introduction to the Botany of Field Crops. Vol.
I. Cereals; Vol. II. Non-cereals. Central News Agency, Ltd.,
Johannesburg. 1127 pp.

1277. HEGEDÜS, A. 1954. Bemerkungen zu den Wurzelspitzengeweben der Krebs-
schere (Stratioites aloides L.). Bot. Közlem. 45:29-34.

1278. HEIL, H. 1923. Die Bedeutung des Haustoriums von Arceuthobium.
Centralb. Bakt. Parasit. Infek. 59:26-55.

1279. -------. 1924. Chamaegigas intrepidus Dtr., eine neue Auferstehungs-
pflanze. Beih. Bot. Centralb. 41:41-50.

1280. -------. 1926. Haustorialstudien an Struthanthus-Arten. Flora
121:40-76.

1281. HEIM, P. 1946. Sur les méristèmes des racines aériennes des Orchidées.
Comp. Rend. Hebd. Séances Acad. Sci. 222:813-815.

1282. HEIM, R. 1945. Sur les racines aériennes de aériennes de Phalaenopsis
schilleriana Rchb. Comp. Rend. Hebd. Séances Acad. Sci. 220:
365-367.

1283. HEIM DE BALSAC, R. 1925. Contributions à l'étude biologique des riz,
Oryza sativa L. Structure des organes, végétatifs du riz, à l'état
adulte. Riz. Rizic. 1:39-46.

1284. ----------------. 1926. Atlas oconographique du cotonnier. Organ-
      isation-structure. Gossypium sp. Contribution à l'étude botanique
      des cotonniers. Coton Cult. Coton. Vol. I., Fasc. 3. pp. 79-87;
      163-169.

1285. HEIMSCH, C. 1949. Ontogenetic changes in vascular tissues of maize
      roots. Amer. Jour. Bot. 36:797. (Abstract)

1286. ----------., RABIDEAU, G.S., and WHALEY, W.G. 1950. Vascular develop-
      ment and differentiation in two maize inbreds and their hybrid.
      Amer. Jour. Bot. 37:84-93.

1287. ----------. 1951. Development of vascular tissues in barley roots.
      Amer. Jour. Bot. 38:523-537.

1288. ----------. 1960. A new aspect of cortical development in roots.
      Amer. Jour. Bot. 47:195-201.

1289. HEINRICHER, E. 1901. Die grünen Halbschmarotzer. III. Bartschia und
      Tozzia. Jahrb. Wiss. Bot. 36:665-752.

1290. -------------. 1902. Zur Kenntnis von Drosera. Zeitschr. Ferd. Tirol.
      Vorarl. 3 Folge 46:1-29.

1291. -------------. 1906. Beiträge zur Kenntnis der Rafflesiaceae. I.
      Denk. Kaiserl. Akad. Wiss. Wien (Math.- Nat. Kl.) 78:57-81.

1292. -------------. 1907. Beiträge zur Kenntnis der Gattung Balanophora.
      Sitz. Akad. Wiss. Wien 116:439-465.

1293. -------------. 1909. Die grünen Halbschmarotzer. V. Melampyrum.
      Jahrb. Wiss. Bot. 46:273-376.

1294. -------------. 1915. Beiträge zur Biologie der Zwergmistel, Arceutho-
      bium oxycedri, besonders zur Kenntnis des anatomischen Baues und der
      Mechanik ihrer explosiven Beeren. Sitz. Kaiser. Akad. Wiss. Wien
      (Math.- Nat. Kl.) Abt. 1. 124:181-230.

1295. -------------. 1915. Die Keimung und Entwicklungsgeschichte der
      Wacholdermistel, Arceuthobium oxycedri, auf Grund durchgeführter
      Kulturen geschildert. Sitz. Kaiser. Akad. Wiss. Wien (Math.- Nat.
      Kl.) Abt. 1. 124:319-352.

1296. -------------. 1917. Warum die Samen anderer Pflanzen auf Mistelschleim
      nicht oder nur schlecht Keimen. Sitz. Akad. Wiss. Wien Abt. 1.
      126:839-892.

1297. -------------. 1921. Das Absorptionssystem von Arceuthobium oxycedri
      (DC.) M. Bieb. Berich. Deutsch. Bot. Gesel. 39:20-25.

1298. --------------. 1924. Das Absorptionssystem der Wacholdermistel Arceuthobium oxycedri (DC.) M. Bieb. mit besondere Berücksichtigung seiner Entwicklung und Leistung. Sitz. Akad. Wiss. Wien (Math.-Nat. Kl.) Abt. 1. 132:143-194.

1299. --------------. 1926. Ueber die Anschlussverhältnisse der Loranth-oideae an die Wirte und die verschiedenartigen Wucherungen (Rosen-bildungen), die gebildet werden. Bot. Archiv 15:299-325.

1300. --------------. 1931. Monographie der Gattung Lathraea. G. Fischer, Jena. 152 pp.

1301. HEJNOWICZ, Z. 1955. Growth distribution and cell arrangement in apical meristems. Acta Soc. Bot. Poloniae 24:583-608.

1302. -------------. 1956. Growth and differentiation in the root of Phleum pratense. I. Growth distribution in the root. Acta Soc. Bot. Poloniae 25:459-478.

1303. -------------. 1956. Growth and differentiation in the root of Phleum pratense. Part II. Distribution of cell divisions in the root. Acta Soc. Bot. Poloniae 25:615-628.

1304. -------------. 1959. Growth and cell division in the apical meristem of wheat roots. Physiol. Plant. 12:124-138.

1305. -------------., and BRODZKI, P. 1960. The growth of root cells as the function of time and their position in the root. Acta Soc. Bot. Poloniae 29:625-644.

1306. HELWIG, B. 1928. Über die Frage der Heterorhizie bei Radix Valerianae officinalis. Berich. Deutsch. Bot. Gesel. 46:595-609.

1307. HEMENWAY, A.F., and BREAZEALE, L. 1935. A study of Neomammillaria macdougali (Rose). Amer. Jour. Bot. 22:493-499.

1308. HENDERSON, M.W. 1919. A comparative study of the structure and sapro-phytism of the Pyrolaceae and Monotropaceae with reference to their derivation from the Ericaceae. Contrib. Bot. Lab. Univ. Penn. 5:42-109.

1309. HENRICI, M. 1929. Structure of the cortex of grass roots in the more arid regions of South Africa. Union So. Africa Dept. Agric. Sci. Bull. No. 85. pp. 3-12. (Also as: Die struktur van die Bas en Graswortels in die Droë Streke van Suid-Africa. Wetensk. Pamflet 85. Dept. van Landbou, Pretoria)

1310. HERBERT, D.A. 1920 (1918-1919). The West Australian Christmas tree. Nuytsia floribunda (The Christmas tree) - its structure and parasitism. Jour. Proc. Roy. Soc. West. Austral. 5:72-88.

1311. ------------. 1920-1921. Parasitism of the quandong. Jour. Proc. Roy. Soc. West. Austral. 7:75-76.

1312. ------------. 1920-1921. Parasitism of the sandalwood. Jour. Proc. Roy. Soc. West. Austral. 7:77-78.

1313. ------------. 1925. The root parasitism of Western Australian Santalaceae. Jour. Roy. Soc. West. Austral. 11:127-149.

1314. HERTZ, M. 1935. The initial development of the root system of the spruce. Acta Forest. Fenn. 41:1-48. (in Finnish with German summary)

1315. HESSE, H. 1904. Beiträge zur Morphologie und Biologie der Wurzelhaare. Inaugural-Dissertation. Grossherzogl. Herzogl. Sächsischen Gesamt-universität Jena. Gebrüder Georgi, Greussen. 61 pp.

1316. HEYDEL, H.R., and GUTTENBERG, H. VON. 1957. Vergleichende Studien über die Entwicklung von Primär-, Seiten- und sprossbürtigen Wurzeln bei einigen Liliaceen. Bot. Studien Heft 7. pp. 40-90.

1317. HIGASHI, J., MIZOBUCHI, K.I., and NAGOSHI, K. 1959. Pharmacognostical studies on "Huang-chi" VI. Jap. Jour. Bot. 34:311-315.

1318. HILF, H.H. 1927. Wurzelstudien an Waldbäumen. Schaper, Hannover. 121 pp.

1319. HILL, A.W. 1906. The morphology and seedling structure of the geophilus species of Peperomia, together with some views on the origin of the monocotyledons. Annals Bot. 20:395-427.

1320. HILL, J.B. 1914. The anatomy of six epiphytic species of Lycopodium. Bot. Gaz. 58:61-85.

1321. ---------. 1919. Anatomy of Lycopodium reflexum. Bot. Gaz. 68:226-231.

1322. HILL, T.G. 1900. The structure and development of Triglochin maritimum L. Annals Bot. 14:83-107.

1323. ---------., and DE FRAINE, E. 1901. On the seedling structure of gymnosperms. II. Annals Bot. 23:189-227.

1324. ---------., and FREEMAN, W.G. 1903. The root-structure of Dioscorea prehensilis. Annals Bot. 17:413-424.

1325. ---------. 1904. The seedling structure of certain Piperaceae. New Phytol. 3:46-47.

1326. ---------. 1906. On the seedling structure of certain Piperales. Annals Bot. 20:161-175.

1327. ----------., and DE FRAINE, E. 1906 (1907). On the seedling structure
    of gymnosperms. Rept. Brit. Assn. Adv. Sci. No. 76. pp. 759-760.

1328. ----------. 1906 (1907). On the seedling structure of certain Centro-
    spermae. Rept. Brit. Assn. Adv. Sci. No. 76. pp. 760-761.

1329. ----------., and DE FRAINE, E. 1908. On the seedling structure of
    gymnosperms. I. Annals Bot. 22:689-712.

1330. ----------- and ------------. 1909. On the seedling structure of
    gymnosperms. III. Annals Bot. 23:433-458.

1331. ----------- and ------------. 1910. On the seedling structure of
    gymnosperms. IV. Annals Bot. 24:319-333.

1332. ----------- and ------------. 1912. On the seedling structure of
    certain Centrospermae. Annals Bot. 26:175-199.

1333. ----------- and ------------. 1912. On the influence of the structure
    of the adult plant upon the seedling. New Phytol. 11:319-332.

1334. ----------- and ------------. 1913. A consideration of the facts
    relating to the structure of seedlings. Annals Bot. 27:257-272.

1335. HILLER, C.H. 1951. A study of the origin and development of callus
    and root primordia of Taxus cuspidata with reference to the effects
    of growth regulators. Master's thesis. Cornell University, Ithaca.
    68 pp.

1336. HIRCE, E.G., and FINNOCHIO, A.F. 1972. Stem and root anatomy of
    Monotropa uniflora. Bull. Torrey Bot. Club 99:89-94.

1337. HOCKAUF, J. 1904. Über als "Enzian" bezeichnete Wurzeln. Chem.
    Zeitschr. 28:1086-1089.

1338. HODGE, H. 1942. Getting down to the roots. Nat. Hist. 49:76-87.

1339. HOECK, A.V. 1914. The anatomy of Megalodonta beckii. Amer. Midl.
    Nat. 3:336-342.

1340. HOFFMAN, C.A. 1933. Developmental morphology of Allium cepa. Bot.
    Gaz. 95:279-299.

1341. HOLDEN, H.S., and BEXON, D. 1918. Observations on the anatomy of
    teratological seedlings. I. On the anatomy of some polycotylous
    seedlings of Cheiranthus cheiri. Annals Bot. 32:513-530.

1342. -----------. 1920. Observations on the anatomy of teratological
    seedlings. III. On the anatomy of some atypical seedlings of
    Impatiens royalei Walp. Annals Bot. 34:321-344.

1343. -----------., and DANIELS, M.E. 1921. Observations on the anatomy of teratological seedlings. III. On the anatomy of some atypical seedlings of _Impatiens royalei_ Walp. Annals Bot. 34:321-344.

1344. -----------., and CHESTERS, A.E. 1925. The seedling anatomy of some species of _Lupinus_. Jour. Linn. Soc. (Bot.) 47:41-53.

1345. -----------., and CLARKE, S.H. 1926. On the seedling structure of _Tilia vulgaris_ Heyne. Jour. Linn. Soc. (Bot.) 47:329-337.

1346. -----------., and KRAUSE, L. 1937. Observations on the root anatomy of the genus _Aletris_. Jour. Linn. Soc. (Bot.) 50:491-505.

1347. HOLLINSHEAD, M.H. 1911. Notes on the seedling of _Commelina communis_ L. Contrib. Bot. Lab. Univ. Penn. 3:275-287.

1348. HOLLOWAY, J.E. 1915. A comparative study of the anatomy of six New Zealand species of _Lycopodium_. Trans. Proc. New Zeal. Inst. 48:253-303.

1349. --------------. 1916. Studies in the New Zealand species of the genus _Lycopodium_. Part I. Trans. Proc. New Zeal. Inst. 49:80-93.

1350. HOLM, T. 1900. _Pogonia ophioglossoides_ Nutt. A morphological and anatomical study. Amer. Jour. Sci. (4th Ser.). 9:13-19.

1351. -------. 1900. Studies in the Cyperaceae. XII. Segregates of _Carex filifolia_ Nutt. Amer. Jour. Sci. (4th Ser.). 9:355-363.

1352. -------. 1901. _Erigenia bulbosa_ Nutt. A morphological and anatomical study. Amer. Jour. Sci. (4th Ser.). 11:63-72.

1353. -------. 1901. Studies in the Cyperaceae. Amer. Jour. Sci. (4th Ser.). 11:205-223.

1354. -------. 1901. _Eriocaulon decangulare_ L.; an anatomical study. Bot. Gaz. 31:17-37.

1355. -------. 1903. _Triadenum virginicum_ (L.) Rafin. A morphological and anatomical study. Amer. Jour. Sci. (Ser. 4) 16:369-376.

1356. -------. 1904. The root-structure of North American terrestrial Orchideae. Amer. Jour. Sci. (Ser. 4) 18:197-212.

1357. -------. 1905. _Claytonia gronov_. A morphological and anatomical study. Mem. Nat. Acad. Sci. 10:27-27.

1358. -------. 1905. _Croomia pauciflora_ Torr. An anatomical study. Amer. Jour. Sci. (Ser. 4) 20:50-54.

1359. ———. 1905. Studies in the Gramineae. VIII. _Munroa squarrosa_
(Nutt.) Torr. Bot. Gaz. 39:123-136.

1360. ———. 1905. _Anemiopsis californica_ (Nutt.) H. et A. An anatomical study. Amer. Jour. Sci. (Ser. 4) 19:76-82.

1361. ———. 1906. _Bartonia_, Muehl. An anatomical study. Annals Bot.
20:441-465.

1362. ———. 1906. The root structure of _Spigelia marilandica_ L.,
Phlox ovata L. and _Ruellia ciliosa_ Pursh. Amer. Jour. Pharm.
78:553-559.

1363. ———. 1906. On the structure of roots. Ottawa Nat. 20:18-22.

1364. ———. 1906. _Ceanothus americanus_ L. and _ovatus_ Desf. A morphological and anatomical study. Amer. Jour. Sci. (Ser. 4)
22:523-530.

1365. ———. 1906. Commelinaceae. Morphological and anatomical studies
of the vegetative organs of some North and Central American species.
Mem. Nat. Acad. Sci. 10:157-192.

1366. ———. 1907. Morphological and anatomical studies of the vegetative
organs of _Rhexia_. Bot. Gaz. 44:22-33.

1367. ———. 1907. _Garcinia cochinchinensis_ Choisy. Merck's Rept.
16:1-4.

1368. ———. 1907. Rubiaceae: Anatomical studies of the North American
representatives of _Cephalanthus_, _Oldenlandia_, _Houstonia_, _Mitchellia_,
Diodia, and _Galium_. Bot. Gaz. 43:153-186.

1369. ———. 1907. _Ruellia_ and _Dianthera_: an anatomical study. Bot. Gaz.
43:308-329.

1370. ———. 1907. _Anemonella thalictroides_ (L.) Spach; an anatomical
study. Amer. Jour. Sci. (Ser. 4) 24:243-248.

1371. ———. 1907. Medicinal plants of North America. Merck's Rept.
1. _Aconitum uncinatum_ L. 16(3):65-67; 2. _Caulophyllum thalictroides_
(L.) Michx. 16(4):94-96; 3. _Jeffersonia diphylla_ (L.) Pers. 16(5):
125-127; 4. _Polygala senega_ L. 16(6):155-157; 5. _Cunila mariana_ L.
16(7):188-189; 6. _Erythronium americanum_ Ker. 16(8):223-225; 7.
Podophyllum peltatum L. 16(9):250-252; 8. _Aristolochia serpentaria_ L.
16(10):276-279; 9. _Phytolacca decandra_ L. 16(11):312-314; 10.
Lobelia inflata L. 16(12):341-343.

1372. ———. 1908. _Sisyrinchium_: anatomical studies of North American
species. Bot. Gaz. 46:179-192.

1373. -------. 1908. Studies in the Gramineae. IX. The Gramineae of
the alpine region of the Rocky Mountains in Colorado. Bot. Gaz.
46:422-444.

1374. -------. 1908. Isopyrum biternatum Torr. and Gray. An anatomical
study. Amer. Jour. Sci. (Ser. 4) 25:133-140.

1375. -------. 1908. Medicinal plants of North America. Merck's Rept.
11. Gaultheria procumbens L. 17:1-3; 12. Liquidambar styraciflua L.
17:31-34; 13. Cypripedium pubescens Willd. 17:60-62; 14. Gelsemium
sempervirens Ait. 17:86-89; 15. Hedeoma pulegioides Pers. 17:115-
117; 16. Medeola virginiana L. 17:147-148; 17. Geranium maculatum L.
17:172-175; 18. Sanguinaria canadensis L. 17:209-212; 19. Gillenia
trifoliata Moench. 17:234-236; 20. Cimifuga racemosa Nutt. 17:
262-265; 21. Baptisia tinctoria R.Br. 17:295-297; 22. Eupatorium
perfoliatum L. 17:326-328.

1376. -------. 1909. Nyssa sylvatica Marsh. Amer. Midl. Nat. 1:128-137.

1377. -------. 1909. Medicinal plants of North America. Merck's Rept.
23. Sassafras officinale Nees. 18:3-6; 24. Cicuta maculata L.
18:35-38; 25. Adiantum pedatum L. 18:62-65; 26. Collinsonia
canadensis L. 18:87-90; 27. Euphorbia ipecacuanha L. 18:115-118;
28. Chimaphila umbellata (L.) Nutt. 18:143-145; 29. Euonymus
americanus L. and E. atropurpureus Jacq. 18:169-171; 30. Liriodendron
tulipifera L. 18:198-201; 31. Diospyros virginiana L. 18:229-231;
32. Sambucus canadensis L. 18:259-262; 33. Prunus serotina Ehrh.
18:287-291; 34. Cornus florida L. 18:318-321.

1378. -------. 1910. Medicinal plants of North America. Merck's Rept.
35. Quercus alba L. 19:2-4; 36. Aletris farinosa L. 19:33-35; 37.
Agropyrum repens (L.) Beauv. 19:65-68; 38. Rhus toxicodendron L.
19:95-98; 39. Euphorbia corollata L. 19:126-128; 40. Convallaria
majalis L. 19:160-162; 41. Glechoma hederacea L. 19:194-196; 42.
Rubus villosus Ait. 19:217-220; 43. Solanum carolinense L. 19:249-
251; 44. Apocynum cannabinum L. 19:277-280; 45. Grindelia squa-
rrosa (Pursh) Dunal 19:310-312; 46. Rhus glabra L. 19:338-340.

1379. -------. 1911. Mollugo verticillata L. Amer. Jour. Sci. (Ser. 4)
31:525-532.

1380. -------. 1911. Medicinal plants of North America. Merck's Rept.
47. Coptis trifolia (L.) Salisb. 20:4-6; 48. Stillingia sylvatica L.
20:36-38; 49. Arisaema triphyllum (L.) Torr. 20:66-69; 50. Arctostaphylos uva-ursi Spreng. 20:95-96; 51. Ilex opaca Ait. 20:124-
126; 52. Monarda punctata L. 20:154-156; 53. Asarum canadense L.
20:185-187; 54. Cephalanthus occidentalis L. 20:216-218; 55.
Scutellaria lateriflora L. 20:247-249; 56. Acorus calamus L. 20:
277-281; 57. Ampelopsis quinquefolia L.C.Rich. 20:309-311; 58.
Magnolia glauca L. 20:336-339.

1381. --------. 1911. Anatomical structure of the olive (Olea europea). U.S.D.A. Bur. Plant Ind. Bull. No. 192. 25:47-53.

1382. --------. 1912. Medicinal plants of North America. Merck's Rept. 59. Hamamelis virginiana L. 21:5-9; 60. Helianthemum canadense (L.) L.C.Rich. 21:38-41; 61. Lycopus virginicus L. 21:68-70; 62. Epiphegus virginiana Bart. 21:129-130; 63. Chenopodium anthelminticum L. and C. ambrosioides L. 21:178-181; 64. Kalmia latifolia L. 21:240-242; 65. Heuchera americana L. 21:267-269; 66. Impatiens fulva Nutt. 21:297-300; 67. Xanthorrhiza apiifolia L'Her. 21:323-326.

1383. --------. 1913. Phryma leptostachya L., a morphological study. Bot. Gaz. 56:306-318.

1384. --------. 1913. Medicinal plants of North America. Merck's Rept. 68. Saponaria officinalis L. 22:9-12; 69. Viburnum prunifolium L. 22:35-37; 70. Leptandra virginica (L.) Nutt. 22:61-64; 71. Datura stramonium L. 22:87-91; 73. Epigaea repens L. 22:144-146; 74. Ranunculus bulbosus L. 22:178-180; 75. Hydrastis canadensis L. 22:202-204; 76. Rhamnus purshiana DC. 22:232-235; 77. Solidago odora Ait. 22:252-254; 78. Menispermum canadense L. 22:281-284; 79. Dioscorea villosa L. 22:311-315.

1385. --------. 1914. Medicinal plants of North America. Merck's Rept. 80. Verbascum thapsus L. 23:4-5; 81. Sabbatia angularis (L.) Pursh 23:110-111; 82. Achillea millefolium L. 23:142-144; 84. Chamaelirium luteum (L.) Gray 23:268-269; 85. Hepatica triloba Chaix. var. americana DC. 23:293-295.

1386. --------. 1915. Medicinal plants of North America. Merck's Rept. 86. Juniperus virginiana L. 24:6-9; 87. Thuja occidentalis L. 24:28-30; 88. Castanea dentata (Marsh) Borkh. and C. pumila (L.) Mill. 24:85-87; 89. Veratrum viride Ait. 24:109-111; 90. Carica papaya L. 24:136-140; 91. Jatropha gossypifolia L. 24:165-167; 92. Ananassa sativa Lindl. 24:192-194; 93. Vanilla planifolia Andrews 24:212-215; 94. Maranta arundinacea L. 24:238-241; 95. Petivera alliacea L. 24:266-269; 96. Coffea arabica L. 24:297-300.

1387. --------. 1916. Medicinal plants of North America. Merck's Rept. 97. Aralia L. and Panax L. 25:11-15; 62-65.

1388. --------. 1917. Medicinal plants of North America. Merck's Rept. 98. Ambrosia artemisiaefolia L. and A. trifida L. 26:62-64; 179-180.

1389. --------. 1918. Medicinal plants of North America. Merck's Rept. 99. Cissampelos pareira L. 27:7-9.

1390. --------. 1918. Medicinal plants of North America. Merck's Rept. 100. Juglans nigra L. and J. cinerea L. 27:115-117; 168-170.

94

1391. -------. 1920. A morphological study of Cicer arietinum. Bot. Gaz. 70:446-452.

1392. -------. 1921. Morphological study of Carya alba and Juglans nigra. Bot. Gaz. 72:375-388.

1393. -------. 1921. Chionophila Benth. A morphological study. Amer. Jour. Sci. (Ser. 5) 1:31-38.

1394. -------. 1921. Dirca palustris L. A morphological study. Amer. Jour. Sci. (Ser. 5) 2:177-182.

1395. -------. 1923. Chenopodium ambrosioides L. A morphological study. Amer. Jour. Sci. (Ser. 5) 6:157-167.

1396. -------. 1923. Chelone glabra L. A morphological study. Amer. Jour. Sci. (Ser. 5) 6:265-270.

1397. -------. 1924. Gratiola L. and Sophronanthe Benth. A morphological study. Amer. Jour. Sci. (Ser. 5) 7:132-140.

1398. -------. 1924. Polypremum procumbens L. A morphological study. Amer. Jour. Sci. (Ser. 5) 7:210-218.

1399. -------. 1924. Ilsanthes, Scrophularia and Linaria. A morphological study. Amer. Jour. Sci. (Ser. 5) 8:395-410.

1400. -------. 1924. Comandra umbellata (L.) Nutt. Amer. Midl. Nat. 9:1-13.

1401. -------. 1924. Apios tuberosa Moench. Amer. Midl. Nat. 9:118-136.

1402. -------. 1924. Hibernation and rejuvenation, exemplified by North American herbs. Amer. Midl. Nat. (Ser. 4) 31:525-532.

1403. -------. 1925. On the development of buds upon roots and leaves. Annals Bot. 39:867-881.

1404. -------. 1925. Leptandra and Veronica - A morphological study. Amer. Jour. Sci. (Ser. 5) 9:460-471.

1405. -------. 1926. Studies in the Compositae. I. Krigia virginica (L.) Willd. Amer. Midl. Nat. 10:1-17.

1406. -------. 1926. Saururus cernuus L. A morphological study. Amer. Jour. Sci. (Ser. 5) 12:162-168.

1407. -------. 1927. Boehmeria cylindrica (L.) Sw. A morphological study. Amer. Jour. Sci. (Ser. 5) 13:115-122.

1408. -------. 1927. Polygonum: sectio tovara. Bot. Gaz. 84:1-26.

1409. -------. 1927. Sciaphilous plant types. Beih. Bot. Centralb. 44:1-88.

1410. -------. 1929. Morphology of North American species of Polygala. Bot. Gaz. 88:167-185.

1411. -------. 1931. The apparent influence of inulin on the meristem in roots of Compositae. Beih. Bot. Centralb. 47:359-377.

1412. -------. 1931. The seedling of Hamamelis virginiana L. Rhodora 33:81-92.

1413. HOLROYD, R. 1912. Morphology and physiology of the axis in Cucurbitaceae. Bot. Gaz. 78:1-45.

1414. HOLZNER-LENDBRADL, I. 1963. Beiträge zur Kenntnis der Histogenese von Baldrianwurzeln unter besonder Berücksichtigung der ölfuhrenden Gewebe. Beitr. Biol. Pflanz. 39:323-366.

1415. HOMÈS, J. 1968. Aspects particuliers de la structure du velamen chez des plantules de Cymbidium Sw. cultivées in vitro. Bull. Soc. Roy. Bot. Belgique 101:257-263.

1416. HOMOYALKO, S.Y. 1961. Spiral root structure in sugar beet. Ukray. Bot. Zhur. 18:55-63. (in Ukrainian with English summary)

1417. HOOK, D.D., BROWN, C.L., and KORMANIK, P.P. 1970. Lenticel and water root development of swamp tupelo under various flooding conditions. Bot. Gaz. 131:217-224.

1418. HORNE, W.T. 1904. An anomalous structure on the leaf of a bean seedling. Bull. Torrey Bot. Club 31:585-588.

1419. HORSLEY, S.B. 1971. Root tip injury and development of the paper birch root system. Forest Sci. 17:341-348.

1420. ------------., and WILSON, B.F. 1971. Development of the woody portion of the root system of Betula papyrifera. Amer. Jour. Bot. 58:141-147.

1421. HOSHIKAWA, K. 1969. Underground organs of the seedlings and systematics of Gramineae. Bot. Gaz. 130:192-203.

1422. HOTTA, M. 1971. Study of the family Araceae - General remarks. Jap. Jour. Bot. 20:269-310.

1423. HOW, J.E. 1942. The mycorrhizal relations of Larch. III. Mycorrhiza formation in nature. Annals Bot. (N.S.) 6:103-129.

1424. HOWARTH, W.O. 1927. The seedling development of Festuca rubra L. var. tenuifolia Mihi, and its bearing on the morphology of the grass embryo. New Phytol. 26:46-57.

1425. HOWE, K.J., and STEWARD, F.C. 1962. Growth, nutrition, and metabolism of Mentha piperita L. Part II. Anatomy and development of Mentha piperita L. Mem. Cornell Univ. Agric. Exp. Sta. No. 379. pp. 11-40.

1426. HOWE, M.D. 1931. Origin of leaf, and adventitious and secondary roots of Ceratopteris thalictroides. Bot. Gaz. 92:326-329.

1427. HUBER, B. 1961. Grundzuge der Pflanzenanatomie. Springer-Verlag, Berlin. 243 pp.

1428. HUFFORD, G.N. 1938. Development and structure of the watermelon seedling. Bot. Gaz. 100:100-122.

1429. HULBARY, R.L. 1948. Three-dimensional cell shape in the tuberous roots of Asparagus and in the leaf of Rhoeo. Amer. Jour. Bot. 35:558-566.

1430. HUNZIKER, A.T. 1946. Raíces gemíferas en algunas plantas lenosas argentinas. Rev. Argentina Agron. 13:47-54.

1431. HURRIER, P., and PERROT, E. 1906. Des falsifications et des succédanés du Gin-seng. Bull. Sci. Pharm. 13:659-669.

1432. HYAKUTAKE, S., and DE SOUZA GROTTA, A. 1965. Contribuição ae estudo morfológico e anatômico de Anemopaegma arvense (Vell.) Stellfeld Var. petiolata Bur., Bignoniaceae. Rev. Fac. Farm. Bioq. Univ. São Paulo 3:51-78.

1433. ------------. 1967. Contribuição ao estudo morfológico e anatômico da espécie Peltastes peltatus (Vell.) Woodson (Echites pellata Vell.), Apocynaceae. Rev. Fac. Farm. Bioq. Univ. São Paulo 5:77-91. (in Portuguese)

1434. ------------. 1969. Contribuição para o estudo botanico de Davilla rugosa Poiret var. rugosa, Dilleniaceae. Rev. Fac. Farm. Bioq. Univ. São Paulo 7:285-293. (in Portuguese)

1435. IANISHEVSKII, D.E. 1937. Morphology of seedlings of cork oak. Bot. Zhur. 22:420-434. (in Russian with French summary)

1436. IKENBERRY, G.J., Jr. 1960. Developmental vegetative morphology of Avena sativa. Diss. Absts. 20(7):2513.

1437. IMAMURA, S.J. 1929. Über Hydrobium japonicum Imamura, eine neue Podostemonacee in Japan. Bot. Mag. (Tokyo) 43:332-339.

1438. INGLE, H.D., and DADSWELL, H.E. 1953. The anatomy of the timbers of the South-west Pacific area. II. Apocynaceae and Annonaceae. Austral. Jour. Bot. 1:1-26.

1439. IRONSIDE, A.F. 1911. The anatomical structure of the New Zealand Piperaceae. Trans. Proc. New Zeal. Inst. 44:339-348.

1440. ISHIHARA, K. 1963. Zonal structure of root apices. Gamma Field Symposium (Japan) No. 2. pp. 13-23.

1441. IVANOV, L.A. 1916. Anatomical structure of root tips in pine. Izv. Imp. Lesn. Inst. (Petrograd) No. 30, Part 2. pp. 151-162. (in Russian)

1442. IYENGAR, M.A., and PENDSE, G.S. 1963. Studies on the pharmacognosy of the root-bark of Streblus asper Lour and its tincture. Indian Jour. Pharm. 25:372-375.

1443. JACCARD, P. 1914. Structure anatomique de racines hypertendues. Rev. Gén. Bot. 25(BIS):359-372.

1444. JACKSON, V.G. 1922. Anatomical structure of the roots of barley. Annals Bot. 36:21-39.

1445. JACOB DE CORDEMOY, H. 1906. Étude sur le développement de l'appareil secreteur de l'Eperua falcata Aublet. Ann. Inst. Colon. Marseille Seite 2. 4:1-22.

1446. --------------------. 1912. Sur la structure de deux Melastomacées épidendres à racines tuberisées de l'Est de Madagascar. Comp. Rend. Hebd. Séances Acad. Sci. 154:1523-1525.

1447. --------------------. 1913. Recherches anatomiques sur les Medineilla de Madagascar. Ann. Sci. Nat., Bot. (Sér. 9) 18:67-145.

1448. JACQUES-FÉLIX, H. 1957. Sur une interprétation de l'embryon des Graminées; la nature adventive des racines séminales. Comp. Rend. Hebd. Séances Acad. Sci. 245:2085-2088.

1449. JAGELS, R. 1963. Gelatinous fibers in the roots of quaking aspen. For. Sci. 9:440-443.

1450. JANASZ, S. 1904. Beschreibung einiger Zuckerrübenrassen. Mitt. Landw. Inst. Königl. Univ. Breslau 2:913-970.

1451. JAMES, L.E. 1950. Studies in the vascular and developmental anatomy of the subgenus Hesperastragalus. Amer. Jour. Bot. 37:373-378.

1452. JANSE, J.M. 1913. Der aufstiegende Strom in der Pflanze. II. Jahrb. Wiss. Bot. 52:509-602.

1453. ----------. 1927. Eine neue Einteilung der Pflanzenbewegungen. Flora (N.F.) 22:1-32.

1454. JEAN, M. 1927. Essai sur l'anatomie comparée du liber interne dans quelques familles de Dicotylédones: Étude des plantules. Le Botaniste 17:225-364.

1455. JEFFREY, E.C. 1903. The comparative anatomy and phylogeny of the Coniferales. Part 1. The genus Sequoia. Mem. Boston Soc. Nat. Hist. 5:441-459.

1456. ------------. 1905. The comparative anatomy and phylogeny of the Coniferales. Part 2. The Abietineae. Mem. Boston Soc. Nat. Hist. 6:1-37.

1457. ------------. 1917. The Anatomy of Woody Plants. University of Chicago Press. 478 pp.

1458. JEFFRIES, T.A. 1916. The vegetative anatomy of Molina caerulea, the purple heath grass. New Phytol. 15:49-71.

1459. JEFFRIES, T.M., EVANS, W.C., and TREASE, G.E. 1967. Anatomy of the root and stem of Rauwolfia exyphylla Stapf. Planta Med. 15:17-29.

1460. JENÍK, J. 1959. Beitrag zur Kenntnis der Heterorrhizie dikotyler Holzpflanzen. Phyton 8:1-7.

1461. --------., and SEN, D.N. 1964. Morphology of root system in trees: A proposal for terminology. In: Abstracts, 10th International Botanical Congress (Edinburgh). pp. 393-394.

1462. --------. 1965. Root pneumatophores in Anthocleista nobilis. Jour. West African Sci. Assn. 10:63.

1463. --------. 1967. Root adaptations in West African trees. Jour. Linn. Soc. (Bot.) 60:25-29.

1464. --------., and MENSAH, K.A.O. 1967. Root system of tropical trees. 1. Ectotrophic mycorrhizae of Afzelia africana Sm. Preslia 39:59-65.

1465. --------., and KUBIKOVA, J. 1969. Root system of tropical trees. 3. The heterorhizis of Aeschynomene elaphroxylon (Guill. et Perr.) Taub. Preslia 41:220-226.

1466. --------. 1970. Root system of tropical trees. 4. The stilted peg-roots of Xylopia staudtii Engl. et Diels. Preslia 42:25-32.

1467. --------. 1970. Root system of tropical trees. 5. The peg-roots and pneumathodes of Laguncularia racemosa Gaertn. Preslia 42:105-113.

1468. --------. 1971. Root system of tropical trees. 6. The aerial roots of Entandrophragma angolense (Welw.) C.D.C. Preslia 43:1-4.

1469. ————————. 1971. Root system of tropical trees. 7. The facultative peg-roots of Anthocleista nobilis G. Don. Preslia 42:105-113.

1470. JENSEN, W.A., and KAVALJIAN, L.G. 1958. An analysis of cell morphology and periodicity of division in the root tip of Allium cepa. Amer. Jour. Bot. 45:365-372.

1471. ————————. 1959. The root apical meristem and its cellular organization. In: Recent Advances in Botany. University of Toronto Press. I:794-795.

1472. ————————. 1959. Cell development and differentiation in root tips. In: Recent Advances in Botany. University of Toronto Press. II: 1269-1272.

1473. ————————. 1966. The problem of cell development in plants. In: Plant Biology Today: Advances and Challenges. (eds.) Jensen, W.A., and Kavaljian, L.G. 2nd ed. Wadsworth Publishing Co., Inc., Belmont, Calif. pp. 11-26.

1474. JESSEN, K. 1912. The structure and biology of arctic flowering plants. 6. Ranunculaceae. Medd. Gronl. 36:335-440.

1475. ————————. 1921 (1913). The structure and biology of arctic flowering plants. 9. Cornaceae. Medd. Gronl. 37:3-126.

1476. JIRÁSEK, V. 1964. Beitrag zur Erkenntnis des histologischen Wurzelbaues der Gräser (Poaceae). Acta Univ. Carol. (Biol. 1) 1964:61-68.

1477. JODIN, H. 1903. Recherches anatomiques sur les Borraginées. Ann. Sci. Nat., Bot. (Sér. 8) 17:263-346.

1478. JOESTING, F. 1902. Beiträge zur Anatomie der Sperguleen, Polycarpeen, Paronychieen, Sclerantheen und Pterantheen. Beih. Bot. Centralb. 12:139-181.

1479. JOHANSEN, D.A. 1941. A proposed new botanical term. Chronica Bot. 6:440.

1480. JOHNSON, D.S. 1914. Studies of the development of the Piperaceae. II. The structure and seed development of Peperomia hispidula. Amer. Jour. Bot. 1:323-339.

1481. ————————. 1933. Structure and development of Pilularia minuta. Duriev manuscript. Bot. Gaz. 95:104-127.

1482. ————————. 1938. Structure and development of Regnellidium diphyllum. Amer. Jour. Bot. 25:141-156.

1483. JOHNSON, M.A. 1928. The pericycle in the root of Equisetum. Proc. Indian Acad. Sci. 38:137-138.

1484. ------------. 1933. Origin and development of tissues in Equisetum scirpoides. Bot. Gaz. 94:469-494.

1485. JONES, W.N. 1925. Root-cap development in Calluna vulgaris. Nature 116:677.

1486. JONES, W.R. 1912. The development of the vascular structure of Dianthera americana. Bot. Gaz. 54:1-30.

1487. JÖNSSON, B. 1902. Die ersten Entwicklungsstadien der Keimpflanze bei den Succulenten. Acta Univ. Lund. 38:1-34.

1488. JOSHI, A.C. 1931. Contributions to the anatomy of the Chenopodiaceae and Amarantaceae. I. Anatomy of Alternanthera sessilis R.Br. Jour. Indian Bot. Soc. 10:213-231.

1489. ----------. 1931. Contributions to the anatomy of the Chenopodiaceae and Amarantaceae. II. Primary vascular system of Achyranthes aspera L., Cyathula prostrata Blume, and Pupalia lappacea Juss. Jour. Indian Bot. Soc. 10:265-292.

1490. ----------. 1935. Secondary thickening in the stem and root of Stellera chamaejasmae Linn. Proc. Indian Acad. Sci. (Sect. B) 2:424-436.

1491. ----------. 1937. Some salient points in the evolution of the secondary vascular cylinder of Amarantaceae and Chenopodiaceae. Amer. Jour. Bot. 24:3-9.

1492. ----------. 1940. A note on the anatomy of the roots of Ophioglossum. Annals Bot. (N.S.) 4:663-664.

1493. JOSHI, P.C. 1935. Development of the anomalous secondary vascular rings in the root of Spergula, L. Proc. Indian Acad. Sci. (Sect. B) 1:729-735.

1494. ----------. 1935. A preliminary note on the occurrence of liane type of structure in the stem and root of Thylacospermum rupifragrum Schrenk. Curr. Sci. 3:300-301.

1495. ----------. 1936. Anatomy of the vegetative parts of two Tibetan Caryophyllaceae - Arenaria musciformis Wall. and Thylacospermum rupifragrum Schrenk. Proc. Indian Acad. Sci. (Sect. B) 4:52-65.

1496. JOST, L. 1932. Die Determinierung der Wurzelstruktur. Bot. Zeit. 25:481-522.

1497. JOYEUX, L. 1929. Recherches anatomiques, systématiques et étholo- giques sur les Asparagus. Mém. Acad. Roy. Belgique (Sér. 2), Science 10:3-45.

1498. ---------. 1930. Nouvelle contribution à l'anatomie et à la systématique des Asparagus. Bull. Acad. Roy. Belgique (Cl. Sci.), Sér. 5. 16:244-263. (Also in: Rec. Quel. Trav. Anat. Vég. Exéc. Liège de 1929 à 1935, Bruxelles. 19 pp. 1936)

1499. ---------. 1936. Recherches anatomiques, systématiques et éthologiques sur les Asparaginées. Rec. Quel. Trav. Anat. Vég. Exéc. Liège de 1929 à 1935, Bruxelles. 42 pp.

1500. JUHÁSZ, G.D. 1966. Analyse der anatomischen Verhältnisse am vegetativen und reproductiven Spross-System von Cornus mas L. Ann. Univ. Sci. Budap. Rol. Eöt. Nom. (Sect. Biol.) 8:121-153.

1501. JULIANO, J.B. 1935. Anatomy and morphology of the bunga Aeginetia indica Linnaeus. Phillip. Jour. Sci. 56:405-451.

1502. -----------. 1937. Morphology of Oryza sativa Linnaeus. Phillip. Agric. 26:1-76.

1503. JUNIPER, B.E., and CLOWES, F.A.L. 1965. Cytoplasmic organeels and cell growth in root caps. Nature 208:864-865.

1504. ------------., and BARLOW, P.W. 1969. The distribution of new plasmodesmata in the root tip of maize. Planta 89:352-360.

1505. JURIŠIĆ, J. 1934. Zur Kenntnis der Drusenhaare an den Wurzeln von Bryophyllum. Anzeiger Akad. Wiss. Wien 71:192-195.

1506. KAASINEN, M. 1959 (1960). Anatomy of the root in diploid and tetraploid barley. Arch. Soc. Zool. Bot. Fenn. "Vanamo" 14:44-46.

1507. KACZMAREK, R.M. 1915. Crocion achlydophyllum (Greene). Amer. Midl. Nat. 4:74-88.

1508. KADEJ, A.R. 1966. Organization and development of apical root meristem in Elodea canadensis (Rich.) Casp. and Elodea densa (Planck) Casp. Acts Soc. Bot. Poloniae 35:143-158.

1509. KADEJ, F. 1956. The course of regeneration of the root tip of Hordeum vulgare. Acta Soc. Bot. Poloniae 25:681-712.

1510. --------. 1960. Regeneration of the apical cells in fern roots. Acta Soc. Bot. Poloniae 29:363-368.

1511. --------. 1963. Interpretation of the pattern of the cell arrangement in the root apical meristem of Cyperus gracilis var. alternifolius. Acta Soc. Bot. Poloniae 32:295-301.

1512. --------. 1964. The creative activity of cells of the constructive center of root apical meristems. Wiad. Bot. 8:131-139. (in Polish)

1513. KADRY, A.E.R., and TEWFIC, H. 1956. A contribution to the morphology and anatomy of seed germination in Orobanche crenata. Bot. Notiser 109:385-399.

1514. ---------------- and ----------. 1956. Seed germination in Orobanche crenata Forssk. Svensk Bot. Tidskr. 50:270-286.

1515. KAINRADL, E. 1927. Beiträge zur Biologie von Hydrolea spinosa L. mit besonderer Berücksichtigung von Fruchtwand und Samenentwicklung. Sitz. Akad. Wiss. Wien (Math.- Nat. Kl.) 136:167-193.

1516. KAMERLING, Z. Sind di Knollen von Batatas edulis Choisy Wurzeln oder Stengel? Berich. Deutsch. Bot. Gesel. 32:352-360.

1517. KAPIL, R.N., and RUSTAGI, P.N. 1966. Anatomy of the aerial and terrestrial roots of Ficus benghalensis L. Phytomorph. 16:382-386.

1518. KARANDIKAR, G.K., and SATAKOPAN, S. 1959. Shankhpushpi - A pharmacognostic study. I. Evolvulus alsinoides Linn. Indian Jour. Pharm. 21:200-203.

1519. ----------------- and ------------. 1959. Shankhpushpi - A pharmacognostic study. II. Convolvulus microphyllus Sieb. Indian Jour. Pharm. 21:204-207.

1520. ----------------- and ------------. 1959. Shankhpushpi - A pharmacognostic study. III. Clitorea ternatea Linn. Indian Jour. Pharm. 21:327-331.

1521. KARTASHEVA, Z.P. 1968. The structure of the embryo and the seedling of Fagopyrum sagittatum Gilib. and F. tartaricum (L.) Gaertn. Vest. Lenin. Univ. No. 3. pp. 66-67. (in Russian with English summary)

1522. KASIPLIGIL, B. 1954. The growth of the root apices in Umbellularia californica Nutt. and Laurus nobilis L. 8th International Congress of Botany (Paris). Rapports et Communications. Sect. 7-8. pp. 263-264.

1523. ------------. 1962. An anatomical study of the secondary tissues in roots and stems of Umbellularia californica Nutt. and Laurus nobilis L. Madroño 16:205-224.

1524. KASSINEN, M. 1959 (1960). Anatomy of the root in diploid and tetraploid barley. Arch. Soc. Zool. Bot. Fenn. "Vanamo" 14:44-46.

1525. KASTORY, A., and NAMYSLOWSKI, B. 1913. Ueber den anatomischen Bau von Actinidia colomicta und arguta. Kosmos 38:1146-1156. (in Polish with German summary)

103

1526. KATAYAMA, T.C. 1966. Anatomical changes in the seminal root of *Oryza sativa* after the removal of adventitious roots. Rept. Nat. Inst. Genetics 16:66-67.

1527. KAUSCH, W. 1967. Die Primärwurzel von *Zea mays* L. Planta 73: 328-332.

1528. KAUSIK, S.B. 1942. Studies in the Proteaceae. VI. Structure and development of the seedling of *Grevillea robusta* Cunn. Jour. Indian Bot. Soc. 21:145-158.

1529. KAUSSMANN, B. 1963. *Pflanzenanatomie*. Gustav Fischer, Jena. 624 pp.

1530. KAWATA, S. 1956. Studies on root formation in certain cultivated plants. III. Relation between primary root formation and canal development in crown roots of rice plants. Proc. Crop Sci. Soc. Japan 24:232-236.

1531. ---------., KAMATA, E., and YAMAZAKI, K. 1962. Studies on vascular elements in rice plants. Proc. Crop Sci. Soc. Japan 30:266-278. (in Japanese with English summary)

1532. ---------., and YAMAZAKI, K. 1964. Studies on vessel elements of crown roots in rice grown on border and in the inner part of paddy fields. Proc. Crop Sci. Soc. Japan 33:423-431.

1533. ---------., and LAI, K.L. 1965. On the meristematic state of the endodermis in the crown roots of rice plants. Proc. Crop Sci. Soc. Japan 34:210-216. (in Japanese)

1534. ----------- and --------. 1966. On the cell wall thickening of the endodermis in the crown roots of rice plants. Proc. Crop Sci. Soc. Japan 34:440-447. (in Japanese with English summary)

1535. ---------., and SHIBAYAMA, H. 1966. Types of branching in lateral roots of rice plants. Proc. Crop Sci. Soc. Japan 35:59-70.

1536. ---------., and LAI, K.L. 1967. On the differentiation of Casperian dots of the endodermis in the crown roots of rice plants. Proc. Crop Sci. Soc. Japan 36:75-84.

1537. KAWATAKE, M. 1955. Comparative studies of the vascular differentiation in some hypogaeous dicotyledons at young stages with special reference to their morphology. Proc. Crop Sci. Soc. Japan 24:12-15.

1538. KEAN, C.I. 1927. Anatomy of the genus *Mesembryanthemum*. I. Root structure. Trans. Proc. Bot. Soc. Edinb. 29:381-388.

1539. ---------. 1929. Seedling anatomy of the genus *Mesembryanthemum*. Trans. Proc. Bot. Soc. Edinb. 30:164-174.

1540. KEIL, G. 1941. Das Wurzelwerk von Taraxacum officinale Weber. Beih. Bot. Centralb. 60:57-96.

1541. KELLER, I.A. 1900. Notes on hyacinth roots. Proc. Acad. Nat. Sci. Phila. Part II. 52:438-440.

1542. KELLEY, A.P. 1950. Mycotrophy in Plants. Chronica Botanica Co., Waltham, Mass. Vol. 22 (N.S.) 206 pp.

1543. ----------. 1960. The root-endings of beech, maple, and dogwood as found in the Eastern USA. Folia Forest. Polon. 4:45-82.

1544. KELLICOTT, W.E. 1904. The daily periodicity of cell division and of elongation in the root of Allium. Bull. Torrey Bot. Club 31:529-550.

1545. KENNEDY, P.B., and CRAFTS, A.S. 1931. The anatomy of Convolvulus arvensis, wild morning-glory or field bindweed. Hilgardia 5:591-622.

1546. KERNER, A., and OLIVER, F.W. 1903. Forms of roots. In: The Natural History of Plants. Gresham Publishing Co., London. 1:749-777.

1547. KESSLER, K.J. 1966. Growth and development of mycorrhizae of sugar maple (Acer saccharum Marsh.). Canad. Jour. Bot. 44:1413-1425.

1548. KHARE, P. 1964. On the morphology and anatomy of Tectaria cicutaria (L.) Copel. Proc. Indian Acad. Sci. (Sect. B) 60:414-423.

1549. KIESSELBACH, T.A. 1949. The structure and reproduction of corn. Res. Bull. Nebr. Agric. Exp. Sta. No. 161. 96 pp.

1550. KIMMEL, A.M. 1936. Anatomical studies of the seedling of Hibiscus trionum. Bot. Gaz. 98:178-189.

1551. KIMURA, Y., and NAGAMACHI, T. 1935. Ueber die japanischen Arznei-drogen, ihren Anbau, ihre Einsammlung und Zubereitung etc. VII. Ueber die japanische Kalmus und Wurzel von Acorus gramineus. Jap. Jour. Bot. 11:58-67. (in Japanese)

1552. KINDERMANN, A. 1928. Haustorialstudien an Cuscuta-Arten. Planta 5:769-783.

1553. KING, E. 1930. Root-stem transition in the axis of Lycopersicum esculentum. Master's thesis. University of Chicago. 15 pp.

1554. KING, L.J. 1966. The growth and development of weeds. In: Weeds of the World. Interscience, New York. Chapter 7.

1555. KINGSLEY, M. 1911. On the anomalous splitting of the rhizome and root of Delphinium scaposum. Bull. Torrey Bot. Club 38:307-317.

1556. KISELEVA, N.S. and SHELUKHIN, N.V. 1969. Atlas on the Anatomy of Plants. "Vysheishaia Shkola," Minsk. 287 pp. (in Russian)

1557. ------------. 1971. Anatomy and Morphology of Plants. "Vysheishaia Shkola," Minsk. 318 pp. (in Russian)

1558. KISSER, J. 1925. Ueber das verhalten von Wurzeln in feuchter Luft. Jahrb. Wiss. Bot. 46:416-439.

1559. ---------. 1930. Untersuchungen über die bei der Keimung geschälter Leguminosensamen auftreteden Wurzel- und Hypokotylkrümmungen. Beitr. Biol. Pflanz. 18:161-184.

1560. KLATT, A. 1909. Über die Entstehung von Seitenwurzeln an gekrümmten Wurzeln. Berich. Deutsch. Bot. Gesel. 27:470-476.

1561. KNIAZEVA, L.A. 1958. Formation of gutta-percha in young roots of European Euonymus. Dokl. Akad. Nauk SSSR (Bot. Sci. Sect.) 119:86-90.

1562. KNOBLOCH, I.W. 1954. Developmental anatomy of chicory. The root. Phytomorph. 4:47-54.

1563. KNUTH, R. 1924. Dioscoreaceae. In: Das Pflanzenreich. (ed.) Engler, A. Verlag von Wilhelm Engelmann, Leipzig. 87:13-15.

1564. KNY, L. 1901. Ueber den Einfluss von Zug und Druck auf die Richtung der Scheidewände in sich theilenden Pflanzellen. Jahrb. Wiss. Bot. 37:55-98.

1565. ------. 1906. Cuscuta trifolii Babington. Bot. Wandt. 10:461-466.

1566. ------. 1908. Über das Dickenwachstum des Holzkörpers der Wurzeln in seiner Beziehung zur Lotlinie. Berich. Deutsch. Bot. Gesel. 26:19-50.

1567. ------. 1911. Scheitelwachstum der Phanerogamenwurzel. Bot. Wandt. Abt. 13. No. 13. pp. 557-563.

1568. KOCH, L. 1911-1912. Pharmacognostischer Atlas. Band 1. Die Rinden, Hölzer und Rhizome. 147 pp.; Band 2. Die Wurzeln, Knollen, Zwiebeln und Kräuter. 183 pp. Gebrüder Borntraeger, Leipzig.

1569. KOERNICKE, M. 1908. Über Rindenwurzeln tropischer Loranthaceen. Naturw. Runds. 23:552-553.

1570. ------------. 1908. Über die Rindenwurzeln tropischer Loranthaceen. Mitt. Gesel. Deutsch. Naturf. Aerzte 80:186-187.

1571. KÖHLER, W.R. (1901-1902) 1903. Über die plastichen und anatomischen Veränderungen bei Keimwurzeln und Luftwurzeln, hervorgerufen durch partielle, mechanische Hemmungen. Naturf. Gesel. 28/29:59-105.

1572. KOJIMA, H. 1928. On the relation between cell-division and elongation in the root of Vicia faba. Jour. Dept. Agric. Kyushu Imp. Univ. 2:75-91.

1573. KOKKONEN, P. 1931. Untersuchungen über die Wurzeln der Getreide-pflanzen. Acta Forest. Fenn. 37:5-144.

1574. -----------. 1931. Roots and root systems of cocksfoot (Dactylis glomerata). Maatal. Aikak. 3:33-57.

1575. KOLESNIKOV, V. 1971. The Root System of Fruit Plants. MIR Publishers, Moscow. 267 pp. (Russian translation by I. Aksenova)

1576. KONAR, R.N. 1963. Anatomical studies on Indian pines with special reference to Pinus roxburghii Sar. Phytomorph. 13:388-402.

1577. ----------., and OBEROI, Y.P. 1970. Anatomical studies on Podocarpus gracilior. Phytomorph. 19:122-133.

1578. KONDRATYEVA-MELVILLE, E.A. 1959. Some anatomical peculiarities of seedling structure of Quercus robur L. Vest. Lenin. Univ. 1:42-47.

1579. ------------------------. 1963. The structure of the embryo and the seedling of Acer platanoides L. Bot. Zhur. 48:199-210.

1580. KORMANIK, P.P., and BROWN, C.L. 1967. Root buds and the development of root suckers in sweetgum. For. Sci. 13:338-345.

1581. KORSMO, E. 1954. Anatomy of Weeds. Kirstes Boktrykkeri, Oslo. 413 pp.

1582. KORTA, J. 1962. Aegopodium podagraria L. I. Analyse anatomique. Acta Biol. Cracov. (Ser. Bot.) 5:63-76.

1583. KOSTLER, J.N., BRÜCKNER, E., and BIBELRIETHER, H. 1968. Die Wurzeln der Waldbaume. Paul Parey, Hamburg. 284 pp.

1584. KOZLOWSKI, T.T. 1971. Growth and Development of Trees. Vol. 1. Seed germination, ontogeny, and shoot growth. Academic Press, New York. 443 pp.

1585. --------------. 1971. Growth and Development of Trees. Vol. 2. Cambial growth, root growth, and reproductive growth. Academic Press, New York. 514 pp.

1586. KOZMA, A. 1967. The histological characteristics of phellum develop-ment in alfalfa (Medicago sativa L.). Bot. Közlem. 54:129-136. (in Hungarian with English summary)

107

1587. KRAEMER, H. 1910. A Textbook of Botany and Pharmacognosy. 4th ed. J.B.Lippincott Co., Philadelphia and London. 888 pp.

1588. ----------. 1910. The histology of the rhizome and roots of Phlox ovata L. Amer. Jour. Pharm. 82:470-475.

1589. ----------. 1914. Applied and Economic Botany. 2nd ed. John Wiley and Sons, New York. 806 pp.

1590. KRAFT, M.M. 1948. Contribution à l'étude des processus de lignification. Bull. Soc. Vaud. Sci. Nat. 64:101-115.

1591. ----------. 1949. Étude histologique de quelques racines aeriennes d'Orchidées. Bull. Soc. Vaud. Sci. Nat. 64:201-211.

1592. KRASILJNIKOVA, A.I., YESYREVA, V.I., and POROSHINA, M.P. 1967. Some data on the structure of roots and tubercles in Alopecurus pratensis L. (Gramineae). Bot. Zhur. 52:686-689.

1593. KRASOVSKAIA, I.V. 1928. Research review on the morphology and physiology of roots. Trudy Prikl. Bot. Genet. Selek. 18:1-121.

1594. KRAUS, C. 1903. Untersuchungen zu den physiologischen grundlagen der Pflanzenkultur. I. Die Wachstumweise der Beta-rüben. Naturw. Zeitschr. Forst-Landw. 1:220-236.

1595. KRAUSS, B.H. 1949. Anatomy of the vegetative organs of the pineapple, Ananas comosus (L.) Merr. III. The root and the cork. Bot. Gaz. 110:550-587.

1596. KRENNER, J.A. 1958. The natural history of the sunflower broomrape (Orobanche cumana Wallr.). Acta Bot. 4:113-144.

1597. KROLL, G.H. 1912. Kritische Studie über die Verwertbarkeit der Wurzelhaubentypen für die Entwicklungsgeschichte. Beih. Bot. Zentralb. 28:134-158.

1598. KROEMER, K. 1903. Wurzelhaut, Hypodermis und Endodermis der Angiospermenwurzel. Biblio. Bot. 12:1-159.

1599. ----------. 1905 (1906). 2. Über die Anatomie der Rubenwurzel. Berich. Königl. Lehr. Wein, - Obst-Garten. Geisen. Rheim pp. 207-222.

1600. ----------. 1906 (1907). Über die Bewurzelung der Rebe. Berich. Königl. Lehr. Wein, - Obst-Garten. Geisen. Rheim pp. 182-201.

1601. KRONENBERG, H.G. 1966. Aerial roots of orchids. Orchideeën (N.S.) 28:119-120. (in Dutch)

1602. KUBÍKOVÁ, J. 1967. Contribution to the classification of root systems of woody plants. Preslia 39:236-243.

1603. ------------. 1968. Contributions to the exodermis in the rootlets of _Fraxinus excelsior_ L. Biol. Plant. 10:455-461.

1604. KUBITZKI, K., and BORCHERT, R. 1964. Morphologischen Studien an _Isoëtes triquetra_ A. Braun und Bemerkungen über das Verhältnis der Gattung _Stylites_ E. Amstutz zur Gattung _Isoëtes_ L. Berich. Deutsch. Bot. Gesel. 77:227-233.

1605. KUDELKA, W. 1926. Die Pfeferminze (_Mentha piperita_). Botanische Studium. Kosmos 51:139-176.

1606. KÜHL, R. 1933. Vergleichend Entwicklungsgeschichtliche Untersuchungen an der Insectivore _Nepenthes_. Beih. Bot. Centralb. 51: 311-334.

1607. KUIJT, J. 1955. Dwarf mistletoes. Bot. Rev. 21:569-626.

1608. --------. 1960. Morphological aspects of parasitism in the dwarf mistletoes (_Arceuthobium_). Univ. Calif. Publ. Bot. 30:337-435.

1609. --------. 1961. Notes on the anatomy of the genus _Oryctanthus_ (Loranthaceae). Canad. Jour. Bot. 39:1809-1816.

1610. --------. 1963. On the ecology and parasitism of the Costa Rican tree mistletoe, _Gaiadendron punctatum_ (Ruiz and Pavon) G. Don. Canad. Jour. Bot. 41:927-938.

1611. --------. 1964. Critical observations on the parasites of New World mistletoes. Canad. Jour. Bot. 42:1243-1278.

1612. --------. 1965. The anatomy of haustoria and related organs of _Gaiadendron_ (Loranthaceae). Canad. Jour. Bot. 43:687-694.

1613. --------. 1965. On the nature and action of the Santalalean haustorium, as exemplified by _Phthirusa_ and _Antidaphne_ (Loranthaceae). Acta Bot. Neerl. 14:278-307.

1614. --------. 1966. Parasitism in _Pholisma_ (Lennoaceae). I. External morphology of subterranean organs. Amer. Jour. Bot. 53:82-86.

1615. --------. 1967. Parasitism in _Pholisma_ (Lennoaceae). II. Anatomical aspects. Canad. Jour. Bot. 45:1155-1162.

1616. --------. 1967. On the structure and origin of the seedling of _Psittacanthus schiedeanus_ (Loranthaceae). Canad. Jour. Bot. 45:1497-1506.

1617. --------. 1969. The Biology of Parasitic Flowering Plants. University of California Press, Berkeley and Los Angeles. 246 pp.

1618. ----------., and DOBBINS, D.R. 1971. Phloem in the haustorium of Castilleja (Scrophulariaceae). Canad. Jour. Bot. 49:1735-1736.

1619. KUKKONEN, I. 1967. Vegetative anatomy of Uncinia (Cyperaceae). Annals Bot. (N.S.) 31:523-544.

1620. KUKLINA, L.A. 1964. Anatomical-morphological structure of corn. Trudȳ Sverdl. Selsk. Inst. 11:199-209.

1621. KULKARNI, A.R., and MUDGAL, P.V. 1969 (1970). Contribution to the anatomy of Pedaliaceae. I. Anatomy of Sesamum laciniatum Klein. Jour. Shivaji Univ. 2/3:123-139.

1622. KUMAZAWA, M. 1930. Studies on the structure of Japanese species of Ranunculus. Jour. Fac. Sci. Imp. Univ. Tokyo (Sect. 3, Botany) 2:297-343.

1623. ------------. 1930. Morphology and biology of Glaucidium palmatum Sieb. et Zucc. with notes on affinities to the allied genera Hydrastis, Podophyllum and Diphylleia. Jour. Fac. Sci. Imp. Univ. Tokyo (Sect. 3, Botany) 2:346-380.

1624. ------------. 1937. Developmental history of the abnormal structure in the geophilous organ of Aconitum. Bot. Mag. (Tokyo) 51:914-925.

1625. ------------. 1937. Ranzania japonica (Berberidaceae). Its morphology, biology and systematic affinities. Jap. Jour. Bot. 9:55-70.

1626. ------------. 1956. Morphology and development of the sinker in Pecteilis radiata (Orchidac.). Bot. Mag. (Tokyo) 69:455-461.

1627. ------------. 1958. The sinker of Platanthera and Perularia. Its morphology and development. Phytomorph. 8:137-145.

1628. ------------. 1958. Vascular connection of the axillary shoot and the adventitious root with their mother axis. Vascular anatomy in maize. VI. Bot. Mag. (Tokyo) 71:70-76. (in Japanese)

1629. ------------. 1961. Studies on the vascular course in maize plant. Phytomorph. 11:128-139.

1630. KUNKEL, G. 1965. Der Standort: Kompetenzfaktor in der Stelwurzel-bildung. Biol. Zentralb. 84:641-651.

1631. KURSCHAT, L. 1931. Untersuchungen über den Strangverlauf in den Wurzeln einiger Liliifloren. Beih. Bot. Centralb. 48:435-450.

1632. KUSANO, S. 1901. On the parasitism of Buckleya quadriala B. et H. (Santalaceae). Bot. Mag. (Tokyo) 15:42-46.

1633. ---------. 1902. Studies on the parasitism of <u>Buckleya quadriala</u>
B. et H., a santalaceous parasite, and on the structure of its
haustorium. Jour. Coll. Sci. Imp. Univ. Tokyo 17(Art. 10):1-42.

1634. ---------. 1903. Notes on <u>Aeginetia indica</u>, Linn. Bot. Mag.
(Tokyo) 17:81-83.

1635. ---------. 1908. On the parasitism of <u>Siphonostegia</u> (Rhinantheae).
Bull. Coll. Agric. Tokyo Imp. Univ. 8:51-57.

1636. ---------. 1909. Further studies on <u>Aeginetia indica</u>. Beih. Bot.
Centralb. 24:286-300.

1637. ---------. 1911. On the root-cotton, a fibrous cork tissue of a
tropical plant. Jour. Coll. Agric. Imp. Univ. Tokyo 4:67-82.

1638. KÜSTER, E. 1925. <u>Pathologische Pflanzenanatomie</u>. 3rd ed. Gustav
Fischer, Jena. 558 pp.

1639. KUTSCHERA, L. 1960. <u>Wurzelatlas: Mitteleuropäischer Ackerunkrauter
und Kulturpflanzen</u>. DLG-Verlags-GMBH, Frankfurt. 574 pp.

1640. KUTTELWASCHER, H. 1964. Entwicklungsanatomie und Vitalfärbe-Studien
an Luftwurzeln einiger tropischer Orchideen. Sitz. Öster. Akad.
Wiss. (Math.- Nat. Kl.) 173:441-483.

1641. -----------------. 1965. Entwicklungsanatomie Untersuchungen an
Orchideen-Luftwurzeln. Berich. Deutsch.Bot. Gesel. 78:307-313.

1642. LABANAUSKAS, C.K., and JAKOBS, J.A. 1957. Cork formation in taproots
and crowns of alfalfa. Agron. Jour. 49:95-97.

1643. LACHMANN, P. 1907. Origine et développement des racines et des
radicelles du <u>Ceratopteris thalictroides</u>. Rev. Gén. Bot. 19:523-556.

1644. LAING, E.V. 1923. Tree roots: their action and development. Trans.
Roy. Scot. Arbor. Soc. 37:6-21.

1645. ----------. 1932. Studies on tree roots. Bull. For. Comm. No. 13.
pp. 3-73.

1646. LAITAKARI, E. 1929. The root system of pine (<u>Pinus sylvestris</u>).
A morphological investigation. Acta Forest. Fenn. 33:1-380. (in
Finnish with English summary)

1647. ------------. 1935. The root system of birch (<u>Betula verrucosa</u> and
<u>odorata</u>). Acta Forest. Fenn. 41:5-216. (in Finnish with English
summary)

1648. LAKSHMINARAYANA, S., and VENTKATESWARLU, V. 1950. Occurrence of
velamen in <u>Eulophia graminea</u>. Sci. Cult. 5:327-328.

1649. LAMONT, B. 1972. The morphology and anatomy of proteoid roots in the genus Hakea. Austral. Jour. Bot. 20:155-174.

1650. ---------. 1972. "Proteoid" roots in the legume Viminaria juncea. Search 3:90-91.

1651. LANG, F.X. 1901. Untersuchungen über Morphologie, Anatomie und Samenentwicklung von Polypompholyx und Byblis gigantea. Flora 88:149-206.

1652. LANG, W.H. 1915. Studies in the morphology of Isoëtes. I. The general morphology of the stock of Isoëtes lacustris. Mem. Proc. Manchester Lit. Phil. Soc. 59:1-28.

1653. ---------. 1915. Studies in the morphology and anatomy of the Ophioglossaceae. III. On the anatomy and branching of the rhizome of Helminthostachys zeylanica. Annals Bot. 29:1-54.

1654. LANGLET, O. 1927. Zur Kenntnis der polysomatischen Zellkerne im Wurzelmeristem. Svensk. Bot. Tidskr. 21:397-442.

1655. LAROCHE, J. 1965. Tracheogénèse dans la racine de la première pousse transitoire chez Equisetum arvense L. Rev. Gén. Bot. 72:615-620.

1656. LARRIVAL, M.T. 1954. Étude des plantules de Bupleurum rotundifolium L., B. fruticosum L. and B. ranunculoides L. Bull. Soc. Hist. Nat. Toulouse 89:8-18.

1657. LA RUE, C.D. 1934. Root grafting in trees. Amer. Jour. Bot. 21:121-126.

1658. LAYKHOUVKIN, A.G., and PETROVA, L.R. 1968. Structural properties of leaves and roots of some lodging and nonlodging varieties of Oryza sativa L. Bot. Zhur. 53:1209-1218.

1659. LEAVITT, R.G. 1901. Predetermined root-hair cells in Azolla and other plants. Science 13:1030-1031.

1660. ------------. 1902. The root-hairs, cap, and sheath of Azolla. Bot. Gaz. 34:414-419.

1661. ------------. 1904. Trichomes of the root in vascular cryptogams and angiosperms. Proc. Boston Soc. Nat. Hist. 31:273-313.

1662. LEBEDENKO, L.A. 1959 (1960). The ontogeny of the wood of the roots and stems of several representatives of Fagales. Dokl. Akad. Nauk SSSR (Bot. Sci. Sect.) 127:193-195.

1663. --------------. 1959. The formation of wood in roots and stems of eastern oak (Quercus macranthera F. et M.). Nauch. Dokl. Vÿss. Shkolÿ, (Biol. Nauki) 2:126-131. (in Russian)

1664. ---------------. 1961. Certain regular patterns in the ontogenesis of the wood in the roots and trunk of <u>Castanea sativa</u>. Moskov. Obshch. Ispyt. Prirody Biul. (Biol. Ser.) 66:66-71.

1665. ---------------. 1962. Comparative anatomical analysis of the mature wood of roots and stems of some woody plants. Akad. Nauk SSSR, Otd., Trudȳ, Inst. Lesa Drev. 51:124-134.

1666. LE CLERG, E.L., and DURRELL, L.W. 1928. Vascular structure and plugging of alfalfa roots. Bull. Colo. State Univ. Agric. Exp. Sta. No. 339. 19 pp.

1667. LECOMPTE, H. 1915. Les tubercule des Balanophoracées. Bull. Soc. Bot. France 62:216-225.

1668. LEDOUX, M. 1909. Sur les variations morphologiques et anatomiques de quelques racines consécutives aux lésions mecaniques. Rev. Gén. Bot. 21:225-240.

1669. LEE, C.L., and CHANG, H.Y. 1958. Morphological studies of <u>Sagittaria sinensis</u>. I. The anatomy of roots. Acta Bot. Sinica 7:71-86. (in Chinese with English summary)

1670. --------. 1958. A note on the structure of <u>Eichhornea crassipes</u>. Acta Bot. Sinica 7:162-164. (in Chinese with English summary)

1671. LEE, E. 1912. Observations on the seedling anatomy of certain Sympetalae. I. Tubiflorae. Annals Bot. 26:727-746.

1672. -------. 1914. Seedling anatomy of certain Sympetalae. II. Compositae. Annals Bot. 28:303-329.

1673. LEECH. J.H., III. 1960. The growth and development of the primary root of <u>Zea</u> <u>mays</u>. Diss. Absts. 21(6):1346.

1674. ------------------, MOLLENHAUER, H. H., and WHALEY, W.G. 1963. Ultra-structural changes in the root apex. Symposia, Society for Experimental Biology No. XVII. Cell differentiation. pp. 74-84.

1675. LEEMANN, A.C. 1927. Contribution à l'étude de l'<u>Asarum europaeum</u> L. avec une étude particulière sur le développement des cellules secretices. Bull. Soc. Bot. Genève (Sér. 2) 19:92-173.

1676. LEGAULT, A. 1908. Recherches anatomiques sur l'appareil végétatif des Geraniacées. Comp. Rend. Hebd. Séances Acad. Sci. 147:382-384.

1677. LEHMANN, C. 1926. Untersuchungen über die Anatomie der Kartoffel-knolle, unter besonderer Berücksichtigung des Dickenwachstums und der Zellgrösse. Planta 2:87-131.

1678. LEHMANN-BAERTS, M. 1967. Les plantules de Gnetum africanum: Organography and anatomy. La Cellule 66:331-342.

1679. LEISER, A.T. 1968. A mucilaginous root sheath in Ericaceae. Amer. Jour. Bot. 55:391-398.

1680. LE MERRE, J., and NÈGRE, R. 1960. Contribution à l'étude morphologique et anatomique de Ferula communis L. Bull. Soc. Hist. Nat. Afrique Nord 51:24-49.

1681. LEMESLE, R. 1947. La constitution anatomique du bois secondaire chez les Ipécacuanhas vrais. Rev. Gén. Bot. 54:138-152.

1682. ----------. 1947. La constitution anatomique du bois secondaire homogène des Ipécacuanhas. Comp. Rend. Hebd. Séances Acad. Sci. 224:144-145.

1683. ----------, GUYOT, M. 1965. Particularités de la structure normale dans le tige et racine chez l'Azorella trifurcata Pers. Comp. Rend. Hebd. Séances Acad. Sci. 261:509-512.

1684. LEMLI, J. 1957. Botanik der Rauwolfia unter besonderer Berücksichtigung der medezinisch wichtigen Arten. Planta Med. 5:135-144.

1685. LENGYEL, G. 1907. Az európai Corispermum és Camphorosma fajok anatomiája. Növény. Közlem. 6:103-129. (in Hungarian)

1686. LENHART, D.Y. 1934. Initial root development of Longleaf pine. Jour. Forestry 32:459-461.

1687. LENOIR, M. 1913. Sur le début de la differenciation vasculaire dans la plantule des Veronica. Comp. Rend. Hebd. Séances Acad. Sci. 156:1084-1085.

1688. ---------. 1920. Évolution du tissue vasculaire chez quelques plantules des Dicotylédones. Ann. Sci. Nat., Bot. (Sér. 10) 2:1-123.

1689. LENZ, F. 1911. Über der durchbruch der Seitenwurzeln. Beitr. Biol. Pflanz. 10:235-264.

1690. LESHEM, B. 1970. Resting roots of Pinus halapensis: structure, function, and reaction to water stress. Bot. Gaz. 131:99-104.

1691. LETOUZEY, R. 1969. Manual de Botanique Forestière. Afrique Tropicale. Vol. I. Botanique générale. Centre Technique Forestier Tropical. Nogents/Marne. Imprimerie Jouve, Paris. 189 pp.

1692. LEWINSKY, E. 1924. Vergleichende Anatomie der Wurzeln und Rhizome einiger pharmakognostich wichtiger Solanacee. Bot. Archiv 6:313-333.

114

1693. LEWIS, F.J. 1900. Formation of an irregular endodermis in the roots of Ruscus. Annals Bot. 14:157-159.

1694. LEWIS, R.F. 1963. A comparative study of the root epidermis within the Gramineae. Master's thesis. Long Island University, New York. 92 pp.

1695. LEZENIUS, E. 1910. Vergleichende pharmakognostische Untersuchung der chinensischen Wurzel Tang-Kui, des aus ihr zubereiteten Fluid-Extraktes und Eumenols Merck. Pharm. Zent. Deutsch. 51:221-233.

1696. LIEBAU, O. 1914. Beiträge zur Anatomie und Morphologie der Mangrove-Pflanzen, inbesondere ihre Wurzelsystems. Beitr. Biol. Pflanz. 12:181-213.

1697. LIEBIG, J. 1931. Ergänzungen zur Entwicklungsgeschichte von Isoetes lacustris L. Flora 125:321-358.

1698. LIERMANN, K. 1926. Beiträge zur vergleichenden Anatomie der Wurzeln einiger pharmazeutisch verwendeter Umbelliferen. Inaugural-Dissertation. Universität Basel. R. Noske, Borna-Leipzig. 99 pp.

1699. LIESE, J. 1924. Beiträge zur Anatomie und Physiologie der Wurzel-holzes der Waldbäume. Berich. Deutsch. Bot. Gesel. 42:(91)-(97).

1700. --------. 1926. Beitrag zur Kenntnis der Wurzelsystems der Kiefer (Pinus sylvestris). Zeitschr. Forst-Jagd. 58:179-181.

1701. LIFE, A.C. 1901. The tuber-like rootlets of Cycas revoluta. Bot. Gaz. 31:265-271.

1702. LIHNELL, D. 1930. Zur Kenntnis der Anatomie von Strychnos tieute. Svensk Bot. Tidskr. 24:26-32.

1703. ----------. 1939. Untersuchungen über die Mykorrhizen und die Wurzel-pilze von Juniperus communis. Symb. Bot. Upsal. 3:9-143.

1704. LINDEMUTH, K. 1924. Beiträge zur Biologie von Vicia hirsuta Koch. und ihre Bedeutung als landwirtschaftliches Unkraut. Bot. Archiv 7:195-251.

1705. LINDINGER, L. 1906. Zur Anatomie und Biologie der Monokotylenwurzel. Beih. Bot. Centralb. 19:321-358.

1706. ------------. 1907. Über den morphologischen Wert der an Wurzeln entstehenden Knollen einiger Dioscorea-Arten. Beih. Bot. Centralb. 21:311-324.

1707. ------------. 1908. Die bewurzelungsverhältnisse grosse Monokotylen-formen und ihre bedeutung für den Gärtner. Gartenflora 57:281-291; 308-318; 367-368.

1708. -----------. 1908. Die sekundären Adventivewurzeln von <u>Dracaena</u> und der morphologische Wert der Stigmarien. Jahrb. Hamburg. Wiss. Anst. 26:59-88.

1709. -----------. 1908. Die struktur von <u>Aloë dichotoma</u> L. mit anschliessenden Betrachtungen. Beih. Bot. Centralb. 24:211-253.

1710. LINDNER, R.C. 1939. Effects of indoleacetic and naphthylacetic acids on development of buds and roots in horseradish. Bot. Gaz. 100:500-527.

1711. LINK, C.B. 1941. An anatomical study of the seedling of <u>Chrysanthemum morifolium</u> Bailey. Abst. Doct. Diss. Ohio State Univ. No. 34. pp. 333-339.

1712. LINK, H.K. 1966. Anatomy and physiology of the daffodil. Amer. Hort. Mag. 45:77-91.

1713. LINSBAUER, K. 1907. Über Wachstum und geotropismus der Aroideen-Luftwurzeln. Flora 97:267-298.

1714. -----------. 1930. Die epidermis. In: <u>Handbuch der Pflanzenanatomie.</u> (ed.) Linsbauer, K. Abt. 1. Teil 2: Histologie. Gebrüder Borntraeger, Berlin. 4:1-283.

1715. LIPSCOMB, H.A. 1969. An anatomical and morphological study of <u>Phryma leptostachya</u> L. with possible systematic applications. Diss. Absts. 29(12):4549-B.

1716. LLOYD, F.E. 1911. Guayule (<u>Parthenium argentatum</u> Gray), a rubber plant of the Chihuahuan desert. Publ. Carn. Inst. Wash. No. 139. 213 pp.

1717. ----------. 1942. <u>The Carnivorous Plants</u>. Chronica Botanica Co., Waltham, Mass. 352 pp.

1718. LLOYD, J.U., and LLOYD, C.G. 1930. Drugs and medicines of North America. Bull. Lloyd Lib. Bot. Pharm. Mat. Med. No. 29. Reproduction Series, No. 9. pp. 5-184.

1719. ---------- and -----------. 1931. Drugs and medicines of North America. Bull. Lloyd Lib. Bot. Pharm. Mat. Med. No. 30. Reproduction Series, No. 9, Part 2. pp. 187-299.

1720. LOBSTEIN, J.E., and GRUMBACH, J. 1932. Étude botanique, chimique et pharmacodynamique de la racine de <u>Stemona tuberosa</u> (drigue vermifuge sino annamite). Bull. Sci. Pharm. 39:26-34.

1721. --------------., and WEILL, A. 1932. La racine de palmier nain, falsification de la salsepareille. Bull. Sci. Pharm. 39:657-663.

1722. LODEWICK, J.E. 1928. Seasonal activity of the cambium in some Northeastern trees. Bull. New York State Coll. For., Univ. Tech. Publ. No. 23. 1:1-87.

1723. LÖFFLER, B. 1923. Beiträge zur Entwicklungsgeschichte der weiblichen Blüte, der Beere und des ersten Saugorgans der Mistel (Viscum album L.). Thar. Forst. Jahrb. 74:49-62.

1724. LONGNECKER, W.M. 1941. Root-stem transition of Opuntia engelmannii Salm-Dyck. Doctoral dissertation. University of Chicago. 9 pp.

1725. LOPRIORE, G. 1904. Verbänderung infolge des Köpfens. Berich. Deutsch. Bot. Gesel. 22:304-312.

1726. -----------. 1904. Kuntslich erzeugte Verbänderung bei Phaseolus multiflorus. Berich. Deutsch. Bot. Gesel. 22:394-396.

1727. -----------., and CONIGLIO, G. 1904. La fasciazione delle radici in rapporto ad azioni traumatiche. Atti Accad. Gioenia Sci. Nat. Catania (Mem. VII). Ser. 4. 17:1-56.

1728. -----------. 1907. Über bandförmige Wurzeln. Abh. Kaiser. Leop. Carol. Deutsch. Akad. Nat. 88:1-114.

1729. -----------. 1908. Zwillingswurzeln. In: Wiesner-Festschrift. (ed.) Linsbauer, K. C. Konegen, Wien pp. 535-547.

1730. -----------. 1908. Homo- und Antitrope in der Bildung von Seiten-wurzeln. Berich. Deutsch. Bot. Gesel. 26:299-312.

1731. -----------. 1922. Stele tubulari di radici nastriformi della Vicia faba. Atti Accad. Gioenia Sci. Nat. Catania (Mem. 16). Ser. 5. 13:1-22.

1732. -----------. 1923. Omo-e anitropa nelle formazione di radici laterali. Studia Mendeliana (Brünae) pp. 364-383.

1733. LOTOVA, L.I., and LYARSKAYA, R.P. 1959. Certain anatomical features of the concrescence of the roots of Cedrus deodara and C. atlantica. Nauch. Dokl. Vyss. Shkoly. (Biol. Nauk) 4:99-104.

1734. LOVEJOY, B. 1913. The anatomy of Gaertneria deltoidea Torr. Kansas Univ. Sci. Bull. 7:277-288.

1735. LOYAL, D.S., and GREWAL, R.K. 1967. Some observations on the morpho-logy and anatomy of Salvinia with particular reference to S. auriculata Aubl. and S. natans All. Res. Bull. Panjab Univ. (N.S.) Science 18:13-28.

117

1736. LUCHININA, A.K. 1967. The structure of seeds and seedlings of licorice species Glycyrrhiza glabra L. and G. uralensis Fisch. Bot. Zhur. 52:1267-1276.

1737. LUCKAN, L. 1917. Ecological morphology of Abutilon theophrasti. Kansas Univ. Sci. Bull. 10:219-228.

1738. LUDWIG, F. 1909. Lemnaceae. In: Lebensgeschichte der Blutenpflanzen Mitteleuropas. (eds.) von Kirchner, O., Loew, E., and Schröter, C. Eugen Ulmer, Stuttgart. 1:57-80.

1739. LUHAN, M. 1951. Zur Wurzelanatomie unserer Alpenpflanzen. I. Primulaceae. Sitz. Öster. Akad. Wiss. (Math.- Nat. Kl.) Abt. 1. 160:481-507.

1740. --------. 1952. Zur Wurzelanatomie unserer Alpenpflanzen. II. Saxifragaceae und Rosaceae. Sitz. Öster. Akad. Wiss. (Math.- Nat. Kl.) Abt. 1. 161:199-237.

1741. --------. 1954. Zur Wurzelanatomie unserer Alpenpflanzen. III. Gentianaceae. Sitz. Öster. Akad. Wiss. (Math.- Nat. Kl.) Abt. 1. 163:89-107.

1742. --------. 1955. Das Abschlusgewebe der Wurzeln unserer Alpenpflanzen. Berich. Deutsch. Bot. Gesel. 68:87-92.

1743. --------. 1959. Zur Wurzelanatomie unserer Alpenpflanzen. IV. Compositae. Sitz. Öster. Akad. Wiss. (Math.- Nat. Kl.) Abt. 1. 168:607-641.

1744. --------. 1959. Neues zur Anatomie der Alpenpflanzen. Berich. Deutsch. Bot. Gesel. 72:262-267.

1745. --------. 1960. The root anatomy of Ranunculus hybridus and Ranunculus brevifolius. Berich. Deutsch. Bot. Gesel. 74:477-480.

1746. LUKÁCSOVICS, E.J. 1966. Some problems in root histogenesis. A. Root apex development. Acta Biol. Debric. (Ser. 2) Biology 4:82-98. (in Hungarian)

1747. LUNAN, M. 1937. Anatomical investigations on Ligusticum scoticum. Trans. Bot. Soc. Edinb. 32:353-363.

1748. LUND, S., and ROSTRUP, E. 1901. The creeping thistle, Cirsium arvense. A monograph. Det Kong. Danske Vid. Sel. Med. (Nat.- Math.) Ser. 6. 10:153-313. (in Danish with French summary)

1749. LUNDGÅRDH, H. 1914. Das Wachstum des Vegetationspunktes. Berich. Deutsch. Bot. Gesel. 32:77-83.

1750. ------------. 1950. Lärobok i Vaxtfysiologi med Växanatomi. Svenska
Bokförlaget Bonniers, Stockholm. 703 pp. (in Swedish)

1751. LUTHRA, J.C. 1921. Striga as root parasite of sugarcane. Agric.
Jour. India 16:519-523.

1752. LUTZ, L. 1910. Sur le mode de formation de la gomme adragante. Bull.
Soc. Bot. France 57:250-257.

1753. LUXOVÁ, M. 1965. Zemědělská Botanika. Vol. 1. Anatomie a morfologie
rostlin. Státní Zemědělske Nakladatelství v Praze 303 pp. (in
Czechoslovakian)

1754. ---------. 1970. Structure and conductivity of the corn root system.
Biol. Plant (Praha) 12:47-57.

1755. LYFORD, W.H., and WILSON, B.F. 1964. Development of the root system
of Acer rubrum. Harvard For. Papers No. 10. Harvard University
Petersham, Mass. pp. 1-17.

1756. LYLE, E.A. 1937. The structure and development of the seedling of
Beta vulgaris. Master's thesis. University of Chicago. 16 pp.

1757. LYNCH, J.J. 1915. Sambucus pubens var. xanthocarpa. Amer. Midl.
Nat. 4:177-181.

1758. LYON, H.L. 1904. The embryogeny of Ginkgo. Minn. Bot. Studies
(Ser. 3) 3:275-290.

1759. LYUBARSKI, E.L. 1965. On the ecological anatomy of the root system
of certain aquatic plants with long roots. Bot. Zhur. 50:119-123.

1760. MACDOUGAL, D.T., and LLOYD, F.E. 1900. The roots and mycorhizas of
some of the Monotropaceae. Bull. New York Bot. Gard. 1:419-429.

1761. ---------------., and CANNON, W.A. 1910. The condition of parasitism
in plants. Publ. Carn. Inst. Wash. No. 129. pp. 1-24.

1762. ---------------. 1914. The determinative action of environic factors
upon Neobeckia aquatica Greene. Flora 106:264-280.

1763. MACFARLANE, J.M. 1911. Cephalotaceae. In: Das Pflanzenreich. (ed.)
Engler, A. Verlag W. Engelmann, Leipzig. 47:IV(116). pp. 1-15.

1764. MACLEOD, D.G. 1961. Some anatomical and physiological observations
on two species of Cuscuta. Trans. Proc. Bot. Soc. Edinb. 39:
302-315.

1765. MAEGDEFRAU, K., and WUTZ, A. 1962. Die Pneumatophoren von Symphonia.
Veröffent. Geobot. Inst. Rübel Zurich 37:183-187.

1766. MAGENC, P. 1914. Les Badamiers. Étude pharmacographique du genre Terminalia L. Ann. Inst. Colon. Marseille (Sér. 3) 2:1-110.

1767. MAGER, H. 1907. Beiträge zur Anatomie der physiologischen Scheiden der Pteridophyten. Biblio. Bot. Heft 66. pp. 1-68. (Same title published by E. Schweitzerbartsche verlagsbuchhandlung - E. Nägele - Stuttgart. 1907. 58 pp.)

1768. --------. 1913. Versuch über die Metakutisierung. Flora 106:42-50.

1769. --------. 1932. Beiträge zur Kenntnis der primaren Wurzelrinde. Planta 16:666-708.

1770. MAGNUS, W. 1911. Mycorrhiza. Bot. Wandt. Abt. 13. pp. 525-551.

1771. MAHABLE, T.S., and UDWADIA, N.N. 1960. Studies on palms: IV. Anatomy of palm roots. Proc. Nat. Inst. Sci. India (Sect. B) Biol. Sci. 26:73-104.

1772. MAHESHWARI, P., and SINGH, B. 1934. The morphology of Ophioglossum fibrosum Schum. Jour. Indian Bot. Soc. 13:103-123.

1773. --------------., and VASIL, V. 1961. Gnetum. Botanical monograph No. 1. Council of Scientific and Industrial Research, New Delhi. 142 pp.

1774. MAHEU, J., and CHARTIER, J. 1927. Faux Ipéca et origine botanique de l'Ipéca strié mineur Manettia ignita Schum. Bull. Sci. Pharm. 34: 347-357.

1775. ---------- and -----------. 1927. Étude de l'herbe dite "a la femme battue" (Tamus communis L.), cause de dermites. Bull. Sci. Pharm. 34:566-575.

1776. MAHLBERG, P.G. 1960. Embryogeny and histogenesis in Nerium oleander L. I. Organization of primary meristematic tissues. Phytomorph. 10: 118-131.

1777. MAHLSTEDE, J.P., and WATSON, D.P. 1952. An anatomical study of adventitious root development in stems of Vaccinium corymbosum. Bot. Gaz. 113:279-285.

1778. MAIGE, A., and GATIN, C.L. 1902. Sur la structure des racines tuberculeuses du Thrincia tuberosa. Comp. Rend. Hebd. Séances Acad. Sci. 134:302-303.

1779. MAILLEFER, A. 1919. L'anatomie de l'Equisetum arvense. Atti Soc. Elvet. Sci. Nat. 100:110.

1780. ------------. 1921. Sur la présence d'une assise dans la racine d'Acorus calamus. Bull. Soc. Vaud. Sci. Nat. 53:77-79.

1781. MAJUMDAR, G.P. 1929. Secondary growth in thickness in the roots of Amorphophallus campanulatus Bl. Abstract. Proc. 16th Indian Sci. Congr. Assn. (Madras) 16:229-230.

1782. ------------. 1932. Heter-archic roots in Enhydra fluctuans Lour. Jour. Indian Bot. Soc. 11:225-227.

1783. MALCOM, W.M., II. 1963. The root-parasitism of Castilleja coccinea. Diss. Absts. 24(2):482-483.

1784. MALINOWSKI, E. 1966. Anatomia Roslin. Panstwowe Wydawn. Naukone, Warszawa. 579 pp. (in Polish)

1785. MALLORY, T.E., CHIANG, S., CUTTER, E.G., and GIFFORD, E.M., Jr. 1970. Sequence and pattern of lateral root formation in five selected species. Amer. Jour. Bot. 57:800-809.

1786. MANEVAL, W.E. 1914. The development of Magnolia and Liriodendron, including a discussion of the primitiveness of the Magnoliaceae. Bot. Gaz. 57:1-31.

1787. MANGENOT, G. 1947. Recherches sur l'organisation d'une Balanophoracée, Thonningia coccinea Vahl. Rev. Gén. Bot. 54:201-244; 271-294.

1788. MANGIN, L. 1910. Introduction à l'étude des mycorhizes des arbres forestiers. Nov. Arch. Mus. Hist. Nat. (Sér. B) 2:245-276.

1789. MANI, A.P. 1962. Air-space tissue in Cyperus. Sci. Cult. 28:39-40.

1790. ---------. 1963. Size-structure correlation in the vascular system of roots in Cyperus. Sci. Cult. 29:357-358.

1791. MANN, A.G. 1921. Observations on the interruption of the endodermis in a secondarily thickened root of Dracaena fruticosa Koch. Proc. Roy. Soc. Edinb. 41:50-59.

1792. MANN, C.E.T., and BALL, E. 1926. Studies in the root and shoot growth of the strawberry. Jour. Pomol. Hort. Sci. 5:149-169.

1793. ------------- and -------. 1927. Studies in the root and shoot development of the strawberry. Jour. Pomol. Hort. Sci. 6:81-112.

1794. -----------. 1930. Studies in the root and shoot growth of the strawberry. V. The origin, development, and function of the roots of the cultivated strawberry (Fragaria virginiana x chiloensis). Annals Bot. 44:55-86.

1795. MANN, L.K. 1952. Anatomy of the garlic bulb and factors affecting bulb development. Hilgardia 21:195-251.

1796. MANNICH, C., and BRANDT, W. 1904. Über die Wurzel von Heteropteris pauciflora Juss., eine neue Verfälschung der Ipecacuanha. Arb. Pharm. Inst. Univ. Berlin 2:132-136.

1797. MANSFIELD, W. 1916. Histology of Medicinal Plants. J. Wiley and Sons, New York. 305 pp.

1798. MANTEUFFEL, A. 1926. Untersuchungen über den Bau und Verlauf der Leitbundel in Cucurbita pepo. Beih. Bot. Centralb. (Abt. 1) 43:153-166.

1799. MARIN, M.A., and VAZQUEZ, M.M. 1959. Estudio Morfologico y Anatomico de los Centenos Españoles. Instituto Nacional de Investigaciones Agronomicas Centro de Cerealicultura, Madrid. 223 pp.

1800. MARINOVIĆ, R., and TATIĆ, B. 1965. Plant Morphology. Univerzitet u Beogradu. Naucna Knjiga, Beograd. 324 pp. (in Yugoslavian)

1801. MARKGRAF, F., and ENDRESS, P. 1967. Die Keimpflanze von Berardia. Viert. Nat. Gesel. Zürich. 112:209-222.

1802. MARKLE, M.S. 1917. Root systems of certain desert plants. Bot. Gaz. 64:177-205.

1803. MAROTI, M. 1950. Die Entwicklung der Wurzel von Stratioites aloides L. Acta Biol. Acad. Sci. Hung. 1:363-370.

1804. MARTENS, P., and LEHMANN-BAERTS, M. 1967. Le sucoir de l'embryon et de la plantule chez Gnetum africanum. La Cellule 66:345-359.

1805. ----------. 1971. Les Gnétophytes. In: Handbuch der Pflanzenanatomie. (eds.) Zimmermann, W., Carlquist, S., Ozenda, P.G., and Wulff, H.D. Spezieller Teil. Band 12. Teil 2. Gebrüder Borntraeger, Berlin. 295 pp.

1806. MARTIN, F.W., and ORTIZ, S. 1963. Origin and anatomy of tubers of Dioscorea floribunda and D. spiculiflora. Bot. Gaz. 124:416-421.

1807. MARTIN, J.N. 1930 (1931). The anatomy of the crowns and roots of the annual and biennial white sweet clover compared. Proc. Iowa Acad. Sci. 37:209-210.

1808. -----------., and HERSHEY, A.L. 1935. The ontogeny of the maize plant and the early differentiation of stem and root structures and their morphological relationships. Iowa State Coll. Jour. Sci. 9:489-503.

1809. MARTIN, J.T., and JUNIPER, B.E. 1970. The Cuticles of Plants. St. Martin's Press, New York. 347 pp.

1810. MARTIN-SANS, E., and LHÉRITIER, G. 1934. La racine de salsepareille indigène. Bull. Sci. Pharm. 41:524-533.

1811. MASUI, K. 1926. A study of the mycorrhiza of <u>Abies firma</u> S. et Z., with special reference to its mycorrhizal fungus, <u>Cantharellus floccosus</u> Schw. Mem. Coll. Sci., Kyoto Imp. Univ. (Ser. B) 2:15-84.

1812. --------. 1926. The compound mycorrhiza of <u>Quercus paucidentata</u> Fr. Mem. Coll. Sci., Kyoto Imp. Univ. (Ser. B) 2:85-92.

1813. --------. 1926. A study of the ectotrophic mycorrhiza of <u>Alnus</u>. Mem. Coll. Sci., Kyoto Imp. Univ. (Ser. B) 2:189-209.

1814. --------. 1927. A study of the ectotrophic mycorrhizas of woody plants. Mem. Coll. Sci., Kyoto Imp. Univ. (Ser. B) 3:149-279.

1815. MATHIESEN, FR. J. 1921. The structure and biology of arctic flowering plants. 11. Primulaceae. Medd. Gronl. 37:361-507.

1816. ----------------. 1921. The structure and biology of arctic flowering plants. 15. Scrophulariaceae. Medd. Gronl. 37:361-507.

1817. MATTE, H. 1908. Sur le développement morphologique et anatomique de germinations des Cycadacées. Mém. Soc. Linné. Norm. 23:37-94.

1818. --------. 1909. Sur la structure de l'embryon et des germinations du genre <u>Zamia</u> L. Bull. Soc. Sci. Méd. Ouest 18:125-145.

1819. MATTHIESEN, F. 1908. Beiträge zur Kenntnis der Podostemaceen. Biblio. Bot. Heft 68 pp. 1-55.

1820. MATTHYSEN, J.O. 1912. Cytologische und anatomische Untersuchungen an <u>Beta vulgaris</u>, nebst einigen Bemerkungen über die Enzyme dieser Pflanze. Zeitschr. Ver. Deutsch. Zucker-Ind. 62:137-151.

1821. MATTICK, F. 1967. Die Wurzelscheibe von <u>Phytolacca dioica</u> und andere Beispiele von Schreibenwurzeln. Bot. Jahrb. 86:38-49.

1822. MAUGINI, E. 1947. Ricerche morfologiche e anatomiche su <u>Mimosa pudica</u> L. I. Lo sviluppo del sistems condutiore nelle plantule in stadi anteriori alla comparasa del cambio. Nuovo Giorn. Bot. Ital. (N.S.) 54:505-519.

1823. -----------. 1947. Contributo all'anatomica dell'apparato vegetativo di <u>Guizotia abyssinica</u> Cass. Nuovo Giorn. Bot. Ital. (N.S.) 54:568-581.

1824. MAYBROOK, A.C. 1917. On the haustoria of <u>Pedicularis vulgaris</u> Tournef. Annals Bot. 31:499-511.

1825. MAYDRECH, B.A.R. 1968. Structure of the seed and developmental anatomy of the seedling of <u>Prunus persica</u> 'OKINAWA.' Diss. Absts. 28(11):70-B.

1826. MAYR, S.I. 1928. Über die Keimung und erste Entwicklung der Riemenmistel (Loranthus europaeus Jacq.). Sitz. Akad. Wiss. Wien (Math.- Nat. Kl.) 137:345-362.

1827. MCARTHUR, I.C.S., and STEEVES, T.A. 1969. On the occurrence of root thorns on a Central American palm. Canad. Jour. Bot. 47: 1377-1382.

1828. MCCALL, M.A. 1934. Developmental anatomy and homologies in wheat. Jour. Agric. Res. 48:283-321.

1829. MCCANN, C. 1935. Some observations on Nymphaea pubescens Willd. Jour. Bombay Nat. Hist. Soc. 37:895-901.

1830. MCCARTHY, J. 1962. The form and development of knee roots in Mitragyna stipulosa. Phytomorph. 12:20-30.

1831. MCCLATCHIE, I. 1917. Observations on the root-system of Impatiens royalei, Walp. Jour. Linn. Soc. (Bot.) 43:493-516.

1832. MCCLURE, F.A. 1966. The Bamboos. Harvard University Press, Cambridge. 347 pp.

1833. MCCORMICK, F. 1916. Notes on the anatomy of the young tuber of Ipomoea batatas Lam. Bot. Gaz. 61:388-398.

1834. MCCRONE, G. 1915. Histology of Malva rotundifolia. Kansas Univ. Sci. Bull. 9:261-267.

1835. MCCUSKER, A. 1971. Knee roots in Avicennia marina (Forsk.) Vierh. Annals Bot. 35:707-712.

1836. MCDOUGALL, W.B. 1914. On the mycorrhizas of forest trees. Amer. Jour. Bot. 1:51-74.

1837. --------------. 1921. Thick-walled root hairs of Gleditsia and related genera. Amer. Jour. Bot. 8:171-175.

1838. MCGAHAN, M.W. 1961. Studies on the seed of banana. II. The anatomy and morphology of the seedling of Musa balbisiana. Amer. Jour. Bot. 48:630-637.

1839. MCKENZIE, D.W. 1961. A note on vessel structure in the xylem of roots of M.XVI apple rootstock. Jour. Hort. Sci. 36:138.

1840. MCLEAN, R.C., and IVIMEY-COOK, W.R. 1951. Textbook of Theoretical Botany. Vol. 1. Longmans, Green and Co., London. 1069 pp.

1841. -------------- and ------------------. 1952. Textbook of Practical Botany. Longmans, Green and Co., London. 476 pp.

124

1842. MCLUCKIE, J. 1922. Studies in symbiosis. I. The mycorhiza of
Dipodium punctatum R.Br. Proc. Linn. Soc. New So. Wales 47:
293-310.

1843. -----------. 1922. Studies in symbiosis. II. The apogeotropic
roots of Macrozamia spiralis and their physiological significance.
Proc. Linn. Soc. New So. Wales 47:319-328.

1844. -----------. 1922. A contribution to the parasitism of Notothixos
incanus (Oliv.) var. subaureus. Proc. Linn. Soc. New So. Wales
47:571-580.

1845. -----------. 1924. Studies in parasitism. I. A contribution to
the physiology of genus Cassytha. Proc. Linn. Soc. New So. Wales
49(Part I):55-78.

1846. MCMARTIN, A. 1933. The root system of Acanthus. Trans. Proc. Bot.
Soc. Edinb. 31:272-297.

1847. MCMINN, R.G. 1963. Characteristics of Douglas-fir root systems.
Canad. Jour. Bot. 41:105-122.

1848. MCMURRY, E.B., and FISK, E.L. 1936. Vascular anatomy of the seedling
of Melilotus alba. Bot. Gaz. 98:121-134.

1849. MCNAIR, J.B. 1921. The morphology and anatomy of Rhus diversiloba.
Amer. Jour. Bot. 8:179-191.

1850. MCNICOLL, I.S. 1929. Notes on strand plants. III. Salsola kali L.
Trans. Proc. Bot. Soc. Edinb. 30:147-156.

1851. MCPHERSON, D.C. 1939. Cortical air spaces in the roots of Zea mays L.
New Phytol. 38:190-202.

1852. MCQUADE, H.A., CUMBIE, B.G., and SHELDRICK, D.L.W. 1970. The wool-
like covering of the roots of Lannea alata. I. The origin and morphology of the wooly coat. Amer. Jour. Bot. 57:1240-1244.

1853. MCQUILKIN, W.E. 1935. Root development of pitch pine, with some
comparative observations on short leaf pine. Jour. Agric. Res.
51:983-1016.

1854. MCVEIGH, I. 1934. Vegetative reproduction in Camptosaurus rhizophyllus.
Bot. Gaz. 95:503-510.

1855. MECHELKE, F. 1951. Über sporadische Polysomatie in Wurzelspitzen bei
Hordeum vulgare L. Öster. Bot. Zeitschr. 98:420-426.

1856. MEDVE, R.J. 1969. Development and morphology of the beaded rootlets,
mycorrhizae, and associated root fan structures of red maple (Acer
rubrum L.). Diss. Absts. 30(1):82-B.

1857. MEDVEDVA, R.G., and KHOVINA, L.A. 1964. Anatomical structure of Rheum wittrocki. II. Underground part of the plant. Akad. Nauk Kazak. SSSR, Izv. Ser. Biol. Nauk 1:44-52.

1858. MEDZMARIASHVILI, I.D. 1962. A contribution to the study of the root tuber in Anemone hortensis. Trudy̅ Sukhum. Bot. Sada 14:85-90.

1859. MEHRA, P.N., and SINGH, H.P. 1955. Observations on the anatomy of Alsophila glabra Hook. Sci. Cult. 21:273.

1860. ----------., and BHATNAGAR, J.K. 1958. A comparative study of 'Satawar' and its supposed botanical source. Indian Jour. Pharm. 20:33-40.

1861. ----------., and SHARMA, O.P. 1963. Anatomy of Eleocharis planta-ginea R.Br. Res. Bull. Panjab Univ. (Science) 14:289-305.

1862. ----------., and JOLLY, S.S. 1963. The identity and pharmacognosy of the adulterant of Nardostachys jatamansi DC. Planta Med. 11:8-15.

1863. ------------ and ----------. 1968. Pharmacognosy of Indian bitters. I. Gentiana kurroo Royle and Picrorhiza kurroa Royle ex Benth. Res. Bull. (Sci.) Panjab Univ. (N.S.) 19:141-156.

1864. ----------., and PURI, H.S. 1968. Pharmacognostic investigations on Radix Aconiti Heterophyllii (Ativisha) and its adulterants. Res. Bull. (Sci.) Panjab Univ. (N.S.) 19:439-449.

1865. ----------., and BHATNAGAR, J.K. 1968. Further studies on the botanical source of "Satawar." Res. Bull. (Sci.) Panjab Univ. (N.S.) 19:131-140.

1866. ----------., and SHARMA, O.P. 1969. Systematic anatomy of Cyperus L. Res. Bull. (Sci.) Panjab Univ. (N.S.) 20:119-137.

1867. ----------., and HANDA, S.S. 1969. Pharmacognosy of Saptrangi - antidiabetic drug of Indian origin. Res. Bull. (Sci.) Panjab Univ. (N.S.) 20:487-502.

1868. MEHTA, K.R. 1934. The root system of Asphodelus tenuifolius Cavan. Jour. Indian Bot. Soc. 13:271-275.

1869. MEJSTŘÍK, V. 1970. The anatomy of roots and mycorrhizae of the orchid Dendrobium cunninghamii Lindl. Biol. Plant. 12:105-109.

1870. MELCHIOR, H. 1921. Über den anatomischen Bau der Saugorgane von Viscum album L. Beitr. Allgem. Bot. 2:55-87.

1871. MELLOR, A.E. 1911. The seedling structure of Dryas octopetala. Naturalist No. 656. pp. 310-312.

126

1872. MENARINI, D. 1934. Differenziazione degli elementi legnosi in
    plantule di Leguminose. Atti Reale Accad. Naz. Lincei 19:741-747.

1873. MENON, K.P.V., DAVIS, T.A., ANANDAN, A.P., and PILLAI, N.G. 1955.
    "Aerial" roots in the coconut palm. Indian Coconut Jour. 8:79-91.

1874. ------------., and PANDALAI, K.M. 1958. The Coconut Palm: A Mono-
    graph. Indian Central Coconut Committee, Ernakulam, India. 384 pp.

1875. MENSAH, K.O.A., and JENIK, J. 1968. Root system of tropical trees.
    2. Features of the root system of Iroko (Chlorophora excelsa Benth.
    et Hook.). Preslia 40:21-27.

1876. MENZ, J. 1910. Beiträge zur vergleichenden Anatomie der Gattung
    Allium nebts einigen Bemerkungen über die anatomischen Beziehungen
    zwischen Allioideae und Amaryllidoideae. Sitz. Akad. Wiss. Wien
    (Math.- Nat. Kl.) Abt. 1. 119:475-533.

1877. -------. 1922. Osservazioni sull'anatomica degli organi vegetativi
    delle specie italiane del genere Allium (Tourn.) L. apparteneti all
    sezione "Molium" G. Don. Sassari R. Univ. Ist. Bot. Bull., Mem. 5.
    1:1-27.

1878. MENZIES, B.P. 1954. Seedling development and haustorial system of
    Loranthus micranthus Hook. f. Phytomorph. 4:397-409.

1879. ------------., and MCKEE, H.S. 1959. Root parasitism in Atkinsonia
    ligustrina (A. Cunn. ex F. Muell.) F. Muell. Proc. Linn. Soc. New
    So. Wales 84:118-127.

1880. MERICLE, L.W., and WHALEY, G.W. 1953. Cell wall structure in apical
    meristems. Bot. Gaz. 114:383-392.

1881. MESSERI, A. 1925. Ricerche sullo sviluppo del sistema vascolare in
    alcune Monocotyledoni. Nuovo Giorn. Bot. Ital. 32:317-362.

1882. ----------. 1929. L'evoluzione del sistema conduttore in Asphodelus
    microcarpus Viv. e il significato dei tipi Anemarrhena nel passagio
    dalla radice al caule. Nuovo Giorn. Bot. Ital. 36:46-56.

1883. ----------. 1930. Lo sviluppo del sistema conduttore di Zamia media
    Jacq. Nuovo Giorn. Bot. Ital. 37:461-509.

1884. ----------. 1931. Lo sviluppo del sistema conduttore di Ginkgo
    biloba L. Nuovo Giorn. Bot. Ital. 38:78-127.

1885. ----------. 1934. Differenziazione (intercalare) in fasci radicolari
    ed ipocotilari di plantule di Conifere. Atti Reale Accad. Naz.
    Lincei 19:178-185.

1886. -----------. 1935. Ricerche sulla morfologia della plantula nelle
        Coniferae. Le Pinaceae. Nuovo Giorn. Bot. Ital. (N.S.) 42:
        267-363.

1887. -----------. 1940. Ricerche sulla morfologia della plantula nelle
        Coniferae. Le Araucarieae. Nuovo Giorn. Bot. Ital. 47:119-154.

1888. METCALFE, C.R. 1931. The "aerenchyma" of Sesbania and Neptunia.
        Kew Bull., Miscl. Info. No. 3. pp. 151-154.

1889. -------------. 1931. The breathing roots of Sonneratia and Bruguiera;
        a review of the recent work by Troll and Dragendorff. Kew Bull.,
        Miscl. Info. No. 10. pp. 465-467. (see listing 2683)

1890. -------------., and CHALK, L. 1950. Anatomy of the Dicotyledons.
        2 Vols. Clarendon Press, Oxford. 1500 pp.

1891. -------------. 1960. Anatomy of the Monocotyledons. Vol. 1.
        Gramineae. Clarendon Press, Oxford. 731 pp.

1892. -------------. 1971. Anatomy of the Monocotyledons. Vol. 5.
        Cyperaceae. Clarendon Press, Oxford. 597 pp.

1893. MEYER, F.J. 1920. Das Leitungssystem von Equisetum arvense. Jahrb.
        Wiss. Bot. 59:263-286.

1894. -----------. 1922. Beiträge zur Kenntnis der Leitbündelanatomie.
        II. Das Leitbundelsystem von Polypodium vulgare. Bot. Archiv 2:
        278-280.

1895. -----------. 1925. Untersuchungen über den Strangverlauf in den
        radialen Leitbundeln der Wurzeln. Jahrb. Wiss. Bot. 65:88-97.

1896. -----------. 1930. Die Leitbundel der Radices filipendulae (Wurzel-
        anschwellungen) von Maranta kerchoveana Morr. Berich. Deutsch.
        Bot. Gesel. 48:51-57.

1897. -----------. 1933. Beiträge zur vergleichenden Anatomie der Typhaceen.
        Beih. Bot. Centralb. 51:335-376.

1898. MEYER, K.I. 1962. Morphology of Corn; Morphology, Anatomy, and
        Embryology. Izd-vo Moskovskogo Universiteta, Moskva. 295 pp. (in
        Russian)

1899. MEYER, L. 1940. Zur Anatomie und Entwicklungsgeschichte der
        Bromeliaceenwurzeln. Planta 31:492-522.

1900. MEYER, R.E., HORTON, H.L., HAAS, R.H., ROBISON, E.D., and RILEY, T.E.
        1971. Morphology and anatomy of Honey Mesquite. U.S.D.A. Tech.
        Bull. No. 1423. 186 pp.

1901. MEYER, R.R., and WALKER, E.R. 1931. The vegetative anatomy of
Impatiens pallida. Trans. Amer. Micro. Soc. 50:1-19.

1902. MIA, A.J. 1959. A comprehensive investigation on the morphology of
Rauwolfia vomitoria Afzl. Doctoral thesis. North Carolina State
University, Raleigh. 202 pp.

1903. MICHEELS, H. 1900. Contribution à l'étude anatomique des organes
végétatifs et floraux chez Carludovica plicata Kl. Arch. Inst.
Bot. Univ. Liège 2:3-86.

1904. MIJAYI, Y., and KOKUBU, T. 1964. Studies on the young tubers of
sweet potatoes Ipomoea batatas Lam., with special reference to
varietal difference in the capacity of accumulating starch. I-III.
Bull. Fac. Agric. Kagosh. Univ. 15:101-125. (in Japanese)

1905. MIKAILOVSKAIA, I.S. 1960. Age changes in the anatomical structure
of the taproot in some species of Leguminosae. Bot. Zhur. 45:
875-880. (in Russian)

1906. ------------------. 1961. Development of the internal structure of
grassy leguminous plant roots. In: Soveschanie po Morfogenezu
Rastenii, 1st, Moscow, 1959. Morfogenez Rastenii. Trudy. (Moskva)
Izd-vo Moskovskogo Universiteta. 1:639-(643). (in Russian)

1907. ------------------. 1965. Change of anatomical structure of roots
of some grains in connection with their ageing. Moskov. Obshch.
Prirody, Biul. (Otd. Biol.). 70:81-86. (in Russian)

1908. ------------------. 1967. On the age of individuals in the population
of Lotus corniculatus L. Bot. Zhur. 52:379-385.

1909. MIKSCHE, J.P. 1960. Developmental vegetative morphology of Glycine
max. Diss. Absts. 20(7):2514.

1910. ------------., and GREENWOOD, M. 1966. Quiescent centre of the primary
root of Glycine max. New Phytol. 65:1-4.

1911. MILANEZ, F.R. 1936. E structura secundaria das raizes de Rhipsalis.
Rodriguésia 2:165-175.

1912. MILHOFER, J. 1933. Botanische und chemische Untersuchungen an
Artemisia caerulescens L. Acta Bot. Inst. Bot. Univ. Zagreb.
8:1-96.

1913. MILLER, H.A., and WETMORE, R.H. 1945. Studies in the developmental
anatomy of Phlox drummondii Hosk. II. The seedling. Amer. Jour.
Bot. 32:628-634.

1914. MILLER, R.H. 1958. Morphology of Humulus lupulus. I. Developmental
anatomy of the primary root. Amer. Jour. Bot. 45:418-431.

1915. -----------. 1958. Morphology of _Humulus lupulus_. II. Secondary growth in the root and seedling vascularization. Amer. Jour. Bot. 46:269-277.

1916. MILLER, W.L. 1952. Root-stem transition in _Mirabilis nyctaginea_ Michx. Proc. So. Dakota Acad. Sci. 31:21-23.

1917. MILLNER, M.E. 1934. Anatomy of _Silene vulgaris_ and _Silene maritima_ as related to ecological and genetical problems. I. Root structure. New Phytol. 33:77-95.

1918. MIRANDE, M. 1901. Recherches physiologiques et anatomiques sur les Cuscutacées. Bull. Sci. France Belgique 34:1-280.

1919. -----------. 1905. Recherches sur l'développement et l'anatomie des Cassythacées. Ann. Sci. Nat., Bot. (Sér. 9) 2:181-285.

1920. MIROV, N.T. 1967. _The Genus Pinus_. University of California Press, Berkeley. 602 pp.

1921. MISRA, R.C. 1966. Morphological studies in _Plantago_. I. Germination of seed and seedling anatomy. Jour. Indian Bot. Soc. 45:116-121.

1922. MITRA, G.C., and BAL, S.N. 1947. Pharmacognosy of the root bark of _Abroma augusta_ Linn. Indian Jour. Pharm. 9:120-125.

1923. MODESTOV, A.P. 1915. Les racines des plantes herbacées. I.N. Kushneret, Moscow. No. 1. pp. 1-138. (in Russian with French summaries)

1924. MOENS, P. 1963. The vascularization of the embryo and the seedling of _Coffea canephora_ Pierre. La Cellule 64:69-126.

1925. MOGENSON, H.L. 1965. A contribution to the embryogeny and developmental anatomy of the seedling in _Quercus_ L. Diss. Absts. 26(6): 3000-3001.

1926. --------------. 1967. A contribution to the developmental anatomy of the root of _Quercus_ L.

1927. MOIR, M.A. 1931. Notes on salt-marsh plants. I. _Glaux maritima_ Linn. Trans. Proc. Bot. Soc. Edinb. 30:304-312.

1928. MOKEEVA, E.A. 1940. A biological and anatomical investigation of Alfalfa; _Medicago sativa_ L. Tashkent, Sel'khozgiz UzRR 123 pp. (in Russian)

1929. MOLISCH, H. 1917. Über das Treiben von Wurzeln. Sitz. Akad. Wiss. Wien Abt. 1. 126:3-12.

130

1930. ----------. 1938. Rote Wurzelspitzen. Berich. Deutsch. Bot. Gesel. 46:311-317.

1931. ----------., and HÖFLER, K. 1961. Anatomie der Pflanze. Gustav Fischer, Jena. 172 pp.

1932. MOLLENHAUER, H.H. 1967. A comparison of root cap cells of epiphytic, terrestrial and aquatic plants. Amer. Jour. Bot. 54:1249-1259.

1933. MOLLIARD, M. 1914. Effets de la compression sur la structure des racines. Rev. Gén. Bot. 25:529-538.

1934. MONDO, A. 1939. Ricerche anatomiche su Mesembryanthemum acinaciforme L. Nuovo Giorn. Bot. Ital. 46:227-258.

1935. MONTESANTOS, N. 1912. Morphologische und biologische Untersuchungen über einige Hydrocharideen. Flora 105:1-32.

1936. MOORE, L.B. 1940. The structure and life-history of the root parasite Dactylanthus taylori Hook. f. New Zeal. Jour. Sci. Tech. 21(4B): 206-224.

1937. MOREAU, C., and MOREAU, M. 1958. Lignification et réactions aux traumatismes de la racine du palmier à huile en pèpiniéres. Oléagineux 13:735-741.

1938. MORELLE, E. 1904. Histologie comparee des Gelsimiées et Spigeliées. Doctoral théses. Université de Paris. Imprimerie Commercienne, Commercy. 162 pp.

1939. MORI, T. 1959. Histological analysis of the thickness of rice roots. Proc. Crop Sci. Soc. Japan 28:12-14.

1940. MORISSET, CH. 1964. Structure et genèse du velamen dans les racines aériennes d'une orchidée epiphyte: le Dendrobium nobile Lindl. Rev. Gén. Bot. 71:529-591.

1941. MORRISON, T.M. 1953. Comparative histology of secondary xylem in buried and exposed roots of dicotyledonous trees. Phytomorph. 3:427-430.

1942. MORSTATT, H. 1903. Beiträge zur Kenntnis der Resedaceen. Inaugural- Dissertation. Ruprecht-Karls-Universität zu Heidelberg. A. Bunz Erben, Stuttgart. 65 pp.

1943. MOSELEY, M.F., JR. 1956. The anatomy of the water storage organ of Ceiba parvifolia. Trop. Woods No. 104. pp. 61-79.

1944. MOSER, F. 1932. Wurzelstudien an einigen Pflanzen aus der Solfatare des Grossen Salak in West-Java. Öster. Bot. Zeitschr. 81:5-30.

1945. MOSS, E.H. 1924. Fasciated roots in Caltha palustris L. Annals Bot. 38:789-791.

1946. ----------. 1926. Parasitism in the genus Comandra. New Phytol. 25:264-276.

1947. ----------., and GORHAM, A.L. 1953. Interxylary cork and fission of stems and roots. Phytomorph. 3:285-294.

1948. MOZINGO, H.N. 1951. Changes in the three dimensional shape during growth and division of living epidermal cells in the apical meristem of Phleum pratense roots. Amer. Jour. Bot. 38:495-511.

1949. MULAY, B.N., and DESHPANDE, B.D. 1953. Presence of velamen in the earth roots of some species of the genus Asparagus (Linn.). Rajas. Univ. Studies (Biol. Sci. and Med.) 1:58-73.

1950. ----------., PANIKKAR, T.K.B., and PRASAD, M.K. 1956. Collateral vascular bundles in some orchid roots. Proc. Rajas. Acad. Sci. 6:70-73.

1951. ----------., and PRASAD, M.K. 1956. On the structure and development of velamen in the roots of some terrestrial orchids. Proc. Indian Sci. Congr. (Agra) Part III. p.246.

1952. ----------., and PANIKKAR, K.B. 1956. Origin, development and structure of velamen in the roots of some species of terrestrial orchids. Proc. Rajas. Acad. Sci. 6:31-48.

1953. ----------., and SARIN, Y.K. 1956. On the histological structure of Satyrium nepalense Don. and other terrestrial orchids. Proc. Indian Sci. Congr. (Agra) Part III. p. 246.

1954. ----------., and SALUJHA, S.K. 1957. Histology of some desert grasses. Jour. Indian Bot. Soc. 36:106-111.

1955. ----------., DESHPANDE, B.D., and WILLIAMS, H.B. 1958. Study of velamen in some epiphytic and terrestrial orchids. Jour. Indian Bot. Soc. 37:123-127.

1956. ----------., and DESHPANDE, B.D. 1959. Velamen in terrestrial monocots. I. Ontogeny and morphology of velamen in the Liliaceae. Jour. Indian Bot. Soc. 38:383-390.

1957. MULLAN, D.P. 1932. Observations on the biology and physiological anatomy of some Indian halophytes. Jour. Indian Bot. Soc. 11:103-118; 285-302.

1958. ----------. 1933. Observations on the biology and physiological anatomy of some Indian halophytes. Jour. Indian Bot. Soc. 12:165-182; 235-253.

1959. ------------. 1936. On the anatomy of _Ipomoea aquatica_ Forsk., with special reference to the development of aerenchyma as a result of injury. Jour. Indian Bot. Soc. 15:39-50.

1960. ------------. 1940. The root structure of _Chlorophytum tuberosum_ Baker. Jour. Indian Bot. Soc. 19:235-239.

1961. ------------. 1940. The anatomy of _Spinifex squarrosus_ Linn. with special reference to the morphology of the leaf-blade. Jour. Indian Bot. Soc. 18:241-246.

1962. ------------. 1940. On the alterations in the tissues of _Melochia corchorifolia_ Linn. and _Corchorus capsularis_ Linn. on a change of environment. Jour. Indian Bot. Soc. 41:877-890.

1963. ------------. 1945. The biology and anatomy of _Scirpus grossus_ Linn. fil. Jour. Bombay Nat. Hist. Soc. 45:402-407.

1964. MULLENDORE, N. 1935. Anatomy of the seedling of _Asparagus officinalis_. Bot. Gaz. 97:356-375.

1965. --------------. 1948. Seedling anatomy of _Brachypodium distachyum_. Bot. Gaz. 109:341-348.

1966. MULLER, C.H. 1946. Root development and ecological relations of Guayule. U.S.D.A. Tech. Bull. No. 923. 114 pp.

1967. MÜLLER, H. 1906. Über die Metakutisierung der Wurzelspitze und über die verkorkten Scheiden in den Achsen der Monokotyledonen. Bot. Zeit. 64:53-84.

1968. MÜLLER, J. 1900. Über die Anatomie der Assimilationswurzeln von _Taeniophyllum zollingeri_. Sitz. Kaiser. Akad. Wiss. (Math.- Nat. Kl.) 90:667-683.

1969. MÜLLER, R. 1908. Radix Senegae und ihre Subsitutionen. Pharm. Praxis 7:309-325.

1970. MURÉN, A. 1934. Untersuchungen über die Wurzeln der Wasserpflanzen. Ann. Bot. Soc. Zoo.- Bot., Fenn. Vanamo No. 8. 5:1-56.

1971. MUROMTSEV, I.A. 1948. Apple tree root hairs. Dokl. Vses. Akad. Selsk. Nauk Imeni V.I. Lenina (Moskva) No. 7. pp. 23-25. (in Russian)

1972. --------------. 1962. Root hairs of fruit plants. Evolution of the root and root cap. Trudy Plodoov. Inst. Michurina 14:25-41. (in Russian)

1973. MURRAY, B.E. 1957. The ontogeny of adventitious stems on roots of creeping-rooted alfalfa. Canad. Jour. Bot. 35:463-475.

1974. MURRAY, J. 1934. Anatomical features in the roots of the genus
Primula. Trans. Proc. Bot. Soc. Edinb. 31:323-326.

1975. MURRAY, S.A. 1971. A comparative analysis of cell morphology in
maturing tissues of roots from shocked and unshocked bulbs of
Allium cepa L. Amer. Jour. Bot. 58:119-123.

1976. MURTY, Y.S. 1960. Studies in the order Piperales. I. Contribution
to the study of vegetative anatomy of some species of Peperomia.
Phytomorph. 10:50-59.

1977. MYERS, G.A., BEASLEY, C.A., and DERSCHEID, L.A. 1964. Anatomical
studies of Euphorbia esula L. Weeds 12:291-295.

1978. MYLIUS, G. 1913. Das Polyderm. Biblio. Bot. 18:1-114.

1979. NAFDAY, U.R. 1966. Studies in the tubiflorae of Nagpur. VII.
Anatomy of Dopatrium junceum Buch.- Ham. ex Benth. Proc. Indian
Acad. Sci. (Sect. B) 64:32-37.

1980. NAGAO, T. 1965. On the vascular patterns of the roots in tobacco
seedlings. Sci. Papers Cent. Res. Inst., Japan. Govt. Monop. Bur.
No. 107. pp. 235-238. (in Japanese)

1981. NAKAJI, M. 1927. Morphological observations of the root of Lygodium
japonicum Sw. Bot. Mag. (Tokyo) 41:452-455. (in Japanese)

1982. NAPP-ZINN, K. 1953. Studien zur Anatomie einiger Luftwurzeln. Öster.
Bot. Zeitschr. 100:322-330.

1983. NARAYANA, H.S. 1963. A contribution to the structure of root nodule
in Cyamopsis tetragonoloba Taub. Jour. Indian Bot. Soc. 42:273-280.

1984. NARAYANASWAMY, S. 1950. Occurrence of velamen and mycorrhiza in the
subterranean roots of the orchid Spiranthes australis Lindl. Curr.
Sci. 19:250-251.

1985. NARDUCCI, A. 1957. Morphology, anatomy and developmental cycle of
Asphodelus racemosus L. var. aestivus Brot. Nuovo Giorn. Bot. Ital.
64:319-346.

1986. NASSONOV, V.A. 1935. The anatomy of the pistachio, Pistacia vera L.
Trudy Prikl. Bot. Genet. Selek. (Ser. 3) 4:113-134. (in Russian
with English summary)

1987. NAST, C.G. 1941. The embryogeny and seedling morphology of Juglans
regia L. Lilloa 6:163-205.

1988. NATIVIDADE, J.V. 1940. Sôbre a existência de raízes aéreas latente
na Oliveira (Olea europaea L.) e os novos aspectos do problema da
propagação vegetativa. Agron. Lusitana 2:25-73. (in Portuguese
with English and French summaries)

1989. ---------------. 1941. O significado ecológico e fisiológico do
sistema radiculare aéreo da Oliveira (Olea europaea L.) e da
Alfarrobeira (Ceratonia siliqua L.). Agron. Lusitana 3:85-91.

1990. ---------------. 1944. Um tipo anómolo de periderme lenticular
em raízes de Quercus suber L. Bol. Soc. Broteriana 19(Ser. 2a,
Part I):311-322. (in Portuguese)

1991. NEGER, F.W. 1922. Beiträge des Baues und der wirkungsweise der
Lentizellen. II. Berich. Deutsch. Bot. Gesel. 40:306-313.

1992. NELSON, P.E., and WILHELM, S. 1952. Strawberry root anatomy with
special reference to black root rot. Phytopath. 42:517.

1993. ------------- and ----------. 1957. Some anatomic aspects of the
strawberry root. Hilgardia 26:631-642.

1994. NĚMEC, B. 1901. Ueber die Wahrnehmung des Schwerkraftreizes bein
den Pflanzen. Jahrb. Wiss. Bot. 36:80-178.

1995. --------. 1908. Einige Regenerationsversuche an Taraxacum-Wurzeln.
In: Wiesner-Festschrift. (ed.) Linsbauer, K. C. Konegen, Wien.
pp. 207-215.

1996. --------. 1966. Die Frequenz der Zellteilungen in der Wurzelspitze
von Phaseolus vulgaris nanus. Biol. Plant. (Praha) 8:5-9.

1997. NEUBAUER, H.F. 1961. Bau und Entwicklung der Luftwurzel von Vanilla
planifolia Andr. Beitr. Biol. Pflanz. 36:239-253.

1998. NEUBER, E. 1904. Beiträge zur vergleichenden Anatomie der Wurzeln
vorwiegend offizineller Pflanzen, mit besonderer Berücksichtigung
der Heterorhizie der Dicotylen. Doctoral dissertation. Adolf
Stenzel, vorm. Brehmer und Minuth., Breslau. 70 pp.

1999. NEUMANN, O. 1939. Über die Bildung der Wurzelhaube bei Juglans,
Mimosa und Lupinus. Planta 30:1-20.

2000. NEWMAN, I.V. 1965. Pattern in the meristems of vascular plants.
III. Pursuing the patterns in the apical meristem where no cell is
a permanent cell. Jour. Linn. Soc. (Bot.) 59:185-214.

2001. NICHIPOROVICH, A.A., and IVANITKAJA, E.F. 1944. On the relation
between leaf development and formation of laticiferous vessels in
roots of kok-saghyz and krym-saghyz. Comp. Rend. (Dokl.) Acad.
Sci. URSS 44:33-36.

2002. NICOLAS, G. 1914. Observations sur la structure des racines du
Ranunculus bullatus L. Bull. Soc. Hist. Nat. Afrique Nord 5:62-65.

2003. ----------. 1916. Remarques sur la structure des organes souterrains du Thrincia tuberosa DC. Bull. Soc. Hist. Nat. Afrique Nord 4:9-16.

2004. ----------. 1918. Effets de la compression sur la structure d'une racine de Dracaena. Bull. Soc. Hist. Nat. Afrique Nord 9:114-116.

2005. NIEUWLAND, J.A. 1910. Notes on the seedlings of bloodroot. Amer. Midl. Nat. 1:199-203.

2006. NIGHTINGALE, G.T. 1935. Effects of temperature on growth, anatomy, and metabolism of apple and peach roots. Bot. Gaz. 96:581-639.

2007. NISHIMURA, M. 1922. Comparative morphology and development of Poa pratensis, Phleum pratense and Setaria italica. Jap. Jour. Bot. 1:55-85.

2008. NIYAZOV, B.N. 1968. Some anatomical data on medicinal soapwort (Saponaria officinalis L.). Uzbek. Biol. Zhur. 12:41-44. (in Russian with Uzbekian summary)

2009. NOBÉCOURT, P. 1921. Les tubercules des Ophrydées. Bull. Soc. Bot. France 68:62-68.

2010. -------------. 1922. Étude sur les organes souterrains de quelques Ophrydées de Java. Bull. Soc. Bot. France 69:226-232.

2011. NOELLE, W. 1910. Studien zur vergleichenden Anatomie und Morphologie der Koniferenwurzeln mit Rücksicht auf die Systemtik. Bot. Zeit. 68:169-266.

2012. NOLL, F. 1900. Über den bestimmenden Einfluss von Wurzelkrümmingen auf Entstehung und Anordnung der Seitenwurzeln. Landw. Jahrb. 29: 361-426.

2013. -------. 1907. Über ein Adventiv-Wurzelsystem bei dikotylen Pflanzen. Sitz. Nat. Verein Preuss. Rhein. Westf. (Zweite halfte) Abt. A. pp. 54-57.

2014. NORDHAUSEN, M. 1907. Über Richtung und Wachsthum der Seitenwurzeln unter dem Einfluss äusserer und innerer Fakturen. Jahrb. Wiss. Bot. 44:557-634.

2015. --------------. 1912. Über kontraktile Luftwurzeln. Flora 105:101-126.

2016. NOZU, Y. 1956. Anatomical and morphological studies of Japanese species of the Ophioglossaceae. II. Rhizome and root. Jap. Jour. Bot. 15: 208-226.

2017. O'BRIEN, T.P., and MCCULLY, M.E. 1969. Plant Structure and Development: A Pictorial Approach. Macmillan Co., London. 114 pp.

2018. OCCHIONI, P. 1943. Da raiz de *Tephrosia toxicaria* Pers., e do seu
aprovietamento no combate ao *Tenthecoris bicolor* Scott. Rodriguésia
7:55-61. (in Portuguese)

2019. OGURA, H. 1958. A note on the lateral root formation of *Eichhornia
crassipes* Solms-Laubach. Ecol. Rev. 14:305-309.

2020. OGURA, T. 1957. Studies of upland rice plants. IV. On morphological
characteristics of seminal root. Proc. Crop Sci. Soc. Japan 25:
154-156. (in Japanese with English summary)

2021. OGURA, Y. 1925. On the structure of *Cyathea spinulosa* Wall. Bot.
Mag. (Tokyo) 39:1-28. (in Japanese)

2022. --------. 1927. Comparative anatomy of Japanese Cyatheaceae. Jour.
Fac. Sci. Tokyo Univ. (Sect. 3, Bot) 1:141-350.

2023. --------. 1927. On the structure of *Diplazium esculentum* (Retz.) Sw.
Bot. Mag. (Tokyo) 41:172-180.

2024. --------. 1938. Anatomy and morphology of *Oleandra wallichii* (Hk.)
Pr., with some notes on the affinities of the genus *Oleandra*.
Jap. Jour. Bot. 9:193-211.

2025. --------. 1938. Anatomie der Vegetationsorgane der Pteridophyten.
In: Handbuch der Pflanzenanatomie. (ed.) Linsbauer, K. Gebrüder
Borntraeger, Berlin. Abt. 2. Band 7. Teil 2. 476 pp.

2026. --------. 1940. New examples of aerial roots in tropical swamp plants.
Bot. Mag. (Tokyo) 54:327-337.

2027. --------. 1940. On the types of abnormal roots in mangrove and swamp
plants. Bot. Mag. (Tokyo) 54:389-404. (in Japanese with English
summary)

2028. --------. 1952. Morphology of the subterranean organs of *Erythronium
japonicum* and its allies. Phytomorph. 2:113-122.

2029. --------. 1953. Anatomy and morphology of the subterranean organs
in some Orchidaceae. Jour. Fac. Sci. Univ. Tokyo (Sect. 3, Bot.)
6:135-137.

2030. OKONKWO, S.N.C. 1966. Studies on *Striga senegalensis* Benth. I. Mode
of host-parasite union and haustorial structure. Phytomorph. 16:
453-463.

2031. --------------. 1970. Studies on *Striga senegalensis*. V. Origin
and development of buds from roots of seedlings reared in vitro.
Phytomorph. 20:144-150.

2032. OLSEN, C. 1921. The structure and biology of arctic flowering plants. 9. Cornaceae. Medd. Grönl. 37:130-150.

2033. ONOFEGHARA, F.A. 1971. Studies on the development and establishment of Tapinanthus bangwensis. Annals Bot. 35:729-743.

2034. ONOFRY, A. 1940. La cana commune (Arundo donax L.). Cremonese Librario Editore Roma XVIII. 166 pp. (in Italian)

2035. OPPENHEIMER, H.R. 1941. Root cushions, root stalagmites and similar structures. Palestine Jour. Bot. 4:11-19.

2036. ----------------. 1957. Further observations on roots penetrating into rocks and their structure. Bull. Res. Council Israel (Sect. D, Bot.) 6:18-31.

2037. OSBORN, T.G.B. 1922. Some observations on Isoetes drummondii, A.Br. Annals Bot. 36:41-54.

2038. OSTERWALDER, R. 1920. Beiträge zur Kenntnis pharmazeutisch wichtiger Gentiana-Wurzeln. Schweiz. Apoth.- Zeit. 58:201-207.

2039. OTIENO, N.C. 1967. The anatomy of common pasture grasses in Kenya. II. Digitaria scalarum Chiov. (African Couch Grass). East African Agric. For. Jour. 33:23-30.

2040. OVCHINNIKOV, N.N., and NIKOLAYESKY, V.G. 1961. Changes in the anatomical structure of maize roots depending on the site of their formation on the plant. Ukry. Bot. Zhur. 18:16-23.

2041. OZENDA, P. 1965. Recherches sur les phanérogames parasites. I. Revue des travaux recents. Phytomorph. 15:311-338.

2042. PABISCH, H. 1905. Ueber die Tuba-Wurzel (Derris elliptica Benth.). Pharm. Central Deutsch. 46:697-706.

2043. PADMANABHAN, D. 1962. The embryology of Avicennia officinalis L. IV. The seedling. Proc. Indian Acad. Sci. (Sect. B) 56:114-122.

2044. ---------------. 1968. Development from zygote to seedling in Tridax procumbens Linn. Jour. Indian Bot. Soc. 47:94-112.

2045. PAIZIEVA, S.A. 1962. The anatomy and morphology of the embryo and seedling in Cousinia umbrosa Bge., C. pseudoarctium Bornm. and Arctium leiospermum Juz. Vest. Lenin. Univ. (Ser. Biol.) No. 21, Part 4. pp. 148-153. (in Russian with English summary)

2046. PAL, B.P. 1966. Wheat. (I.C.A.R. Cereal Crop Series No. 4) Indian Council Agricultural Research, New Delhi. 370 pp.

2047. PAL, N., and PAL, S. 1962. Studies on the morphology and affinity of the Parkeriaceae. I. Morphological observations of Ceratopteris thalictroides. Bot. Gaz. 124:132-143.

2048. PALAZZO, A.M. 1929. Le lenticelle della radici del Morus. Atti Reale Accad. Sci. Toreno 64:197-207.

2049. PALIWAL, G.S., and KAVATHEKAR, A.K. 1971. Anatomy of vegetative food storage organs. I. Roots. Acta Agron. Acad. Sci. Hung. 20:261-270.

2050. PALLADIN, V.I. 1914. Pflanzenanatomie. 15th ed. (ed.) Tschulok, S. B.G.Teubner, Leipzig. 195 pp.

2051. PAMMEL, L.H. 1909. The underground organs of a few weeds. Proc. Iowa Acad. Sci. 16:31-40.

2052. PANDÉ, S.K. 1935. Notes on the anatomy of a xerophyte fern Niphobolus adnascens from the Malay peninsula. Proc. Indian Acad. Sci. (Sect. B) 1:556-564.

2053. PANIZZA, S., and DE SOUZA GROTTA, A. 1965. Contribuição ao estudo morfológico e anatômico de Solidago microglossa DC. Compositae. Rev. Fac. Farm. Bioq. Univ. São Paulo 3:27-50. (in Portuguese)

2054. ----------. 1967. Contribuição ao estudo morfológico e anatômico da Jacaranda caroba (Velloso) DC. Bignoniaceae. Rev. Fac. Farm. Bioq. Univ. São Paulo 5:93-106. (in Portuguese)

2055. PANKOW, H., and GUTTENBERG, H. VON. 1957. Vergleichenden Studien über die Entwicklung monokotyler Embryonen und Keimpflanzen. Bot. Studien Heft 7. pp. 1-39.

2056. ----------- and -------------------. 1963. The structure of the root apex of Casuarina suberosa Otto and Dietr. Öster. Bot. Zeitschr. 110:132-136.

2057. PANT, D.D. 1943. On the morphology and anatomy of the root system in Asphodelus tenuifolius Cavan. Jour. Indian Bot. Soc. 22:1-26.

2058. ----------., and MEHRA, B. 1961. Occurrence of intra-cortical roots in Bambusa. Curr. Sci. 30:308.

2059. PANTANELLI, E. 1900. Anatomia fisiologia delle Zygophyllaceae. Atti Soc. Nat. Med. (Ser. 4) 2:93-181.

2060. PAOLILLO, D.J., JR. 1963. The developmental anatomy of Isoetes. Illinois Bot. Monographs 31. University of Illinois Press, Urbana. 130 pp.

2061. PARIJA, P., and MISRA, P. 1933. The "root-thorn" of Bridelia pubescens Kurz. Jour. Indian Bot. Soc. 12:227-233.

2062. PARK, J. 1931. Notes on salt marsh plants. III. Triglochin maritimum Linn. Trans. Bot. Soc. Edinb. 30:320-325.

2063. PARKER, K.W., and SAMPSON, A.W. 1930. Influence of leafage removal on anatomical structure of roots of Stipa pulchra and Bromus hordeaceous. Plant Phys. 5:543-553.

2064. PARSA, A. 1934. Contribution à l'étude structurale de quelques dicotylédones xérophiles de l'Iran. Doctoral thèses. Université de Poitiers. Société Française d'Imprimerie et de Libraire, Poitiers. 92 pp.

2065. PARTHSARATHY, M.V. 1966. Studies on metaphloem in petioles and roots of Palmae. Diss. Absts. (Sect. B) 27(3):706B.

2066. -----------------. 1968. Observations on metaphloem in the vegetative parts of palms. Amer. Jour. Bot. 55:1140-1168.

2067. PASCHER, A. 1942. Über Wurzeldimorphismus (Körbchenwurzeln) bei Gagea. Beih. Bot. Centralb. 61:437-461.

2068. PASHKOV, G.D. 1951. Concerning the morphological character of the coleorhiza in Gramineae. Bot. Zhur. 36:597-606. (in Russian)

2069. PASSERINI, N. 1935. Di alcune caratteri differenziali nelle plantule di Cuscuta epithymum Mürr. e di Cuscuta pentagona Eng. Nuovo Giorn. Bot. Ital. (N.S.) 42:508-513.

2070. PATEL, J.S. 1938. The Coconut: A Monograph. Superintendent, Government Press. Madras, India. 313 pp.

2071. PATEL, M.B., and ROWSON, J.M. 1964. Investigations of certain Nigerian medicinal plants. Part IV. The anatomy of the root bark and stem bark of Pleiocarpa pycnantha (K. Schum.) Stapf var. tubicina (Stapf) Pichon. Planta Med. 13:11-22.

2072. ------------ and -----------. 1965. Investigations of certain Nigerian medicinal plants. Part VI. The anatomy of the root and rhizome of Hederanthera barteri (Hook. f.) Pichon. Planta Med. 13:270-279.

2073. PATEL, R.N. 1964. On the occurrence of gelatinous fibres with special reference to root wood. Jour. Inst. Wood Sci. 12:67-80.

2074. -----------. 1965. A comparison of the anatomy of the secondary xylem in roots and stems. Holzforschung 19:72-79.

2075. PAUPARDIN, C. 1962. Nature and origin of the "fibrosity" of sugar beet roots. Comp. Rend. Hebd. Séances Acad. Agric. France 48:597-601.

2076. PAVLU, J. 1924. Contribution à l'histologie expérimental de la beterave (Beta vulgaris). Publ. Biol. École Hautes Études Vet. 3(B46):173-199.

2077. PAVOLINI, A.F. 1909. La Stangeria paradoxa Th. Moor. Nuovo Giorn. Bot. Ital. 16:335-351

2078. PAYNE, M.A. 1933. Morphology and anatomy of Mollugo verticillata L. Univ. Kans. Sci. Bull. 21:399-419.

2079. PECKET, R.C. 1957. The initiation and development of lateral meristems in the pea root. I. The effect of young and mature tissues. Jour. Exp. Bot. 8:172-180.

2080. PEE-LABY, E. 1904. La Passiflore parasite sur les racines du fusain. Rev. Gén. Bot. 16(192):453-457.

2081. PEIRCE, G.J. 1905. The dissemination and germination of Arceuthobium occidentale Eng. Annals Bot. 19:99-113.

2082. PEKLO, J. 1908. Die epiphytischen Mykorrhizen nach neuen Untersuchungen. Acad. Sci. Bohême, Bull. Int'l. (Sci. Math., Nat. Med. Cl.) 13:87-107.

2083. PELOURDE, F. 1906. Recherches anatomiques sur la classification des fougères de France. Ann. Sci. Nat., Bot. (Sér. 9) 4:281-372.

2084. PENFOUND, W.T. 1934. Comparative structure of the wood in the "knees," swollen bases, and normal trunks of the tupelo gum (Nyssa aquatica L.). Amer. Jour. Bot. 21:623-631.

2085. PENZIG, O. 1901. Beiträge zur Kenntnis der Gattung Epirrhizanthes Bl. Ann. Jard. Bot. Buitenz. 17:142-170.

2086. PERCIVAL, J. 1921. The Wheat Plant. A monograph. Duckworth and Co., London. 463 pp.

2087. PERDRIGEAT, C.- A. 1900. Anatomie comparée des Polygonées et ses rapports avec la morphologie et la classification. Actes Soc. Linn. Bord. (Sér. 6) 5:1-91.

2088. PEROL, C. 1967. Sur la structure des racines de l'Hydrostachys maxima Perr. Comp. Rend. Hebd. Séances Acad. Sci. (Sér. D) 264:2459-2462.

2089. PERRAUD, J. 1934. Les caracters anatomiques du xérophytisme dans la végétation Mauritaienne. Doctoral thèses. Université d'Aix, Marseille. Bosc Frères, M. et L. Riou, Lyon. 278 pp.

2090. PERROT, E. 1902. Sur le Ksopo. Poison des Sakalaves (Menabea venenata H. Bn.). Rev. Cult. Colon. 10:105-113.

2091. ----------., and MOREL, F. 1913. Quelques remarques sur l'anatomie
des Ombellifères. Bull. Soc. Bot. France 60:99-106.

2092. PERSIDSKY, D. 1928. Polar structure in the cells of a root's growth-
zones. Bull. Jardin Bot. Kieff No. 7 - 8. pp. 52-56. (in Russian
with English conclusions)

2093. PERSINOS, G.J., and QUIMBY, M.W. 1968. Studies on Nigerian plants.
V. Comparative anatomy of Lophira lanceolata and Lophira alata.
Econ. Bot. 22:206-220.

2094. PERVOVA, Y.A. 1964. On the quantitative anatomy of the roots of
Erysimum canescens Roth and E. silvaticum M.B. depending on the
habit. Ukray. Bot. Zhur. 21:41-48. (in Russian with English
summary)

2095. PETCH, T. 1930. Buttress roots. Ann. Roy. Bot. Gard. Peradeniya
11:277-285.

2096. PETER, A. 1903. Beiträge zur Anatomie der Vegetationsorgane von
Boswellia carteri Birdw. Sitz. Kaiser Akad. Wiss. Wien (Math.- Nat.
Kl.) 112:511-534.

2097. PETERSEN, E. 1908. Zur vergleichenden Anatomie der Zentralzylinder
der Papilionaceen-Wurzel. Beih. Bot. Centralb. 24:20-44.

2098. PETERSON, R.L. 1967. Differentiation and maturation of primary tissues
in white mustard root tips. Canad. Jour. Bot. 45:319-331.

2099. --------------. 1973. Control of cambial activity in roots of turnip
(Brassica rapa). Canad. Jour. Bot. 51:475-480.

2100. PETRIE, P.S. 1938. Osservazioni sull'anatomia della Beta vulgaris L.
Indus. Sacc. Ital. 31:224-232.

2101. ------------., MAZZI, S., and STRIGOLI, P. 1960. Considerations on the
formation of adventitious roots with special reference to: Cucurbita
pepo, Nerium oleander, Menyanthes trifoliata, and Solanum lycopersicum.
Nuovo Giorn. Bot. Ital. 67:131-175.

2102. PETRY, L.C. 1914. The anatomy of Ophioglossum pendulum. Bot. Gaz.
57:169-192.

2103. PFEIFFER, H. 1922/1923. Neue Untersuchungen über abnormen Dickenwachstum einheimischer Pflanzen. I. Über das Dickenwachstum der
Wurzeln von Raphanus sativus L. prol. radicula Pers. (Radieschen)
und anderer Cruciferen. Separat-Atabdruck aus Mikrobiologie. Naturw.
Monatsh. Biol. Chem. Geogr. Unter. Band XII. Heft 2. (Ser. 3) 4 pp.

2104. ------------. 1923. Über die Entstehung der rübenförmigen Wurzelverdickungen der Dikotyledon. Aus der Heimat 36:121-124.

2105. ----------. 1926. Das abnorme Dickenwachstum. In: Handbuch der Pflanzenanatomie. (ed.) Linsbauer, K. Band 9. Gebrüder Borntraeger, Berlin. 268 pp.

2106. PFEIFFER, N.E. 1914. Morphology of Thismia americana. Bot. Gaz. 57:122-134.

2107. -------------. 1931. A morphological study of gladiolus. Contr. Boyce Thomp. Inst. Plant Res. 3:173-195.

2108. PFEIFFER, W.M. 1912. The morphology of Leitneria floridana. Bot. Gaz. 53:189-203.

2109. PFITZER, E., and KRÄNZLIN, F. 1907. Orchidaceae-Monandrae-Coleogyninae. In: Das Pflanzenreich. (ed.) Engler, A. Verlag W. Engelmann, Leipzig. p. 7.

2110. PHILIPP, M. 1923. Über die verkorkten Abschlussgewebe der Monokotylen. Biblio. Bot. Heft 92. pp. 1-27.

2111. ----------. 1923. Über die verkorkten Abschlussgewebe der Monokotylen. E. Schweitzerbart'sche Verlagsbuchhandlung, Stuttgart. 28 pp.

2112. PHILIPSON, W.R. 1937. A revision of the British species of the genus Agrostis Linn. Jour. Linn. Soc. (Bot.) 51:73-151.

2113. --------------. 1959. Some observations on root-parasitism in New Zealand. Trans. Roy. Soc. New Zeal. 87:1-3.

2114. --------------., WARD, J.M., and BUTTERFIELD, B.G. 1971. The Vascular Cambium. Chapman and Hall Ltd., London. 182 pp.

2115. PHILLIPS, W.S. 1937. Seedling anatomy of Cynara scolymus. Bot. Gaz. 98:711-724.

2116. PICKETT, F.L. 1915. A contribution to our knowledge of Arisaema triphyllum. Mem. Torrey Bot. Club 16:1-55.

2117. PIEHL, M.A. 1962. The parasitic behavior of Melampyrum lineare and a note on its seed color. Rhodora 64:15-23.

2118. ----------. 1962. The parasitic behavior of Dasistoma macrophylla. Rhodora 64:331-336.

2119. ----------. 1963. The natural history and taxonomy of the genus Comandra (Santalaceae). Diss. Absts. 23(9):3094.

2120. ----------. 1963. Mode of attachment, haustorium structure, and hosts of Pedicularis canadensis. Amer. Jour. Bot. 50:978-985.

2121. ----------. 1965. Studies of root parasitism in _Pedicularis_
       lanceolata. Mich. Bot. 4:75-81.

2122. PIENAAR, K.J. 1968. An anatomical and ontogenetic study of the
       roots of South African Liliaceae. I. Structure of the meristem-
       atic root apex and origin of the differentiation of the primary
       tissues. Jour. So. African Bot. 34:37-60. (in Dutch with
       English summary)

2123. ------------. 1968. An anatomical and ontogenetic study of the
       roots of South African Liliaceae. II. Anatomy of the adventitious
       roots at primary growth. Jour. So. African Bot. 34:91-110. (in
       Dutch with English summary)

2124. ------------. 1968. An anatomical and ontogenetic study of the
       roots of South African Liliaceae. III. The ontogeny and morphology
       of the velamen and exodermis. Jour. So. African Bot. 34:113-125.
       (in Dutch with English summary)

2125. ------------. 1968. An anatomical and ontogenetic study of the
       roots of South African Liliaceae. IV. The genesis and development
       of the contractile roots. Jour. So. African Bot. 34:203-214. (in
       Dutch with English summary)

2126. PIERPAOLI, I. 1916. Ricerche anatomiche, istologiche ed embriolo-
       giche sulla _Putoria_ _calabrica_ Pers. Annali Bot. 14:83-100.

2127. PILLAI, A., and PILLAI, S.K. 1962. Air-spaces in the root of some
       monocotyledons. Proc. Indian Acad. Sci. (Sect. B) 55:296-301.

2128. ----------. 1963. Root apical organization in gymnosperms -- some
       cycads and _Ginkgo_ _biloba_. Proc. Indian Acad. Sci. (Sect. B)
       57:211-222.

2129. ----------. 1964. Root apical organization in gymnosperms. -- Some
       conifers. Bull. Torrey Bot. Club 91:1-13.

2130. ----------. 1966. Root apical organization in gymnosperms; root
       apex of _Ephedra_ _foliata_, with a suggestion on the possible evolu-
       tionary trend of root apical structures in gymnosperms. Planta
       70:26-33.

2131. PILLAI, S.K., and SACHDEVE, S. 1960. Effect of some surgical
       excisions on the regeneration of the root apex of _Sorghum_ _vulgare_
       Pers. Curr. Sci. 29:233-234.

2132. ------------., and PILLAI, A. 1961. Root apical organization in
       monocotyledons -- Palmae. Proc. Indian Acad. Sci. (Sect. B)
       54:218-233.

2133. ----------, ------------., and SACHDEVA, S. 1961. Root apical
organization in monocotyledons -- Zingiberaceae. Proc. Indian
Acad. Sci. (Sect. B) 54:240-256.

2134. ---------- and ----------. 1961. Root apical organization in
monocotyledons -- Xyridaceae. Proc. Indian Acad. Sci. (Sect. B)
54:234-240.

2135. ---------- and ----------. 1961. Root apical organization in
monocotyledons -- Marantaceae. Proc. Indian Acad. Sci. (Sect. B)
54:302-317.

2136. ---------- and ----------. 1961. Root apical organization in
monocotyledons -- Musaceae. Jour. Indian Bot. Soc. 40:444-455.

2137. ---------- and ----------. 1961. Root apical organization in
monocotyledons -- Cannaceae. Jour. Indian Bot. Soc. 40:645-656.

2138. ----------, ------------., and GIRIJAMMA, P. 1962. Apical organization of the roots of dicotyledons. I. Root apices of some
members of the Ranunculaceae, Malvaceae, Bombacaceae and Euphorbiaceae. Proc. Rajas. Acad. Sci. 8:43-59.

2139. ---------- and ----------. 1962. Reactions of root apices to some
surgical experiments. Jour. Indian Bot. Soc. 41:148-155.

2140. ----------. 1963. A tentative suggestion on the evolutionary trend
in the root apical structures in some members of the Scitaminales.
Phyton 10:253-258.

2141. ----------., VIJAYALEKSHMI, P., and GEORGE, O.M. 1965. Apical
organization of the roots of dicotyledons. II. Root apices of some
members of Proteaceae, Cruciferae, Piperaceae, Amaranthaceae,
Onagraceae, Gentianaceae and Scrophulariaceae. Proc. Indian Acad.
Sci. (Sect. B) 61:267-276.

2142. ----------., GRORGE, O.M., and VIJAYALEKSHMI, P. 1965. Apical
organization in the roots of dicotyledons. III. Root apices of some
members of the Compositae. Proc. Indian Acad. Sci. (Sect. B)
61:296-308.

2143. ----------, and SUKUMARAN, K. 1969. Histogenesis, apical meristems,
and anatomy of Cyamopsis tetragonoloba. Phytomorph. 19:303-312.

2144. ----------. 1969. Root apical organization of certain dicotyledons.
Abstract. In: Seminar on Morphology, Anatomy and Embryology of Land
Plants. (eds.) Johri, B.M., et al. Centre of Advanced Study of
Botany. University of Delhi, India.

2145. PINKERTON, M.E. 1936. Secondary root hairs. Bot. Gaz. 98:147-158.

2146. PIROGOV, V.S. 1962. Specific anatomical features of the roots of Gypsophila sp. abounding on cliffs. Bot. Zhur. 46:1152-1161.

2147. ------------. 1968. The specialization of the roots of lithophytes belonging to the genus Draba and of some other rock-dwelling plants. Bot. Zhur. 53:350-357. (in Russian with English summary)

2148. PIROTTA, R. 1902. Origine e differenziazione degli elementi vasculari primari nella radici delle Monocotiledoni. Atti Reale Accad. Lincei I. 11:49-52; 158-162.

2149. ----------. 1903-1904. Ricerche ed osservazioni intorno alla origine ed alla differenziazione degli elementi vascolari primari nella radici delle Monocotiledoni. Annali Bot. 1:43-48; 345-357.

2150. PIRWITZ, K. 1931. Physiologische und anatomische Untersuchungen an Speichertracheiden und Velamina. Planta 14:19-76.

2151. PITOT, A. 1945. Remarques sur l'évolution du bois primaire dans le système radiculaire de Pinus halapensis. La Rev. Sci. Année 83 3244:237-238.

2152. --------. 1958. Rhizophores et racines chez Rhizophora sp. Bull. Inst. Franc. Afrique Noire. (Sér. A) 20:1103-1138.

2153. PIXLEY, E.Y. 1968. A study of the ontogeny of the primary xylem in the roots of Lycopodium. Bot. Gaz. 129:156-160.

2154. PIZZONI, P. 1906. Contribuzione alla conoscenz degli austori dell' Osyris alba. Annali Bot. 4:79-98.

2155. PLANTEFOL, L. 1947. Hélices foliares, point végétif et stéle chez les Dicotyledonés. La notion d'anneau initial. Rev. Gén. Bot. 54:49-80.

2156. PLAUT, M. 1910. Untersuchungen zur Kenntnis der physiologischen Scheiden bei den Gymnospermen, Equisetaceen und Bryophyten. Jahrb. Wiss. Bot. 47:121-185.

2157. --------. 1910. Epiblem, Hypodermis und Endodermis der Zuckerrübe. Mitt. Kaiser-Wilhelm-Inst. Landw. Bromberg 3:63-68.

2158. --------. 1910. Über der Veranderungen im anatomischen Bau der Wurzel während des Winters. Jahrb. Wiss. Bot. 48:142-154.

2159. --------. 1918. Über die morphologischen und mikroskopischen Merkmale der Periodizität der Wurzel, sowie über die Verbreitung der Metakutisierung der Wurzelhaube im Pflanzenreich. Festschrift zur Feier des 100-jährigen Bestehens der Kgl. Württemberg Landwirtschaftlichen Hochschule Hohenheim, Stuttgart. pp. 129-151.

2160. PLAVSIC, S. 1936. Anatomische Untersuchungen über <u>Picea</u> <u>omorica</u>.
Beih. Bot. Centralb. 54:429-493.

2161. PLOWMAN, A.B. 1906. The comparative anatomy and phylogeny of the
Cyperaceae. Annals Bot. 20:1-34.

2162. ------------. 1915. Is the Box-Elder a Maple? A study of comparative anatomy of <u>Negundo</u>. Bot. Gaz. 60:169-192.

2163. POHL, F. 1926. Vergleichende Anatomie von Drainagezöpfen, Land-
und Wasserwurzeln. Beih. Bot. Centralb. 42:229-262.

2164. -------. 1927. Ein Beitrag zur Abhängigkeit der Gefässweite des
Wurzelholzes von äussern Faktoren. Forstw. Zentralb. 49:271-275.

2165. POLLACK, B.M., GOODWIN, R.H.and GREENE, S. 1954. Studies on roots.
II. Effects of coumarin, scopoletin and other substances on growth.
Amer. Jour. Bot. 41:521-529.

2166. POLLACK, J.B. 1902. The relation of the fibro-vascular bundles in
the root and hypocotyl in <u>Echinocystis</u> <u>lobata</u> Torr. and Gray. Third
Rept., Michigan Academy of Science, Lansing. pp. 40-42.

2167. POND, R.H. 1908. Emergence of lateral roots. Bot. Gaz. 46:410-421.

2168. POPESCO, ST. 1926. Recherches sur la région absorbante de la racine.
Bulet. Agric. 4:59-189.

2169. POPHAM, R.A. 1941. The developmental anatomy of <u>Jatropha</u> <u>cordata</u>
(Orteg.) Muell. Abst. Doct. Diss. Ohio State Univ. No. 34. pp.
485-497.

2170. -----------. 1947. Developmental anatomy of seedling of <u>Jatropha</u>
<u>cordata</u>. Ohio Jour. Sci. 47:1-19.

2171. -----------. 1952. <u>**Developmental** Plant Anatomy: A Workbook for Use
in a General Course</u>. Long's College Book Co., Columbus, Ohio.
361 pp.

2172. -----------. 1955. Zonation of primary and lateral root apices of
<u>Pisum</u> <u>sativum</u>. Amer. Jour. Bot. 42:267-273.

2173. -----------. 1955. Levels of tissue differentiation in primary roots
of <u>Pisum</u> <u>sativum</u>. Amer. Jour. Bot. 42:529-540.

2174. -----------., and HENRY, R.D. 1955. Multicellular root hairs on
adventitious roots of <u>Kalanchoe</u> <u>fedtschenkoi</u>. Ohio Jour. Sci. 55:
301-307.

2175. -----------. 1966. Laboratory Manual for Plant Anatomy. C.V. Mosby Co., St. Louis. 228 pp.

2176. PORODKO, T.M. 1927. Ein eigenartiger Wachstumsmodus der Hauptwurzeln bei Lupinus albus. Planta 4:710-725.

2177. ------------. 1928. Neue Längenwachstumstypen der Hauptwurzeln. Planta 6:234-254.

2178. ------------. 1929. Die Ursachen des anomalen Längenwachstums der Hauptwurzeln. Planta 8:625-641.

2179. PORSCH, O. 1908. Orchidaceae section In: Ergebnisse der botanischen expedition der Kaiserlichen Akademie der Wissenschaften nach Südbrasilien 1901. (eds.) Wettstein, R. von, und Schiffner, V. Band 1. Pteridophyta und Anthophyta. Denk. Akad. Wiss. (Wien) (Math.- Nat. Kl.) 79(1 Halb.):92-167.

2180. ----------. 1931. Araceae. I. Die Anatomie der Nähr- und Haftwurzeln von Philodendron selloum C. Koch. Denk. Akad. Wiss. (Wien) (Math.- Nat. Kl.) 79(2 Halb.):389-454.

2181. POTTIER, J. 1927. Recherches sur l'anatomie comparée des espèces dans la famille des Elatinacées et sur le développement de la tige et de la racine dans le genre Elatine. Imprimerie de l'Est, Besançon. 157 pp.

2182. -----------. 1934. Contribution à l'étude du développement de la racine, de la tige et de la feuille des phanerogames angiospermes. Contribution à l'étude des phanerogames angiospermes. Imprimerie de l'Est, Besançon. 125 pp.

2183. POULSEN, V.A. 1902. Nogle anatomiske studier. Viden. Medd. Nat. Foren. Kjøbh. 54:231-248. (in Danish)

2184. ------------. 1905. Die Stutzwurzeln von Rhizophora. Viden. Medd. Nat. Foren. Kjøbh. 57:153-165. (in Danish)

2185. ------------. 1911. Bidrag til Rodens Anatomi. Biologiska arbejder tilegnede Eugenius Warming. H. Hagerup, København. 8 pp. (in Danish)

2186. POWELL, D. 1925. The development and distribution of chlorophyll in roots of flowering plants grown in the light. Annals Bot. 39:503-513

2187. POWERS, E., and GUARD, A.T. 1950. A study of the primary tissues of the apple root. Amer. Jour. Bot. 37:666. (Abstract)

2188. PRANKERD, T.L. 1911. On the structure and biology of the genus Hottoni. Annals Bot. 25:253-267.

2189. PRASAD, M.K. 1960. Velamen in some terrestrial orchids. Jour. Biol. Sci. (Bombay). 3:48-51.

2190. PRASAD, S. 1945. A histological study of the root of Saussurea lappa C.B. Clarke. Indian Jour. Pharm. 7:81-88.

2191. ---------., LUTHRA, S.P., GUPTA, P.K., and BHATTACHARYA, I.C. 1959. Pharmacognostical studies on Withania somnifera Dunal and W. coagulans Dunal. Indian Jour. Pharm. 21:189-194.

2192. ---------., and WAHI, S.P. 1965. Pharmacognostical investigation on Indian sarsaparilla. I. Root and rootstock of Hemidesmus indicus R. Br. Indian Jour. Pharm. 27:35-39.

2193. ----------, --------------., and KHOSA, R.L. 1969. Pharmacognostical studies on roots of Plumbago zeylandica and P. rosea. In: Recent Advances in the Anatomy of Tropical Seed Plants. (ed.) Chowdhury, K.A. Proceedings, 1st National Symposium, Muslim University, Aligarh, India (Dec. 1966). Hindustan Publishing Corp. (India), Delhi. pp. 197-206.

2194. PRAT, H. 1926. Étude des mycorhizes du Taxus baccata. Ann. Sci. Nat., Bot. (Sér. 10) 8:141-162.

2195. PRATT, A. 1929. Notes on strand plants. IV. Arenaria peploides. Trans. Proc. Bot. Soc. Edinb. 30:157-163.

2196. PRATT, D.J. 1917. An anatomical study of Cycloloma atriplicifolium. Kansas Univ. Sci. Bull. 10:87-120.

2197. PRESTON, R.J., JR. 1943. Anatomical studies of the roots of juvenile lodgepole pine. Bot. Gaz. 104:443-448.

2198. PREVOT, P., and STEWARD, F.C. 1936. Salient features of the root system relative to the problem of salt absorption. Plant Phys. 11: 509-534.

2199. PRICE, S.R. 1911. The roots of some North African desert-grasses. New Phytol. 10:328-340.

2200. PRIESTLEY, J.H. 1920. The mechanism of root pressure. New Phytol. 19:189-200.

2201. ---------------. 1922. Further observations upon the mechanism of root pressure. New Phytol. 21:41-47.

2202. ---------------., and NORTH, E.E. 1922. Physiological studies in plant anatomy. III. The structure of the endodermis in relation to its function. New Phytol. 21:113-139.

149

2203. --------------., and RADCLIFFE, F.M. 1924. A study of the endodermis in the Filicineae. New Phytol. 23:161-193.

2204. --------------. 1928. The meristematic tissues of the plant. Biol.
Rev. 3:1-20.

2205. --------------., and SWINGLE, C.F. 1929. Vegetative propagation
from the standpoint of plant anatomy. U.S.D.A. Tech. Bull. 151.
98 pp.

2206. PRINTZ, H. 1921/1922 (1923). Über den Bau des vegetativen Sprosses
bei Phelipaea lanuginosa C.A. Meyer. Det Kong. Norske Vid. Se.
Skrif. No. 2. pp. 3-49.

2207. PRODINGER, M. 1909. Das Periderm der Rosaceen in systematischer
Beziehung. Denk. Kaiserl. Akad. Wiss. Wien (Math.- Nat. Kl.)
84:329-383.

2208. PROZINA, M.N., and PRIBURA, YU.N. 1962. A comparative anatomical
study of different types of roots of corn. In: Morfologiia Kukuruzy.
(ed.) Meier, K.I. (Moskva) Izd-vo Moskovskogo Universiteta. pp.
196-217. (in Russian)

2209. PRYOR, L.D. 1937. Some observations on the roots of Pinus radiata in
relation to wind resistance. Austral. For. 2:37-40.

2210. PURER, E.A. 1942. Anatomy and ecology of Ammophila arenaria Link.
Madroño 6:167-171.

2211. PURI, Y.P., STANTON, C.R., ASSESAVESNA, S., and FISHLER, D.W. 1966.
Anatomical and agronomic studies of Phormium in western Oregon.
U.S.D.A. Prod. Res. Rept. No. 93. 43 pp.

2212. PURNELL, H.M. 1961. Studies of the family Proteaceae. I. Anatomy
and morphology of the roots of some Victorean species. Austral.
Jour. Bot. 8:38-50.

2213. PURVIS, C. 1956. The root system of the oil palm: its distribution,
morphology and anatomy. Jour. West African Inst. Oil Palm Res.
No. 4. pp. 60-82.

2214. QUER, P.F. 1960. The Anatomy of Plants. Arrow Science Series.
Arrow Books Ltd., London. 128 pp.

2215. QUEVA, C. 1903. Structure des radicelles de la Mâcre. Comp. Rend.
Hebd. Séances Acad. Sci. 136:826-827.

2216. --------. 1903. Les radicelles de la Mâcre et les exceptions aux
définitions des membres des plantes vasculaires. Bull. Soc. Hist.
Nat. Autun 16:99-108.

2217. --------. 1907. Contributions à l'anatomie des Monocotylédonées. II. Les Uvulariées rhizomateuses. Beih. Bot. Centralb. 22:30-77.

2218. --------. 1909. Observations anatomiques sur le _Trapa natans_ L. Comp. Rend. Assoc. Franç. Avan. Sci. 38e Session, Lille 38: 512-517.

2219. --------. 1909. Le _Monotropa hypopitys_ L. anatomie et biologie. Bull. Soc. Hist. Nat. Autun 22:39-50.

2220. --------. 1910. L'_Azolla filiculoides_ Lam. Étude anatomique. Bull. Soc. Hist. Nat. Autun 23:233-256.

2221. RADKEVICH, O.N. 1934. Anatomy of _Scorzonera tau-saghyz_ Lipsch et Bosse. Bot. Zhur. 19:467-494. (in Russian with German summary)

2222. RAJU, M.V.S., STEEVES, T.A., and COUPLAND, R.T. 1963. Developmental studies on _Euphorbia esula_ L.: Morphology of the root system. Canad. Jour. Bot. 41:579-589.

2223. ------------, -------------, and NAYLOR, J.M. 1964. Developmental studies on _Euphorbia esula_ L.: Apices of long and short roots. Canad. Jour. Bot. 42:1615-1628.

2224. ------------, COUPLAND, R.T., and STEEVES, T.A. 1966. On the occurrence of root buds on perennial plants in Saskatchewan. Canad. Jour. Bot. 44:33-37.

2225. RAMSBOTTOM, J. 1922. Orchid mycorrhiza. British Mycol. Soc. Trans. 8:28-61.

2226. RANG, K.H. 1929. An anatomical study of _Monarda fistulosa_ L. Doctoral thesis. University of Wisconsin, Madison. 19 pp.

2227. RANGAN, T.S., and RANGASWAMY, N.S. 1969. Morphogenic investigations on parasitic angiosperms. III. _Cassytha filiformis_ (Lauraceae). Phytomorph. 19:292-300.

2228. RAO, A.N. 1966. Developmental anatomy of natural root grafts in _Ficus globosa_. Austral. Jour. Bot. 14:269-276.

2229. RAO, A.P., and KHARE, P. 1964. Contribution to our knowledge of the sporophyte of _Tectaria amplifolia_ (V.A.V.R.) Christensen. Proc. Indian Acad. Sci. (Sect. B) 59:328-339.

2230. RAO, A.R., and SHARMA, U. 1963. Studies on Indian Hymenophyllaceae. III. Contributions to our knowledge of _Meringium edentulum_ (v.d.b. Copeland). Proc. Indian Acad. Sci. (Sect. B) 57:300-306.

2231. --------., and SRIVASTAVA, P. 1970. Studies on Indian Hymenophyllaceae. X. Contributions to our knowledge of Mecodium polyanthos
(SW.) Copeland. Proc. Indian Acad. Sci. (Part B, Biol. Sci.)
36:248-253.

2232. RAO, J.S. 1953. Role of velamen tissue in roots of orchids. Sci.
Cult. 19:97-99.

2233. RAO, L.N. 1942. Parasitism in the Santalaceae. Annals Bot. (N.S.)
6:131-150.

2234. --------. 1971. Swollen roots of Cycas circinalis L. Curr. Sci.
40:117-118.

2235. RAO, M.R. 1903. Root-parasitism of the sandal tree. Indian For.
29:386-389.

2236. RAO, T.A., GOVINDU, H.C., and THIRUMALACHAR, M.J. 1950. Aerial
roots in some tropical rain forest plants. Jour. Indian Bot. Soc.
29:224-226.

2237. RASCH, W. 1915. Über den anatomischen Bau der Wurzelhaube einiger
Glumifloren und seine Beziehungen zur Beschaffenheit des Bodens.
Beitr. Allgem. Bot. 1:80-114.

2238. RATERA, E.L., and CRISTIANI, L.Q. 1961. Anatomical study of toxic
plants. The "sunchillo" (Wedelia glauca). Rev. Inst. Munic. Bot.
(Buenos Aires) 1:25-37. (in Portuguese)

2239. RATHFELDER, O. 1956. Anatomische Untersuchungen an Pulsatilla. I.
Bot. Jahrb. 77:25-51.

2240. RAUH, W. 1937. Die Bildung von Hypokotyl- und Wurzelsprossen und
ihre Bedeutung für die Wuchsformen der Pflanzen. Nova Acta Leop.
(N.F.) 4:395-553.

2241. --------. 1950. Morphologie der Nutzpflanzen. Quelle und Meyer,
Heidelberg. 290 pp.

2242. --------. 1956. Morphologische, entwicklungsgeschichtliche, histogenetische und anatomische Untersuchungen an den Sprossen der
Didiereaceen. Akad. Wiss. Lit. Jahrb. (Math.- Nat. Kl.) No. 6.
pp. 345-444.

2243. --------., and WEBERLING, F. 1959. Morphologische und anatomische
Untersuchungen an der Valerianaceen Gattung Stangea Graebner. Akad.
Wiss. Lit. Jahrb. (Math.- Nat. Kl.) No. 10. pp. 799-839.

2244. --------., and FALK, H. 1959. Stylites E. Amstutz, eine neue Isoëtacee
aus den Hochanden Perus. Sitz. Heidelb. Akad. Wiss. (Math.- Nat.
Kl.) 1959:3-83.

2245. -------. 1961. Weitere Untersuchungen an Didiereaceen. Sitz. Heidelb. Akad. Wiss. (Math.- Nat. Kl.) 1960/1961(7):185-300.

2246. RAYNER, M.C. 1926. Mycorrhiza. New Phytol. 25:1-50; 65-108; 171-190; 248-263; 338-372.

2247. -----------. 1927. Mycorrhiza. New Phytol. 26:22-45; 85-114.

2248. RAZDORSKIĬ, V.F. 1949. Anatomy of Plants. Soviet Science, Moscow. 524 pp. (in Russian)

2249. RÉAUBOURG, G. 1906. Les Holboellia de la Chine central. Bull. Soc. Bot. France 53:451-461.

2250. REDMOND, D.R. 1957. Observations on rootlet development in yellow birch. For. Chron. 33:208-212.

2251. REED, E.L. 1924. Anatomy, embryology, and ecology of Arachis hypogaea. Bot. Gaz. 78:289-310.

2252. REEDER, J.R., and VON MALTZAHN, K. 1953. Taxonomic significance of root-hair development in the Gramineae. Proc. Nat. Acad. Sci. U.S.A. 39:593-598.

2253. REEVE, R.M. 1948. Late embryogeny and histogenesis in Pisum. Amer. Jour. Bot. 35:591-601.

2254. REICHE, K. 1907. Bau und Leben der hemiparasitischen Phrygilanthus - Arten Chiles. Flora 97:375-400.

2255. ---------. 1908. Zur Kenntnis der Dioscoreaceen - Gattung Epipetrum Phil. Bot. Jahrb. 42:178-190.

2256. ---------. 1921. Die physiologische Bedeutung des anatomischen Baues der Crassulaceen. Flora 114:249-261.

2257. ---------. 1922. Beiträge zur Kenntnis der Gattung Fouquiera. Bot. Jahrb. 57:287-301.

2258. ---------. 1923. Entwicklung, Bau und Leben der Euphorbia radians Benth., einer knollentragenden Art. Flora 116:259-263.

2259. REINDERS, E. 1955. "Morphologie" van het wortel stelsel. In: De Plantenwortel in de Landbouw. Nederlands genootschap voor landbouw- wetenschap. Staatsdrukkerij-en Uitgeverijbedriff, 's-Gravenhage. pp. 18-30. (in Dutch)

2260. REINHARD, E. 1956. Ein Vergleich zwischen diarchen and triarchen Wurzeln von Sinapsis alba. Zeitschr. Bot. 44:505-514.

2261. ----------. 1959. Leitbündelgewebe - Differenzierung in der Wurzel. In: Recent Advances in Botany. Vol. 1. University of Toronto Press, Toronto. pp. 800-805.

2262. REINERT, J. 1958. Untersuchungen über die Morphogenese an Gewebekulturen. Berich. Deutsch Bot. Gesel. 71:15.

2263. RENNERT, R.J. 1902. Seed and seedlings of Arisaema triphyllum and Arisaema dracontium. Bull. Torrey Bot. Club 29:37-54.

2264. RESCH, A. 1961. Zur Frage nach den Geleitzellen im Protophloem der Wurzel. Zeitschr. Bot. 49:82-95.

2265. RESVOLL, T.R. 1900. Nogle arktiske ranunklers morfologi og anatomi. Nyt Mag. Natur. 38:343-369. (in Danish)

2266. ------------. 1929. Rubus chamaemorus L. A morphological-biological study. Nyt Mag. Natur. 67:55-129.

2267. REZNIK, M.A. 1934. Étude anatomique de la plantule de sorgho. Rev. Gén Bot. 46:385-419.

2268. RHODE, H. 1928. Über die kontraktilen Wurzeln einiger Oxalidaceen. Bot. Archiv 22:463-532. (Same as Listing 2304)

2269. RICARD, L. 1931. Sur l'insertion vasculaires des radicelles. Comp. Rend. Hebd. Séances Acad. Sci. 193:874-876.

2270. RICHARDS, H.M. 1904. A case of irregular secondary thickening. Torreya 4:181-184.

2271. RICHARDSON, S.D. 1955. The influence of rooting medium on the structure and development of the root cap in seedlings of Acer saccharinum L. New Phytol. 54:336-337.

2272. RICHTER, A. 1901. Physiologisch-anatomische Untersuchungen über Luftwurzeln mit besonderer Berücksichtigung der Wurzelhaube. Biblio. Bot. 10:1-50.

2273. RICÔME, H. 1904. Passage de la racine à la tige chez l'Auricule. Comp. Rend. Hebd. Séances Acad. Sci. 139:468-470.

2274. ---------. 1922. Sur l'elongation des racines. Comp. Rend. Hebd. Séances Acad. Sci. 174:880-881.

2275. RIDGWAY, C.S. 1913. The occurrence of callose in root hairs. Plant World 16:116-122; 188.

2276. RIEDHART, J.M., and GUARD, A. 1957. On the anatomy of the roots of apple seedlings. Bot. Gaz. 118:191-194.

2277. RIEDL, H. 1937. Bau und Leistungen des Wurzelholzes. Jahrb. Wiss. Bot. 85:1-75.

2278. RIMBACH, A. 1902. Physiological observations on the subterranean organs of some Californian Liliaceae. Bot. Gaz. 33:401-420.

2279. ----------. 1921. Über Wurzelverkürzung bei dikotylen Holzgewächsen. Berich. Deutsch Bot. Gesel. 39:281-284.

2280. ----------. 1921. Über die Wachstumsweise der Wurzel von Incarvillea delavayi. Berich. Deutsch. Bot. Gesel. 39:288-290.

2281. ----------. 1922. Die Wurzelverkürzung bei den grossen Monokotylen-formen. Berich. Deutsch. Bot. Gesel. 40:196-202.

2282. ----------. 1922. Die Wirkung der Wurzelverkürzung bei einigen Nutz- und Zierpflanzen. Angew. Bot. 4:81-90.

2283. ----------. 1924. Die Bewurzelung der Speisezwiebeln. Angew. Bot. 6:458-463.

2284. ----------. 1926. Die Grösse der Wurzelverkürzung. Berich. Deutsch. Bot. Gesel. 44:328-334.

2285. ----------. 1927. Die Geschwindikeit und Dauer der Wurzelverkürzung. Berich. Deutsch. Bot. Gesel. 45:127-130.

2286. ----------. 1928. Endodermiswellung und Casparyscher Punkt. Berich. Deutsch. Bot. Gesel. 46:424-433.

2287. ----------. 1929. Die verbreitung der Wurzelkürzung im Pflanzenreich. Berich. Deutsch. Bot. Gesel. 47:22-31.

2288. ----------. 1932. Nachträgliche Dickenzunahme kontraktiler Monokot-ylen-Wurzeln. Berich. Deutsch. Bot. Gesel. 50:215-219.

2289. RIOLLE, Y.T. 1914. Recherches morphologiques et biologiques sur les radis cultivées. Berger-Levrault, Nancy. 244 pp.

2290. RIOPEL, J.L., and STEEVES, T.A. 1964. Studies on the roots of Musa acuminata cv. Gros Michel. I. The anatomy and development of main roots. Annals Bot. (N.S.) 28:475-490.

2291. RITTER, N. 1917. Histology of Astragalus mollissimus. Kansas Univ. Sci. Bull. 10:197-208.

2292. RIVETT, M.F. 1924. The root-tubercles in Arbutus unedo. Annals Bot. 38:661-677.

2293. RIVIÈRE, S. 1940. Étude anatomique de quelques plantes à évolution vasculaire tres accélérée. Rev. Gén. Bot. 52:49-74.

155

2294. ----------. 1956. Une fausse bistélie de la racine de Tradescantia virginica. Bull. Soc. Bot. France 103:596-598.

2295. RIZZINI, C.T., and HERINGER, E.P. 1961. Underground organs of plants from some southern Brazilian savannas, with special reference to the xylopodium. Phytomorph. 17:105-124.

2296. ROBERTS, E.A. 1916. The epidermal cells of roots. Bot. Gaz. 62: 488-506.

2297. ------------., and HERTY, S.D. 1934. Lycopodium complanatum var. flabelliforme Fernald: Its anatomy and a method of vegetative propagation. Amer. Jour. Bot. 21:688-697.

2298. ROBERTSON, A. 1902. Notes on the anatomy of Macrozamia heteromera Moore. Proc. Cambr. Philos. Soc. 12:1-14.

2299. ------------. 1906. The "droppers" of Tulipa and Erythronium. Annals Bot. 20:429-440.

2300. ROBERTSON, N.F. 1954. Studies on the mycorrhiza of Pinus sylvestris. I. The pattern of development of mycorrhizal roots and its significance for experimental studies. New Phytol. 53:253-283.

2301. RODRIGUEZ, R.L. 1957. Systematic anatomical studies on Myrrhidendron and other woody Umbellales. Univ. Calif. Publ. Bot. 1959. 29: 145-292.

2302. ROGERS, W.E., and NELSON, R.R. 1962. Penetration and nutrition of Striga asiatica. Phytopath. 52:1064-1070.

2303. ROGERS, W.S., and HEAD, G.C. 1966. The roots of fruit plants. Jour. Roy. Hort. Soc. 91:198-205.

2304. ROHDE, H. 1928. Ueber die kontraktilen Wurzeln einiger Oxalidaceen. Bot. Archiv 22:463-532. (Same as Listing 2268)

2305. ROMBERGER, J.A. 1963. Meristems, Growth, and Development in Woody Plants. U.S.D.A. Forest Service, Tech. Bull. No. 1293. 214 pp.

2306. ROMM, H.J. 1953. The development and structure of the vegetative and reproductive organs of kudzu, Pueraria thunbergiana (Sieb. and Zucc.) Benth. Iowa State Coll. Jour. Sci. 27:407-419.

2307. RORDAM, A.M. 1966. Secretory ducts in roots of several Umbelliferae: primary structure. Dansk Tidssk. Farmaci 40:131-139. (in Dutch with German summary)

2308. ROSENDAHL, C.O. 1911. Observations on the morphology of the underground stems of Symplocarpus and Lysichiton, together with some notes on geographical distribution and relationships. Minn. Bot. Studies (Part 2) 4:137-152.

2309. ROSENTHALER, L., and STADLER, P. 1908. Ein Beitrag zur Anatomie von Cnicus benedictus L. Arch. Pharm. 246:436-466.

2310. --------------., and KIENE, K. 1914. Über eine chinesische Rhabarberwurzel. Berich. Deutsch. **Pharm. Gesel.** 24:234-243.

2311. ROSS, H. 1908. Der anatomische Bau der mexicanischen Kautschukpflanze "Guayule," Parthenium argentatum Gray. Berich. Deutsch. Bot. Gesel. 26:248-263.

2312. ROSSO, S.W. 1964. The vegetative anatomy of the Cypripedioideae (Orchidaceae). Diss. Absts. 24:3984.

2313. ----------. 1966. The vegetative anatomy of the Cypripedioideae (Orchidaceae). Jour. Linn. Soc. (Bot.). 59:309-341.

2314. ROSTOVTSEV, S.I. 1948. Handbook of Plant Anatomy. Soviet Science, Moscow. 276 pp. (in Russian)

2315. ROTH, I. 1966. Anatomia de las Plantas Superiors. Ediciones de la Biblioteca. No. 30. Universidad de Venezuela, Caracas. 357 pp.

2316. ROTHWELL, N.V. 1966. Evidence for diverse cell types in the apical region of the root epidermis of Panicum virgatum. Amer. Jour. Bot.

2317. ROVENSKÁ, B. 1968. Anatomical Wheat Atlas. Ceskoslovenské Akademie Ved, Praha. 157 pp. (in Czechoslovakian)

2318. ROW, H.C., and REEDER, J.R. 1957. Root-hair development as evidence of relationships among genera of Gramineae. Amer. Jour. Bot. 44: 596-601.

2319. RUDENSKAYA, S.J. 1938. Development of the latex vessels system as a factor of rubber accumulation in kok-saghyz roots. Comp. Rend. (Dokl.) Acad. Sci. URSS (N.S.) 20:399-403.

2320. RUDENSKAYA, V.J. 1938. Anatomical premises for breeding large-rooted forms of kok-saghyz. Comp. Rend. (Dokl.) Acad. Sci. URSS (N.S.) 20:617-620.

2321. RUDOLPH, K. 1906. Psaronien und Marattiaceen. Vergleichend anatomische Untersuchung. Denk. Kaiserl. Akad. Wiss. Wien (Math.- Nat. Kl.) 78:165-201.

2322. RUER, P. 1967. Morphologie et anatomie du système radiculaire du Palmier à huile. Oléagineux 22:595-599.

2323. RÜGGEBERG, H. (1911) 1912. Beitrag zur Anatomie des Zuckerrübenkeimlings. Jahrb. Ver. Angew. Bot. 9:52-57.

2324. ------------. 1912. Beiträge zur Anatomie der Zuckerrübe. Mitt. Kaiser-Wilh.-Inst. Landw. Bromberg 4:399-415.

2325. RUGGERI, C. 1963. Variations in the root anatomy of some Euramerican poplar hybrids. Pubbl. Centro Sperim. Agric. For. 6:133-139. (1962-63) (in Italian)

2326. RUMPF, G. 1904. Rhizodermis, Hypodermis und Endodermis der Farnwurzel. Biblio. Bot. 13:1-48.

2327. RUNNER, D.K. 1950. The structure and development of the storage root of Humulus lupulus L. Master's thesis. Oregon State College, Corvallis. 37 pp.

2328. RUSSELL, A.M. 1919. A comparative study of Floerkia proserpinacoides and allies. Contr. Bot. Lab. Univ. Penn. 4:401-418.

2329. RUSSELL, M.W. 1934. Origines et particularités des racines adventives caulinaires de quelques Aeschynomene. Rev. Bot. Appl. Agric. Trop. 14:407-411.

2330. RUSSELL, W. 1937. Note sur la structure de la racine de Derris elliptica. Rev. Bot. Appl. Agric. Trop. 17:539-540.

2331. RYWOSCH, S. 1909. Untersuchungen über die Entwicklungsgeschichte der Seitenwurzeln der Monokotylen. Zeitschr. Bot. 1:253-283.

2332. SAHNI, B. 1920. On the structure and affinities of Acmopyle pancheri Pilger. Philos. Trans. Roy. Soc. London (Series B) 210: 253-310.

2333. SAKHARAMA, R.J. 1953. Role of velamen tissue in roots of orchids. Sci. Cult. 19:97.

2334. SAKSENA, R.K., and MATHUR, L.N. 1924/1925. "On an Indian Ophioglossum." Jour. Indian Bot. Soc. 4:307-311.

2335. SALIAEV, R.K. 1958. Anatomical structure of root-tips in the adult pine (Pinus sylvestris L.) and the formation of mycorhiza. Bot. Zhur. 43:869-876. (in Russian)

2336. ------------. 1962. The structure of the root-tips and the formation of mycorhiza of Betula verrucosa Ehrh. In: Akadamiya Nauk SSSR. Laboratoriya Lesovedeniya. Fiziologiya Drevesnykh Rastenii. Ref. 1962. pp. 289-293. (in Russian)

2337. SALTER, T.M. 1952. Notes on the process of forming contractile roots and the lowering of the first bulbils by seedlings of the South African Oxalis which produce endospermous seeds. Jour. So. African Bot. 17:189-194.

2338. SAMPSON, H.C. 1923. The Coconut Palm. John Bale, Sons and Danielsson, Ltd., London. 262 pp.

2339. SANTRA, D.K. 1960. Structure of the root of Datura innoxia Miller. Indian Jour. Pharm. 22:11-13.

2340. SARGANT, E. 1900. A new type of transition from stem to root in the vascular system of seedlings. Annals Bot. 14:633-638.

2341. ----------. 1902. The origin of the seed-leaf in monocotyledons. New Phytol. 1:107-113.

2342. ----------. 1903. A theory of the origin of monocotyledons founded on the structure of their seedlings. Annals Bot. 17:1-92.

2343. ----------., and ARBER, A. 1915. The comparative morphology of the embryo and seedling in the Gramineae. Annals Bot. 114:161-222.

2344. SARIN, Y.K., and KAPOOR, L.D. 1965. Pharmacognostic study of Trillium govanianum Wall. Indian Jour. Pharm. 27:6-10.

2345. SARTON, A. 1905. Recherches expérimentales sur l'anatomie des plantes affines. Ann. Sci. Nat., Bot. (Sér. 9) 2:1-117.

2346. SASS, J.E. 1955. Vegetative morphology. In: Corn and Corn Improvement. (ed.) Sprague, G.F. Academic Press, New York. pp. 63-87.

2347. SATCHITHANANDAM, P. 1957. The apical organization and development of the vegetative and reproductive apices of Oryza sativa L. Ceylon Jour. Sci., Biol. Sci. (N.S.) 1:23-44.

2348. SAVCHENKO, N.L. 1940. Entwicklung und Anordnung des Milchsaftgefässsystems bei Taraxacum kok-saghyz. Comp. Rend. (Dokl.) Acad. Sci. URSS 27:1052-1055.

2349. SAXELBY, E.M. 1908. The origin of the roots in Lycopodium selago. Annals Bot. 22:21-23.

2350. SCARLETTE, W.D.B. 1964. The developmental morphology and vascular anatomy of Coronilla varia L. cv. Penngift. Diss. Absts. 24(12, Part 1):4946-4947.

2351. SCAVONE, O. 1955 (1956). Contribuição ao estudo morfológico e anatômico do Senecio brasiliensis Less. var. tripartitus Baker. Anais Fac. Farm. Odont. Univ. São Paulo 13:15-31. (in Portuguese)

2352. ----------. 1964. Contribuição ao estudo morfológico e anatômico de Cissus gongyloides Burch. Rev. Fac. Farm. Bioq. Univ. São Paulo 2:107-128. (in Portuguese)

2353. ----------. 1965. Contribuição ao estudo morfológico e anatômico de <u>Coleus</u> <u>barbatus</u> Benth. Labiatae. Rev. Fac. Farm. Bioq. Univ. São Paulo 3:249-270. (in Portuguese)

2354. ----------. 1966. Contribuição ao estudo morfológico e anatômico do <u>Spartium</u> <u>junceum</u> Linn. Rev. Fac. Farm. Bioq. Univ. São Paulo 4:311-328. (in Portuguese)

2355. SCHADE, C., and GUTTENBERG, H. VON. 1951. Über die Entwicklung des Wurzelvegetationspunktes der Monocotyledonen. Planta 40:170-198.

2356. SCHAEDE, R. 1945. Über die Korallenwurzeln der Cycadeen und ihre Symbiose. Planta 34:98-124.

2357. SCHAEFER, H. 1942. Ein Gras mit Speicherwurzeln. Berich. Deutsch. Bot. Gesel. 60:284-291.

2358. SCHAFFNER, J.H. 1903. Ohio plants with contractile roots. Ohio Nat. 3:410.

2359. SCHAFFSTEIN, G. 1932. Untersuchungen an ungegliederten Milchröhren. Beih. Bot. Centralb. 49:197-220.

2360. SCHEIRER, D.C., and HILLSON, C.J. 1973. The vascular transition region of <u>Helianthus</u> <u>annuus</u>. I. Bilateral and unilateral patterns of differentiation. Amer. Jour. Bot. 60:242-246.

2361. SCHENCK, H. 1918. Verbänderungen und Gabelungen an Wurzeln. Flora 111/112:503-525.

2362. SCHERER, P.E. 1904. Studien über Gefassbündeltypen und Gefässformen. Beih. Bot. Centralb. 16:67-110.

2363. SCHINDLER, A.K. 1904. Die Abtrennung der Hippuridaceen von den Halorhagaceen. Bot. Jahrb. No. 77. 34:1-77.

2364. SCHLECHTER, R. 1970. <u>Die Orchideen</u>. 3rd ed. Paul Parey, Berlin und Hamburg. Vol. I. pp. 1-64.

2365. SCHLOSS, H. 1913. Zur Morphologie und Anatomie von <u>Hydrostachys</u> <u>natalensis</u> Wedd. Sitz. Akad. Wiss. Wien Abt. I. 122:339-359.

2366. SCHMID, W. 1925. Morphologische, anatomische und entwicklungs-geschichtliche Untersuchungen an <u>Mesembrianthemum</u> <u>pseudotruncatellum</u> Berger. Viert. Nat. Gesel. Zürich 70:1-96.

2367. ----------. 1928. Untersuchungen über den Bau der Wurzel und Sproß-achse der Amarantacee. Viert. Nat. Gesel. Zürich 73:217-297.

2368. ----------. 1932. Beiträge zur Kenntnis von Sarcocaulon rigidum Schinz. Viert. Nat. Gesel. Zürich 77:36-77.

2369. SCHMIDT, O. CHR. 1935. Aristolochiaceae. In: Die Naturlichen Pflanzenfamilien. (eds.) Engler, H.G.A., and Prantl, K.A.E. Aufl. 2. Band 16b. Wilhelm Engelmann, Leipzig. pp. 204-242.

2370. SCHMUCKER, T. 1923. Zur morphologie und Biologie geophiler Pflanzen. Bot. Archiv 4:201-248.

2371. ------------. 1926. Beiträge zur Kenntnis einer merkwurdigen Orchidee, Haemaria discolor Lindl. Flora 121:157-171.

2372. SCHNEGG, H. 1902. Beiträge zur Kenntnis der Gattung Gunnera. Flora 90:161-208.

2373. SCHNEIDER, F. 1913. Beiträge zur Entwicklungsgeschichte der Marsiliaceen. Flora 105:347-369.

2374. ------------. 1968. Technologie des Zuckers. M. and H. Schaper, Hannover. 1067 pp.

2375. SCHNEIDER, H. 1968. The anatomy of Citrus. In: The Citrus Industry. (eds.) Reuther, W., Batchelor, L.D., and Webber, H.J. (Rev. ed.) University of California, Division of Agricultural Sciences. 2:1-85.

2376. SCHNEIDER, J. 1901. Über die Histologie der Zuckerrübe. Zeitschr. Zucker. Böhmen 25:305-326.

2377. SCHOENICHEN, W. 1924. Biologie der Blutenpflanzen. T. Fisher, Freiburg im Breisgau. 216 pp.

2378. ---------------. 1924. Winke für das botanische praktikum: Die Pilzwurzel der Orchideen. Naturforscher (Berlin) 1:277-279.

2379. SCHOLZ, E. 1900-1901. Entwicklungsgeschichte und Anatomie von Asparagus officinale L. Festschrift zum 50 Jahres-Bericht der Schottenfelder k.k. Staats-Realschule im VII Bezirke in Wien pp. 137-151.

2380. SCHOUTE, J.C. 1902. Über Zellteilungsvorgänge im Cambium. Verh. Konink. Akad. Wetens. 9:1-60.

2381. ------------. 1903. Die Stelär-Theorie. Gustav Fischer, Jena 175 pp.

2382. ------------. 1910. Die Pneumatophoren von Pandanus. Ann. Jard. Bot. Buitenz. Suppl. 3, Part 1. pp. 216-220.

2383. ------------. 1938. Morphology; Anatomy. In: Manual of Pteridology. (ed.) Verdoorn, Fr. Martinus Nijhoff, Hague. pp. 1-104.

2384. SCHRÖDER, D. 1926. Unterscheidungsmerkmale der Wurzeln unserer
Wiesen-Weidenpflanzen. Landw. Jahrb. 64:41-64.

2385. SCHUBERT, O. 1913. Bedingungen zur Stecklingsbildung und Propfung
von Monokotylen. Centralb. Bakt. Parasit. Infek. (Abt. 2) 38:
309-443.

2386. SCHÜEPP, O. 1916. Beiträge zur Theorie des Vegetationspunktes.
Berich. Deutsch. Bot. Gesel. 34:847-857.

2387. ----------. 1917. Untersuchungen über Wachstum und Formwechsel von
Vegetationspunkten. Jahrb. Wiss. Bot. 57:17-79.

2388. ----------. 1926. Meristeme. In: Handbuch der Pflanzenanatomie.
(ed.) Linsbauer, K. Abt. I, Teil 2. Histologie. Band 4. Gebrüder
Borntraeger, Berlin. 114 pp.

2389. SCHULLE, H. 1933. Zur Entwicklungsgeschichte von Thesium montanum
Ehrh. Flora 127:140-184.

2390. SCHULMAN, E. 1945. Root growth-rings and chronology. Tree-ring
Bull. 12:2-5.

2391. SCHULZ, E. 1930. Beiträge zur physiologischen und phylogenetischen
Anatomie der vegetativen Organe der Bromeliaceen. Bot. Archiv
29:122-209.

2392. SCHULZ, G. 1926. Scopolia carnioloica Jacquin. Eine botanisch-
pharmacognostische Studie. Bot. Archiv 13:433-448.

2393. SCHULZE, B. 1911. Wurzelatlas. Teil I. Darstellung natürlicher
Wurzelbilder der Halmfrüchte in verscheidenen Studien der Entwicklung.
Paul Parey, Berlin. pp. 5-36.

2394. ----------. 1911. Wurzelatlas. Teil II. Darstellung natürlicher
Wurzelbildung von Leguminosen und Raps in verscheidenen Stadien der
Entwicklung. Paul Parey, Berlin. pp. 5-42.

2395. SCHUMACHER, W. 1934. Die Absorptionorgane von Cuscuta odorata und der
Stoffübertritt aus den Siebröhren der Wirtspflanze. Jahrb. Wiss.
Bot. 80:74-91.

2396. SCHÜTZE, W. 1906. Zur physiologischen Anatomie einiger tropischer
Farne, besonders der Baumfarne. Beitr. Wiss. Bot. 5:329-376.

2397. SCHWARTZ, E.J. 1912. Observations on Asarum europaeum and its
mycorhiza. Annals Bot. 26:769-776.

2398. SCHWEIGER, J. 1909. Vergleichende Untersuchungen über Sarracenia und
Cephalotus folicularis betreffs ihrer etwaigen systematischen Ver-
wandschaft. Beih. Bot. Zentralb. 25:490-539.

2399. SCHWEITZER, J. 1910 (1911). Beiträge zur Anatomie und Entwicklungsgeschichte der Gattung Dipsacus. (Jahrb.) Egyet. Termész. Szövet.
(Budapest) pp. 1-32.

2400. SCHWERIN, F. VON. 1913. Beobachtungen über die Bildung von Luftwurzeln. Gartenflora 62:276-278.

2401. SCOTT, D.G. 1906. The apical meristems of the roots of certain aquatic
monocotyledons. New Phytol. 5:119-129.

2402. SCOTT, D.H., and HILL, T.G. 1900. The structure of Isoetes hystrix.
Annals Bot. 14:413-454.

2403. ----------., and BROOKS, F.T. 1927. An Introduction to Structural
Botany. Part I. Flowering plants. A. and C. Black, Ltd., London.
308 pp.

2404. SCOTT, F.M. 1932. Some features of the anatomy of Fouquiera splendens.
Amer. Jour. Bot. 19:673-678.

2405. ----------., and SHARSMITH, H. 1933. The transition region in the
seedling of Ricinus communis: a physiological interpretation. Amer.
Jour. Bot. 20:176-187.

2406. ----------. 1935. Contribution to the causal anatomy of the desert
willow, Chilopsis linearis. Amer. Jour. Bot. 22:333-343.

2407. ----------. 1935. The anatomy of Cercidium torreyanum and Parkinsonia
microphylla. Madroño 3:33-41.

2408. ----------., and LEWIS, M. 1953. Pits, intercellular spaces, and
internal "suberization" in the apical meristems of Ricinus communis
and other plants. Bot. Gaz. 114:253-264.

2409. ----------. 1963. Root hair zone of soil-grown roots. Nature 199:
1009-1110.

2410. ----------., BYSTROM, B.G., and BOWLER, E. 1963. Root hairs, cuticle,
and pits. Science 140:63-64.

2411. ----------. 1965. The anatomy of plant roots. In: Ecology of Soil-
borne Plant Pathogens. (eds.) Baker, K.F., et al. University of
California Press, Berkeley. pp. 145-151.

2412. SCOTT, L.I., and PRIESTLEY, J.H. 1928. The root as an absorbing
organ. I. A reconsideration of the entry of water and salts in the
absorbing region. New Phytol. 27:125-140.

2413. ----------. 1928. The root as an absorbing organ. II. The delimit-
ation of the absorbing organ. New Phytol. 27:141-174.

2414. ----------., and WHITWORTH, A.B. 1928. On a structural peculiarity of the exodermis of the root of Pelargonium. Proc. Leeds Philos. Lit. Soc. 1:312-317.

2415. SCOTT, R. 1908. The contractile roots of the aroid Sauromatum guttatum. Rep. Brit. Assn. Adv. Sci., Dublin pp. 910-911.

2416. SCULTHORPE, C.D. 1967. The Biology of Aquatic Flowering Plants. St. Martin's Press, New York. 610 pp.

2417. SEAGO, J.L. 1966. Apical organization in roots of the Convolvulaceae. Master's thesis. Miami University. Oxford, Ohio. 61 pp.

2418. ----------. 1969. Developmental anatomy of primary and lateral roots of Ipomoea purpurea. Diss. Absts. Inter. 30(7):3053B.

2419. ----------., and HEIMSCH, C. 1969. Apical organization in roots of the Convolvulaceae. Amer. Jour. Bot. 56:131-138.

2420. ----------. 1971. Developmental anatomy in roots of Ipomoea purpurea. I. Radicle and primary root. Amer. Jour. Bot. 58:604-615.

2421. ----------. 1971. Origin of secondary roots in Ipomoea purpurea seedlings. Amer. Jour. Bot. 58(5, Part 2):456. (Abstract)

2422. SEELIGER, R. 1919. Untersuchungen über das Dickenwachstum der Zuckerrübe (Beta vulgaris L. var. rapa Dum.). Arb. Biol. Reichs. Land-Forst. 10:149-194.

2423. SEIDELIN, A. 1912. The structure and biology of arctic flowering plants. 5. Hippuridaceae, Halorrhagidaceae and Callitrichaceae. Medd. Grönl. 36:297-332.

2424. SELIBER, G. 1905. Variationen von Jussieua repens mit besonderer berücksichtigung des bei der Wasserform vorkommenden Aërenchyms. Nova Acta Abh. Kaiser. Leop.- Carol. Deutsch. Akad. Nat. 84:147-198.

2425. SEN, D.N. 1961. Root ecology of Tilia europaea L. I. On the morphology of mycorrhizal roots. Preslia 33:341-350.

2426. --------., and JENIK, J. 1962. Root ecology of Tilia europaea L. Anatomy of mycorrhizal roots. Nature 193:1101-1102.

2427. --------. 1964. Root system and root ecology of Tilia europaea L. Acta Univ. Carol. (Biol. I Suppl.) 1964:5-83.

2428. SEN, U. 1968. Morphology and anatomy of Ophioglossum reticulatum. Canad. Jour. Bot. 46:957-968.

2429. SENFT, E. 1906. Über einiger medizinisch verwendete Pflanzen aus Familie der Ranunculaceen. Pharm. Praxis 5:1-11.

2430. SENN, G. 1923. Ueber die Ursachen der Brettwurzel-Bildung bei der Pyramiden-Pappel. Verh. Nat. Gesel. Basel 35:405-435.

2431. SERGUÉEFF, M. 1907. Contribution à la morphologie et la biologie des Aponogétonacées. Dcotoral thèses. Université de Genève. W. Kundig et Fils, Genève. 132 pp.

2432. SERVETTAZ, G. 1909. Monographie des Eléagnacées. Beih. Bot. Zentralb. 25:1-420.

2433. SESARMA, P. 1963. On seedling anatomy of <u>Lagenaria</u> <u>vulgaris</u> Seringe and <u>Luffa</u> <u>acutangula</u> Roxb. Sci. Cult. 29:138-139.

2434. SESHARDI, C.R., RAO, M.B., and MUHAMED, S.V. 1958. Studies on root development in groundnut. Indian Jour. Agric. Sci. 28:211-215.

2435. SEVERIN, C.F. 1932. Origin and structure of the secondary root of <u>Sagittaria</u>. Bot. Gaz. 93:93-99.

2436. SEVERINI, G. 1909. Particolarità morfologiche ed anatomiche nelle radici dell'<u>Hedysarum</u> <u>coronarium</u> L. Annali Bot. 7:75-82.

2437. SEWARD, A.C., and GOWAN, J. 1900. The maidenhair tree (<u>Ginkgo</u> <u>biloba</u> L.). Annals Bot. 14:109-154.

2438. -----------., and FORD, S.O. 1902. The anatomy of <u>Todea</u>, with notes on the geological history and affinities of the Osmundaceae. Trans. Linn. Soc. London (Ser. 2) 6:237-260.

2439. ------------- and ---------. 1906. IX. The Araucarieae, recent and extinct. Philos. Trans. Roy. Soc. London (Ser. B) 198:305-409.

2440. SHAH, C.S., and AGHARA, L.P. 1958. Pharmacognosy of Tuticorin Yellow Root - a substitute of Senega. Indian Jour. Pharm. 19:97.

2441. ---------., and VYAS, L.S. 1958. Pharmacognosy of <u>Polygala</u> <u>chinensis</u> Linn. and comparison with Chinensis I.P. Indian Jour. Pharm. 19: 224-227.

2442. ---------., and SUKKAWALA, V.M. 1959. Pharmacognostic study of <u>Mollugo</u> <u>hirta</u> Thunb. (<u>Glinus</u> <u>lotoides</u> Linne). Indian Jour. Pharm. 21:64-68.

2443. ----------, -----------------., and VYAS, L.S. 1959. Pharmacognosy of <u>Withania</u> <u>somnifera</u> Dunal. Indian Jour. Pharm. 21:195-199.

2444. ----------., and PANDHE, M.K. 1960. Pharmacognostic study of <u>Polygala</u> <u>abyssinica</u> R. Br. Indian Jour. Pharm. 22:66-68.

2445. SHAN, REN-HWA. 1944. Anatomical study of certain Umbelliferous seedlings. Sinesia 15:105-118.

2446. SHANNON, E.L. 1953. The production of root hairs by aquatic plants. Amer. Midl. Nat. 50:474-479.

2447. SHARMA, M.R. 1962. Morphological and anatomical investigations on Artocarpus Forst. I. Vegetative organs. Proc. Indian Acad. Sci. (Sect. B) Ref. Oct. 1962. 56:243-258.

2448. SHARMA, O.P., and MEHRA, P.N. 1970. Comparative anatomy of Kobresia Willd. (Cyperaceae). Res. Bull. (Sci.) Panjab Univ. (N.S.) 21: 119-128.

2449. ------------- and ----------. 1972. Systematic anatomy of Fimbristylis Vahl (Cyperaceae). Bot. Gaz. 133:87-95.

2450. SHARMA, U. 1963. Studies on Indian Hymenophyllaceae. Part II. Contributions to our knowledge of Hymenophyllum simonsianum Hooker. Proc. Nat. Inst. Sci. India (Part B, Biol. Sci.) 29:93-105.

2451. SHARMAN, B.C. 1939. The development of the sinker of Orchis mascula. Jour. Linn. Soc. (Bot.) 52:145-158.

2452. SHAW, A.C., LAZELL, S.K., and FOSTER, G.N. 1965. Photomicrographs of the Flowering Plant. Longmans, London. 79 pp.

2453. SHAW, F.J.F. 1909. The seedling structure of Araucaria bidwillii. Annals Bot. 23:321-334.

2454. SHIBATA, K. 1900. Beiträge zur Wachstumsgeschichte der Bambusgewächse. Jour. Coll. Sci. Imp. Univ. Tokyo 8:427-496.

2455. ----------. 1900. On the anatomical structure of vegetative organs of bamboo-plants. Bot. Mag. (Tokyo) 14:206-218; 231-241.

2456. SHIMABUKU, K. 1960. Observation on the apical meristem of rice roots. Bot. Mag. (Tokyo) 73:22-28.

2457. SHUKHOBODSKIĬ, B.A. 1965. On the formation of gutta and localization of gutta receptacles in the young roots of Euonymus L. Rast. Resursy 2:258-266. (in Russian)

2458. SHUSHAN, S. 1959. Developmental anatomy of an orchid, Cattleya x Trimos. In: The Orchids. (ed.) Withner, C.L. Ronald Press, New York. pp. 45-72.

2459. SIEBE, M. 1903. Ueber den anatomischen Bau der Apostasiinae. Inagural-Dissertation. Universität Heidelberg. Heidelb. Verlangsanst., Heidelberg. 63 pp.

2460. SIEBERT, A. 1920. Ergrünungsfähigkeit von Wurzeln. Beih. Bot. Centralb. 37:185-216.

2461. SIEGLER, A.E., and BOWMAN, J.J. 1939. Anatomical studies of root and shoot primordia in 1-year apple roots. Jour. Agric. Res. 58:795-803.

2462. SIFTON, H.B. 1925. Poison canals of Cicuta maculata. Bot. Gaz. 80:319-324.

2463. ------------. 1944. Developmental morphology of vascular plants. New Phytol. 43:87-129.

2464. ------------. 1945. Air-space tissue in plants. Bot. Rev. 11:108-143.

2465. SIIM-JENSEN, J. 1901. Beiträge zur botanischen und pharmacognostichen Kenntnis von Hyoscyamus niger L. Biblio. Bot. Heft 51. pp. 1-89.

2466. SILER, M.B. 1931. The transition from root to stem in Helianthus annus L. and Arctium minus Bernh. Amer. Midl. Nat. 12:425-487.

2467. SILVA, B.L.T. DE. 1936. Secondary thickening in the roots of Dracaena. Ceylon Jour. Sci. (Sect. A, Bot.) 12:127-135.

2468. SIMMLER, G. 1910. Monographie der Gattung Saponaria. Denk. Kaiserl. Akad. Wiss. Wien (Math.- Nat. Kl.) 85:433-529.

2469. SIMON, S. 1904. Untersuchungen über die Regeneration der Wurzelspitze. Jahrb. Wiss. Bot. 40:103-142.

2470. SIMONDS, A.O. 1935. Histological studies of the development of the root and crown of alfalfa. Iowa State Coll. Jour. Sci. 9:641-659.

2471. --------------. 1938. The anatomical development of Lepidium draba. Jour. Agric. Res. 57:917-928.

2472. SIMPSON, P.G., and FINERAN, B.A. 1970. Structure and development of the haustorium in Mida salicifolia. Phytomorph. 20:236-248.

2473. SINGH, B. 1943. A contribution to the anatomy of Salvadora persica L. with special reference to the origin of the included phloem. Jour. Indian Bot. Soc. 23:71-78.

2474. ---------. 1945. An anatomical study of Tiliacora acuminata Miers. Jour. Indian Bot. Soc. 24:135-146.

2475. ---------. 1954. Studies in the family Loranthaceae. Jour. Res. Agra Univ. (Science) 3:301-315.

2476. SINGH, P. 1956. Pharmacognosy of root and rhizome of Cocculus villosus DC. Indian Jour. Pharm. 18:393-396.

167

2477. --------. 1963. Pharmacognostic study of root of Streblus asper
Lour. Planta Med. 11:191-197.

2478. --------. 1966. On the root of Limonia acidissima Auct. Quart.
Jour. Crude Drug Res. 6:916-919.

2479. SINGH, S.N. 1947. Aerial roots in the sponge gourd, Luffa sp.
Jour. Bombay Nat. Hist. Soc. 47:397-398.

2480. SINGH, T.C.N. 1963. An anatomical and ecological study of some
ferns from Mussoori (North-western Himalayas). Jour. Indian Bot.
Soc. 42:475-544.

2481. SINGH, V. 1965. Morphological and anatomical studies in Helobiae.
IV. Vegetative and floral anatomy of Aponogetonaceae. Proc.
Indian Acad. Sci. (Sect. B) 61:147-159.

2482. SINNOTT, E.W. 1939. Growth and differentiation in living plant
meristems. Proc. Nat. Acad. Sci. U.S.A. 25:55-58.

2483. -------------., and BLOCH, R. 1939. Changes in intercellular
relationships during growth and differentiation of living plant
tissues. Amer. Jour. Bot. 26:625-634.

2484. -------------- and --------. 1939. Cell polarity and the differentiation of root hairs. Proc. Nat. Acad. Sci. U.S.A. 25:248-252.

2485. -------------- and --------. 1946. Comparative differentiation in
the air roots of Monstera deliciosa. Amer. Jour. Bot. 33:587-590.

2486. ------------. 1960. Plant Morphogenesis. McGraw-Hill Book Co.,
New York. 550 pp.

2487. SITTE, P. 1958. Die ultrastruktur von Wurzelmeristemzellen der Erbse
(Pisum sativum). Protoplasma 49:447-522.

2488. SKORIĆ, V. 1923. Contributions to the anatomy of the genus Daphne.
Glas. Hrvat. Pirod. Druš. 34:(55-65) 1-11. (in Croatian with
German summary)

2489. SKOTTSBERG, C. 1947. Peperomia barteroana Miq. and P. tristanensis
Christoph. an interesting case of disjunction. Acat Hort. Berg.
16:251-288.

2490. SKUTCH, A.F. 1928. Origin of endodermis in ferns. Bot. Gaz.
86:113-114.

2491. ------------. 1930. Repeated fission of stem and root in Mertensia
maritima - a study in ecological anatomy. Ann. New York Acad. Sci.
32:1-52.

2492. -----------. 1932. Anatomy of the axis of the banana. Bot. Gaz. 93:233-258.

2493. SMITH, A.C. 1901. The structure and parasitism of Aphyllon uniflorum Gray. Contr. Bot. Lab. Univ. Penn. 2:111-121. (Also in Trans. Proc. Bot. Soc. Penn. 1898. 1:111-121)

2494. SMITH, A.I. 1936. Adventitious roots in stem cuttings of Begonia maculata and B. semperflorens. Amer. Jour. Bot. 23:511-515.

2495. -----------. 1942. Adventitious roots in stem cuttings of Tropaeolum majus L. Amer. Jour. Bot. 29:192-194.

2496. SMITH, C.M. 1931. Development of Dionaea muscipula. II. Germination of seed and development of seedling to maturity. Bot. Gaz. 91: 377-394.

2497. SMITH, F.H. 1930. The corm and contractile roots of Brodiaea lactea. Amer. Jour. Bot. 17:916-927.

2498. -----------. 1930. Notes on the contractile roots of Brodiaea lactea (Lindl.) Wats. Northw. Sci. 4:18-19.

2499. SMITH, G.F., and KERSTEN, H. 1942. The relation between xylem thickenings in primary roots of Vicia faba seedlings and elongation, as shown by soft X-ray irradiation. Bull. Torrey Bot. Club 69:221-234.

2500. SMITH, G.M. 1938. Cryptogamic Botany. 2nd ed. Vol. 2. McGraw-Hill Book Co., New York. 399 pp.

2501. SMITH, P.M. 1916. The development of the embryo and seedling of Dioscorea villosa. Bull. Torrey Bot. Club 43:545-558.

2502. SMITH, W. 1909. The anatomy of some Sapotaceous seedlings. Trans. Linn. Soc. London (Bot.) Ser. 2. 7:189-200.

2503. SNOW. L.M. 1905. The development of root hairs. Bot. Gaz. 40:12-48.

2504. SNYDER, W.E. 1962. Plant anatomy as related to the rooting of cuttings. Proc. Plant Prop. Soc. pp. 43-47.

2505. SOEDING, H. 1924. Anatomie der Wurzel, - Stengel- und Rübenbildung von Oelraps und Steckrübe (Brassica napus L. var. oleifera und var. napobrassica). Bot. Archiv 7:41-69.

2506. SOEDJONO, E.H. 1965. Comparative anatomical study of three species of Microseris. Diss. Absts. 25(9):4931.

2507. SOLEREDER, H., and MEYER, F.J. 1928. Systematische Anatomie der Monokotyledonen. Heft 2. Glumiflorae. Gebrüder Borntraeger, Berlin.

169

2508. ------------ and ----------. 1928. Systematische Anatomie der
Monokotyledonen. Heft 3. Principes-Synanthae-Spathiflorae.
Gebrüder Borntraeger, Berlin. pp. 1-175.

2509. ------------ and ----------. 1928. Systematische Anatomie der
Monokotyledonen. Heft 5. Liliiflorae. Gebrüder Borntraeger,
Berlin.

2510. ------------ and ----------. 1929. Systematische Anatomie der
Monokotyledonen. Heft 4. Farinosae. Gebrüder Borntraeger, Berlin.
pp. 1-176.

2511. ------------ and ----------. 1930. Systematische Anatomie der
Monokotyledonen. Heft 6. Scitamineae-Microspermae. Gebrüder
Borntraeger, Berlin. pp. 1-242.

2512. ------------ and ----------. 1933. Systematische Anatomie der
Monokotyledonen. Heft 1. Pandanales, Helobiae, Triuridales. Teil
1. Typhaceae - Scheuzeriaceae. Gebrüder Borntraeger, Berlin. pp.
1-155.

2513. SOLOMON, R. 1931. The anatomy of caudex and root of Eriocaulon
septangulare. Jour. Indian Bot. Soc. 10:139-144.

2514. SOPER, K. 1959. Root anatomy of grasses and clovers. New Zeal. Jour.
Agric. Res. 2:329-341.

2515. SOUÈGES, R. 1931. L'embryon chez le Sagittaria sagittaefolia L. Ann.
Sci. Nat., Bot. (Sér. 10) 13:353-402.

2516. SOUTH, F.W., and COMPTON, R.H. 1908. On the anatomy of Dioon edule
Lindl. New Phytol. 7:222-229.

2517. SPANN, L. 1933. Morphology and anatomy of Lygodesmia juncea (Pursh)
D. Don. Univ. Kans. Sci. Bull. 21:421-438.

2518. SPERLICH, A. 1902. Beiträge zur Kenntnis der Inhaltstoffe in den
Saugorganen der grunen Rhinanthaceen. Beih. Bot. Centralb. 11:
437-485.

2519. ------------. 1925. Die vegetationsorgane der Anthophyten. Organe
besondere physiologischer dignität. A. Die absorptionsorgane der
parasitischen Samenpflanzen. In: Handbuch der Pflanzenanatomie.
(ed.) Linsbauer, K. Gebrüder Borntraeger, Berlin. 9(2):iv-52.

2520. SPESSARD, E.A. 1928. Anatomy of Lycopodium seedling. Bot. Gaz.
85:323-333.

2521. SPIETH, A.M. 1933. Anatomy of the transition region in Gossypium.
Bot. Gaz. 95:338-347.

2522. SPORNE, K.R. 1966. The Morphology of Pteridophytes. Hutchinson and Co., Ltd., London. 192 pp.

2523. SPRAGUE, E.F. 1962. Parasitism in Pedicularis. Madroño 16:192-200.

2524. SPRATT, A.V. 1920. Some anomalies in Monocotyledonous roots. Annals Bot. 34:99-105.

2525. SPRATT, E.R. 1912. The morphology of the root tubercles of Alnus and Elaeagnus, and the polymorphism of the organism causing their formation. Annals Bot. 26:119-128.

2526. -----------. 1912. The formation and physiological significance of root nodules in the Podocarpineae. Annals Bot. 26:801-814.

2527. SPRECHER, A. 1919. Étude sur la sémence et la germination du Garcinia mangostana L. Rev. Gén. Bot. 31:513-531.

2528. SPRENGER, M. 1904. Ueber den anatomischen Bau der Bolbophyllinae. Inaugural-Dissertation. Ruprecht-Karls-Universität zu Heidelberg. J. Hörning, Heidelberg. 61 pp.

2529. SRIVASTAVA, L.M., and ESAU, K. 1961. Relation of dwarf mistletoe (Arceuthobium) to the xylem tissue in conifers. I. Anatomy of the parasite sinkers and their connection with host xylem. Amer. Jour. Bot. 48:159-167.

2530. STABER, M.J. 1909. Notes on the anatomy of Sesbania macrocarpa Muhl. Bull. Torrey Bot. Club 36:625-633.

2531. STAESCHE, K. 1968. Vergleichende anatomische Untersuchungen an Drogen. Apoth.- Zeit. 108:329-332.

2532. STANGLER, B.B. 1956. Origin and development of adventitious roots in stem cuttings of chrysanthemum, carnation, and rose. Mem. Cornell Univ. Agric. Exp. Sta. No. 342. pp. 3-24.

2533. STANT, M.Y. 1964. Anatomy of the Alismataceae. Jour. Linn. Soc. (Bot.) 59:1-42.

2534. ----------. 1967. Anatomy of the Butomaceae. Jour. Linn. Soc. (Bot.) 60:31-60.

2535. ----------. 1970. Anatomy of Petrosavia stellaris Becc., a saprophytic monocotyledon. In: New Research in Plant Anatomy. (eds.) Robson, N.K.B., Cutler, D.F., and Gregory, M. Academic Press, New York. pp. 147-161.

2536. STAPF, O. 1905. The Aconites of India: a monograph. Ann. Roy. Bot. Garden, Calcutta 10:115-197.

2537. STARR, A.M. 1912. Comparative anatomy of dune plants. Bot. Gaz. 54:265-305.

2538. STEEVES, T.A., and SUSSEX, I.M. 1972. Patterns in Plant Development. Prentice-Hall, Inc., Englewood Cliffs, New Jersey. 302 pp.

2539. STEIGER, E. 1920. Beiträge zur Morphologie der Polygala senega L. Berich. Deutsch. Pharm. Gesel. 30:43-116.

2540. STEPHENS, E.L. 1912. The structure and development of the haustorium of Striga lutea. Annals Bot. 26:1067-1076.

2541. --------------. 1912. Note on the anatomy of Striga lutea Lour. Annals Bot. 26:1125-1126.

2542. STERCKX, R. 1900. Recherches anatomiques sur l'embryon et les plantules dans la famille des Renonculacées. Mém. Soc. Roy. Soc. Liège (Sér. 3) 2:3-112. (Also in Arch. Inst. Bot. Univ. Liège 1900 2:3-117)

2543. STERLING, C.M. 1912. Krameria canescens Gray. Kansas Univ. Sci. Bull. 6:363-372.

2544. STERN, K. 1916. Beiträge zur Kenntnis der Nepenthaceen. Flora 109: 213-282.

2545. STEVENS, O.A. 1966. Rhizomes, stolons and roots. Castanea 31:140-145.

2546. STEVENS, W.C. 1916. Plant Anatomy. Blakiston, Philadelphia. 399 pp.

2547. STEVENSON, D.W., and POPHAM, R.A. 1971. The seedling anatomy of Bougainvillea spectabilis. Amer. Jour. Bot. 58(5, Part 2):461. (Abstract)

2548. STEWART, D.F. 1931. Notes on salt-marsh plants. II. Plantago maritima Linn. Trans. Proc. Bot. Soc. Edinb. 30:313-319.

2549. STEWART, W.N. 1948. A comparative study of stigmarian appendages and Isoetes roots. Amer. Jour. Bot. 34:315-324.

2550. STILES, W. 1912. The Podocarpeae. Annals Bot. 26:443-514.

2551. STOCKING, C.R. 1956. Histology and development of the root. In: Handbuch der Pflanzenphysiologie. (ed.) Ruhland, W. Springer-Verlag, Berlin. 3:173-187.

2552. STOJANOW, N. 1916. Über die vegetativ Fortpflanzung der Ophrydineen. Flora 109:1-39.

2553. STOKEY, A.G. 1907. The roots of Lycopodium pithyoides. Bot. Gaz. 44:57-63.

2554. -----------. 1909. The anatomy of Isoetes. Bot. Gaz. 47:311-335.

2555. STOVER, E.L. 1928. The roots of wild rice _Zizania aquatica_ L. Ohio Jour. Sci. 28:43-49.

2556. -----------. 1951. An Introduction to the Anatomy of Seed Plants. D.C. Heath and Co., Boston. 274 pp.

2557. STRASBURGER, E. 1906. Über die Verdickungsweise der Stamme von Palmen und Schraubenbaumen. Jahrb. Wiss. Bot. 43:580-628.

2558. STREET, H.E. 1967. The ageing of root meristems. In: Symposia of the Society for Experimental Biology. Aspects of the biology of ageing. No. 21. pp. 517-542.

2559. -----------., ÖPIK, H., and JAMES, F.E.L. 1968. Fine structure of the main axis meristems of cultured tomato roots. Phytomorph. 17:391-401.

2560. STRIGL, M. 1908. Der Thallus von _Balanophora_, anatomisch-physiologisch geschildert. Sitz. Kaiser. Akad. Wiss. Wien (Math.- Nat. Kl.) 117 (II Halfb.):1127-1175.

2561. STURM, K. 1910. Monographische Studien über _Adoxa moschatellina_ L. Viert. Nat. Gesel. Zürich 55:391-462.

2562. SUESSENGUTH, K. 1927. Über die Gattung _Lennoa_. Ein beitrag zur Kenntnis exotischer Parasiten. Flora (N.F.) 122:264-305.

2563. SUKKAWALA, V.M., and SHAH, C.S. 1960. Pharmacognostic study of _Merremia emarginata_, Hallier. Indian Jour. Pharm. 22:147.

2564. SUN, C.N. 1955. Growth and development of primary tissues in aerated and non-aerated roots of soybean. Bull. Torrey Bot. Club 82:491-502.

2565. --------. 1957. Zonation and organization of root apical meristems of _Glycine max_. Bull. Torrey Bot. Club 84:69-78.

2566. --------. 1959. Zonation and differentiation of tissues in the primary root of soybean. Diss. Abasts. 19(7):1534-1535.

2567. --------. 1962. Fine structure of root cells of _Phaseolus vulgaris_. I. Structure of the meristematic cell. Cytologia 27:204-211.

2568. SURANGE, K.R. 1949. A contribution to the morphology andanatomy of the Cyclanthaceae. Trans. Nat. Inst. Sci. India 3:159-209.

2569. SUSSEX, I.M., and STEEVES, T.A. 1968. Apical initials and the concept of promeristem. Phytomorph. 17:387-391.

2570. SUTHERLAND, G.K., and EASTWOOD, A. 1916. The physiological anatomy of _Spartina townsendii_. Annals Bot. 30:333-351.

2571. SVEDELIUS, N. 1904. On the life-history of Enhalus acoroides. Ann. Roy. Bot. Gard., Perideniya 2:267-297.

2572. SWARTLEY, J.C. 1942. Adventitious root initiation in Forsythia, Ribes and Caragana. Master's thesis. Ohio State University, Columbus. 55 pp.

2573. SWINGLE, C.F. 1930. The anatomy of Euphorbia intisy. Jour. Agric. Res. 40:615-625.

2574. -------------. 1940. Regenerative and vegetative propagation. Bot. Rev. 6:301-355.

2575. -------------. 1952. Regenerative and vegetative propagation. II. Bot. Rev. 18:1-13.

2576. SYKES, M.G. 1910. The anatomy of Welwitschia mirabilis Hook f., in the seedling and adult states. Trans. Linn. Soc. London (Ser. 2, Bot.) 7:327-354.

2577. SZABÓ, Z. 1910. Entwicklungsgeschichtliche Beobachtungen an Knautia-Arten. Bot. Közlem. 8:100-101; (44). (in Hungarian with German summary)

2578. --------. 1910. Nouvelles observations concernant l'histologie et le développement des organes sur les espéces du genere Knautia. Bot. Közlem. 9:133-148.

2579. SZENTPÉTERY, G.B., and SÁRKÁNY, S. 1963. Beobachtungen hinsichtlich der Ontogenese und Organisation der einheimischen Arzneibaldriane. Ann. Univ. Sci. Budap. Rol. Eöt. Nom. (Sect. Biol.) 6:13-41.

2580. ------------------- and ----------. 1964. Histological analysis of the fully developed root in some Valeriana species. Ann. Univ. Sci. Budap. Rol. Eöt. Nom. (Biol.) 7:213-228.

2581. ----------------. 1965. Histogenetische Untersuchung der sprossbürtigen Wurzel bei den Arten Valeriana officinalis L.und Valeriana collina Wallr. I. Bildung der sprossbürtigen Wurzel. Acta Bot. Acad. Sci. Hung. 11:405-419.

2582. ----------------., and SÁRKÁNY, S. 1966. Initiierung und anfangliche Organisation des Wurzelvegetationskegels in den Wurzelgeweben. I. Valeriana officinalis L. und V. collina Wallr. Ann. Univ. Sci. Budap. Rol. Eöt. Nom. (Biol.) 8:243-252.

2583. ----------------. 1967. Histogenetische Untersuchungen an sprossbürtige Wurzeln von Valeriana officinalis L. und Valeriana collina Wallr. II. Entwicklung der Struktur der Wurzelspitze nach Austritt aus der Rhizom Acta Bot. Acad. Sci. Hung. 13:311-324.

2584. ----------------. 1968. Histogenetische Untersuchungen an spross-
bürtigen Wurzeln von Valeriana officinalis L. und Valeriana collina
Wallr. III. Determination, Differenzierung, und Stabilisation der
Plerom-Gewebe. Acta Bot. Acad. Sci. Hung. 14:175-195.

2585. SZKLARSKA, J. 1966. A study on the variability of anatomical structure
of chicory roots (Cichorium intybus L. var. sativum Bisch.). In:
Adv. Front. Plant Sci. (ed.) Chandra, L. 16:74-76.

2586. ------------. 1966. Causes for the formation of cavities in the roots
of chicory. Hod. Ros. Aklim. Nas. 10:69-87. (in Polish)

2587. SZYNAL, T. 1934. La structure de la racine du blé et le climat. Acta
Soc. Bot. Poloniae 11:267-270.

2588. TAINTER, F.H. 1971. The ultrastructure of Arceuthobium pusillum.
Canad. Jour. Bot. 49:1615-1622.

2589. TAKANO, T. 1966. Studies on the pithiness of radish. IV. On the
process of pithy tissue formation in the radish root. Jour. Jap.
Soc. Hort. Sci. 35:152-157.

2590. TAKENOUCHI, Y. 1931. Systematisch-vergleichende Morphologie und
Anatomie der Vegetationsorgane der japanischen Bambus-Arten. Mem.
Fac. Sci. Agric. Taihoku Imp. Univ. 3:1-60.

2591. TANAKA, N. 1968. Studies on root system formation in leguminous crop
plants. Proc. Crop Sci. Soc. Japan 37:424-429.

2592. TANSLEY, A.G. 1902. "Reduction" in descent. New Phytol. 1:131-133.

2593. ------------., and CHICK, E. 1903. On the structure of Schizaea
malaccana. Annals Bot. 17:493-510.

2594. ------------., and THOMAS, E.N. 1904. Root-structure in the central
cylinder of the hypocotyl. New Phytol. 3:104-106.

2595. ------------., and LULHAM, R.B.J. 1905. A study of the vascular system
of Matonia pectinata. Annals Bot. 19:475-519.

2596. ------------., and THOMAS E.R. 1906 (1907). The phylogenetic value
of the vascular structure of spermophytic hypocotyls. Rept. Brit.
Assn. Adv. Sci. No. 76. pp. 761-763.

2597. TATE, P. 1925. On the anatomy of Orobanche hederae Duby and its
attachment to its host. New Phytol. 24:284-293.

2598. TAYLOR, A.R.A. 1957. Studies of the development of Zostera marina L.
I. The embryo and seed. Canad. Jour. Bot. 35:477-499.

2599. ------------. 1957. Studies of the development of *Zostera marina* L. II. Germination and seedling development. Canad. Jour. Bot. 35: 681-695.

2600. TAYLOR, G. 1926. The origin of adventitious growths in *Acanthus montanus*. Trans. Proc. Bot. Soc. Edinb. 29:291-296.

2601. TELANG, S.W., CHAUHAN, G.S., and GUAR, S.K.S. 1969. Anatomy of *Chenopodium* and *Beta benghalensis*. Labdev. Jour. Sci. Tech. (Part B) 7-B:277-279.

2602. TELLEFSON, M.A. 1922. The relation of age to size in certain root cells and in vein-islets of the leaves of *Salix nigra* Marsh. Amer. Jour. Bot. 9:121-139.

2603. TELLINI, G. 1939 (1940). Ricerche anatomiche su *Dianthus arboreus* L. versus (= *D*. *aciphyllus* Sieb.). Nuovo Giron. Bot. Ital. 46:615-642.

2604. TEN EYCK, A.M. 1905. The roots of plants. Bull. Kansas Agric. Exp. Sta. No. 127. pp. 199-252.

2605. TERNETZ, C. 1902. Morphologie und Anatomie der *Azorella selago* Hook f. Bot. Zeit. 60:1-20.

2606. TERRAS, J.A. 1900. The relation between the lenticels and adventitious roots of *Solanum dulcamara*. Trans. Proc. Bot. Soc. Edinb. 21:341-352.

2607. ------------. 1905. Notes on the origin of lenticels, with special reference to those occurring in roots. Trans. Proc. Bot. Soc. Edinb. 22:450-457.

2608. TETLEY, U. 1925. The secretory system of the roots of the Compositae. New Phytol. 24:138-162.

2609. The contractile roots of bulbs. (Anonymous author) Gard. Chron. (Ser. 3) 89:251.

2610. THEODOROV, A., KIRPICZNIKOV, M., and ARTJUSCHENKO, Z. 1962. *Illustrated Organography of Vascular Plants: Stems and Roots*. Academy of Science URSS, Moscow. 350 pp. (in Russian)

2611. *The Structure and Biology of Arctic Flowering Plants*. 1912. Medd. Grön. Part I. Vol. 36. 481 pp. (Author ?)

2612. *The Structure and Biology of Arctic Flowering Plants*. 1921. Medd. Grön. Part II. Vol. 37. 507 pp. (Author ?)

2613. THIBAULT, M. 1946. Contribution à l'étude des radicelles de carotte. Rev. Gén. Bot. 53:434-460.

176

2614. THIEL, A.F. 1931. Anatomy of the primary axis of Solanum melongena. Bot. Gaz. 92:407-419.

2615. ----------. 1933. Vascular anatomy of the transition region of certain solanaceous plants. Bot. Gaz. 94:598-604.

2616. ----------. 1934. Anatomy of the transition region of Helianthus annuus. Jour. Eli. Mitchell Sci. Soc. 50:268-274.

2617. THIESSEN, R. 1908. The vascular anatomy of the seedling of Dioon edule. Bot. Gaz. 46:357-380.

2618. THIRUMALACHAR, M.J., SWAMY, B.G.L., and KHAN, K.B.A. 1942. A note on the epiphytism in Heptapleurum venulosum Seem. Jour. Bombay Nat. Hist. Soc. 43:276-277.

2619. THODAY, D. 1926. The contractile roots of Oxalis incarnata. Annals Bot. 40:571-583.

2620. ----------., and JOHNSON, E.T. 1930. On Arceuthobium pusillum Peck. I. The endophytic system. Annals Bot. 44:393-413.

2621. ----------., and DAVEY, A.J. 1932. Contractile roots. II. On the mechanism of root contraction in Oxalis incarnata. Annals Bot. 46:993-1005.

2622. ----------. 1951. The haustorial system of Viscum album. Jour. Exp. Bot. 2:1-19.

2623. ----------. 1956. Modes of union and interaction between parasite and host in the Loranthaceae. I. Viscoideae, not including Phoradendreae. Proc. Roy. Soc. London (Ser. B) 145:531-548.

2624. ----------. 1957. Modes of union and interaction between parasite and host in the Loranthaceae. II. Phoradendreae. Proc. Roy. Soc. London (Ser. B) 146:320-338.

2625. ----------. 1958. Modes of union and interaction between parasite and host in the Loranthaceae. III. Further observations on Viscum and Korthalsella. Proc. Roy. Soc. London (Ser. B) 148:188-206.

2626. ----------. 1958. Modes of union and interaction between parasite and host in the Loranthaceae. IV. Viscum obscurum on Euphorbia polygona. Proc. Roy. Soc. London (Ser. B) 149:42-57.

2627. THODAY, M.G. 1911. On the histological relations between Cuscuta and its host. Annals Bot. 25:655-682.

2628. THOENES, H. 1929. Morphologie und Anatomie von Cynosurus cristatus und die Erscheinungen der Viviparie bei ihm. Bot. Archiv 25:284-346.

2629. THOMAS, E.N. 1905. Some points on the anatomy of <u>Achrosticum</u> <u>aureum</u>. New Phytol. 4:175-189.

2630. -----------. 1907. A theory of the double leaf-trace founded on seedling structure. New Phytol. 6:77-91.

2631. -----------. 1914. Seedling anatomy of Ranales, Rhoedales, and Rosales. Annals Bot. 28:695-733.

2632. -----------. 1923. Observations on the seedling anatomy of the genus <u>Ricinus</u>. Proc. Linn. Soc. London 135:49-50.

2633. -----------. 1923. Observations on the seedling anatomy of the Ebenales. Rept. Brit. Assn. Adv. Sci. 91st Meeting p. 491.

2634. -----------., and HOLMES, L.E. 1930. The development and structure of the seedling and young plant of the pineapple (<u>Ananas</u> <u>sativus</u>). New Phytol. 29:199-226.

2635. THOMPSON, C.B. 1904. The structure and development of internal phloem in <u>Gelsemium</u> <u>sempervirens</u> Ait. Contrib. Bot. Lab. Univ. Penn. (N.S. 5) 2:41-53. (Also in Trans. Proc. Bot. Soc. Penn. 1898 1:41-53)

2636. THOMPSON, J., and CLOWES, F.A.L. 1968. The quiescent centre and rates of mitosis in the root meristem of <u>Allium</u> <u>sativum</u>. Annals Bot. (N.S.) 32:1-13.

2637. THOMPSON, S.H. 1960. Cellular development and morphogeny of the root tip of <u>Trillium</u> <u>grandiflorum</u>. Bot. Gaz. 121:215-220.

2638. THOMPSON, W.P. 1912. The anatomy and relationships of the Gnetales. I. The genus <u>Ephedra</u>. Annals Bot. 26:1077-1104.

2639. THOMSON, R.B. 1913. On the comparative anatomy and affinities of the Araucarineae. Philos. Trans. Roy. Soc. London (Ser. b) 204:1-50.

2640. -----------. 1934. A seed-plant feature of the roots of Marattiaceae. New Phytol. 33:96-100.

2641. TICHOMIROW, W.A. 1913 (1912). Zur Kenntnis der Wurzelbaues von <u>Smilax</u> <u>excelsa</u> L., der Transkaukasiens-Sarsaparilla, Ekale der Iberier, mit <u>Smilax</u> <u>aspera</u> L. vergelichen. Bull. Soc. Imp. Nat. Moscov (N.S.) 26:401-421.

2642. TIEGS, E. 1913. Beiträge zur Kenntnis der Entstehung und des Wachstums der Wurzelhauben einiger Leguminosen. Jahrb. Wiss. Bot. 52:622-646.

2643. TIMMEL, H. 1927. Über die Bildung anomaler Tracheiden im Phloem. Flora 122:203-241.

2644. TISCHLER, G. 1905. Über das Vorkommen von Statolithen bei wenig oder gar nicht geotropischen Wurzeln. Flora 94:1-67.

2645. -----------. 1910. Untersuchungen an Mangrove- und Orchideen-Wurzeln mit spezieller Beziehung auf die Statolithen-Theorie des Geotropismus. Ann. Jard. Bot. Buitenz. Suppl. 3, Part I. pp. 131-186.

2646. TODA, R., and SATOO, S. 1948. The development of roots arising from callus tissues in young seedling cuttings of pine. Jour. Jap. For. Soc. 30:20-25.

2647. TOMLINSON, P.B. 1956. Studies in the systematic anatomy of the Zingiberaceae. Jour. Linn. Soc. (Bot.) 55:547-592.

2648. --------------. 1959. An anatomical study of the classification of the Musaceae. Jour. Linn. Soc. (Bot.) 55:779-809.

2649. --------------. 1960. The anatomy of Phenakospermum (Musaceae). Jour. Arnold Arb. 41:287-297.

2650. --------------. 1961. Palmae. In: Anatomy of the Monocotyledons. (ed.) Metcalfe, C.R. Vol. 2. Clarendon Press, Oxford. 453 pp.

2651. --------------. 1961. The anatomy of Canna. Jour. Linn. Soc. (Bot.) 56:467-476.

2652. --------------. 1961. Morphological and anatomical characteristics of the Marantaceae. Jour. Linn. Soc. (Bot.) 58:55-78.

2653. --------------. 1962. Phylogeny of the Scitamineae - morphological and anatomical considerations. Evolution 16:192-213.

2654. --------------. 1962. Essays on the morphology of palms. VIII. The root. Principes 6:122-124.

2655. --------------. 1964. Notes on the anatomy of Triceratella (Commelinaceae). Kirkia 4:207-212.

2656. --------------. 1965. Notes on the anatomy of Aphyllanthus (Liliaceae) and comparison with Eriocaulaceae. Jour. Linn. Soc. (Bot.) 59: 163-173.

2657. --------------., and AYENSU, E.S. 1968. Morphology and anatomy of Croomia pauciflora (Stemonaceae). Jour. Arnold Arb. 49:260-275.

2658. --------------., and ZIMMERMANN, M.H. 1968. Anatomy of the palm Rhapis excelsa, VI. Root and branch insertion. Jour. Arnold Arb. 49:307-316.

2659. --------------. 1969. Notes on the vegetative morphology and anatomy of the Petermanniaceae (Monocotyledons). Jour. Linn. Soc. (Bot.) 62:17-26.

2660. --------------. 1969. Commelinales-Zingiberales. In: Anatomy of the Monocotyledons. (ed.) Metcalfe, C.R. Vol. 3. Oxford University Press. 446 pp.

2661. --------------. 1969. The anatomy of the vegetative organs of Juania australis (Palmae). Gentes Herb. 10:412-424.

2662. --------------., and ZIMMERMANN, M.H. 1969. Vascular anatomy of monocotyledons, with secondary growth - an introduction. Jour. Arnold Arb. 50:159-179.

2663. --------------., and FISHER, J.B. 1971. Morphological studies in Cordyline (Agavaceae). I. Introduction and general morphology. Jour. Arnold Arb. 50:159-179.

2664. TONZIG, S. 1948. Elementi di Botanica. Vol. I Casa Editrice Ambrosiana, Milano. 1166 pp.

2665. TORREY, J.G. 1951. Cambial formation in isolated pea roots following decapitation. Amer. Jour. Bot. 38:596-604.

2666. -----------. 1955. On the determination of vascular patterns during tissue differentiation in excised pea roots. Amer. Jour. Bot. 42: 183-198.

2667. -----------. 1956. Physiology of root elongation. Ann. Rev. Plant Phys. 7:237-266.

2668. -----------. 1959. Experimental modification of development in the root. In: Cell, Organism and Milieu. (ed.) Rudnick, D. Ronald Press, New York. pp. 189-222.

2669. -----------. 1961. The initiation of lateral roots. Rec. Adv. Bot. 1:808-812.

2670. -----------. 1963. Cellular patterns in developing roots. Symp. Soc. Exp. Biol. No. 17. Cell differentiation. pp. 285-314.

2671. -----------. 1967. Development in Flowering Plants. The Macmillan Co., New York. 184 pp.

2672. -----------., and LOOMIS, R.S. 1968. Ontogenetic studies of vascular cambium formation in excised roots of Raphanus sativus L. Phytomorph. 17:401-409.

2673. TOUMEY, J.W. 1929. Initial root habit in American trees and its bearing on regeneration. Proceedings, International Congress Plant Science. 4th (1926) Ithaca, New York. 1:713-728.

2674. TOURNEAUX, C. 1910. Recherches sur la structure des plantules chez les Viciées. Le Botaniste 11:313-330.

2675. TOZUN, B. 1961. The anatomy and morphology of Digitalis davisiana Hey. Istanbul Univ. Fen Fak. Mecmuasi (Ser. B) 26:41-47.

2676. --------. 1961. Morphological and anatomical characteristics of Digitalis viridiflora Lindl. Istanbul Univ. Fen Fak. Mecmuasi (Ser. B) 26:111-123.

2677. TREASE, G.E., and EVANS, W.C. 1966. A Textbook of Pharmacognosy. 9th ed. Bailliere, Tindall and Cassell, London. 821 pp.

2678. TRENDELENBURG, R., and MAYER-WEGELIN, H. 1955. Das Holz als Rohstoff. C. Hanser, Munich. 541 pp.

2679. TRIVEDI, B.S., and SINGH, D.K. 1969. Structure and Reproduction of the Gymnosperms. Shashidar Malaviya Prakashan, Lucknow. 198 pp.

2680. TRIVEDI, M.L. 1967. Anatomy of normal and tetracot seedlings of Prosopis juliflora DC. Proc. Indian Acad. Sci. (Sect. B) 65: 114-118.

2681. TROCHAIN, J., and DULAU, L. 1942. Quelques particularités anatomiques d'Avicennia nitida (Verbenaceae) de la mangrove oest-africaine. Trav. Lab. For. Toulouse No. 3. Article XIX. 11 pp.

2682. TROLL, W. 1930. Über die sogennanten Atemwurzeln der Mangroven. Berich. Deutsch. Bot. Gesel. 48:81-99.

2683. --------., and DRAGENDORFF, O. 1931. Ueber die Luftwurzeln von Sonneratia Linn. f. und ihre biologische Bedeutung. Planta 13:311-473. (see Listing 1889)

2684. --------. 1935. Wurzel. In: Handwörterbuch der Naturwissenschaften. Auflage 2. Gustav Fischer, Jena. 10:682-702.

2685. --------. 1937. Vergleichende Morphologie der Höhern Pflanzen. Band 1. Vegetationsorgane. Teil 1. Gebrüder Borntraeger, Berlin. 955 pp.

2686. --------. 1937. Morphologie einschliesslich Anatomie. I. Vegetationsorgane. 4. Wurzel. Forts. Bot. 6:26-27.

2687. --------. 1941. Morphologie einschliesslich Anatomie. III. Wurzel. Forts. Bot. 11:29-37.

2688. --------. 1943. Vergleichende Morphologie der Höhern Pflanzen. Band 1. Vegetationsorgane. Teil 3. Wurzel und Wirzelsystems. Gebrüder Borntraeger, Berlin. pp. 2007-2736.

2689. --------. 1944. Morphologie einschliesslich Anatomie. III. Wurzel. Forts. Bot. 18:27-32.

2690. --------. 1949. Über die Grundbegriffe der Wurzelmorphologie. Öster. Bot. Zeitschr. 96:444-452.

2691. --------., and WEBER, H. 1949. Morphologie einschliesslich Anatomie. III. Wurzel. Forts. Bot. 12:32-34.

2692. ---------- and --------. 1949. Morphologische und anatomische Studien an höhern Pflanzen. Sitz. Heidelb. Akad. Wiss. (Math.- Nat. Kl.) 6:3-83.

2693. ---------- and --------. 1951. Morphologie einschliesslich Anatomie. III. Wurzel. Forts. Bot. 13:50-54.

2694. --------., and WETTER, C. 1952. Beiträge zur Kenntnis der Radikations- verhältnisse von Farnen. Abh. Akad. Wiss. Lit. Mainz (Math.- Nat. Kl.) 1:1-84.

2695. --------. 1952. Über Wuchsform und Wurzelbildung von Asplenium nidus L. Abh. Akad. Wiss. Lit. Mainz (Math.- Nat. Kl.) 1952(1):1-84.

2696. --------., and WEBER, H. 1953. Morphologie einschliesslich Anatomie. III. Wurzel. Forts. Bot. 14:29-34.

2697. --------. 1954, 1957. Praktische Einführung in die Pflanzenmorphologie. Teil 1. Der vegetativ Aufbau. Teil 2. Die blühende Pflanze. Gustav Fischer, Jena. 285 pp.

2698. --------., and WEBER, H. 1954. Morphologie einschliesslich Anatomie. III. Wurzel. Forts. Bot. 16:38-45.

2699. ---------- and --------. 1955. Morphologie einschliesslich Anatomie. III. Wurzel. Forts. Bot. 17:32-36.

2700. ---------- and --------. 1956. Morphologie einschliesslich Anatomie. III. Wurzel. Forts. Bot. 18:27-32.

2701. ---------- and --------. 1959. Morphologie einschliesslich Anatomie. III. Wurzel. Forts. Bot. 21:20-33.

2702. --------. 1961. Cochliostema odoratissimum Lem., Organisation und Lebensweise. Nebst vergleichenden Ausblicken auf andere Commelin- aceen. Beitr. Biol. Pflanz. 36:325-389.

2703. --------., and WEBER, H. 1961. Morphologie einschliesslich Anatomie. III. Wurzel. Forts. Bot. 23:22-28.

2704. ---------- and --------. 1963. Morphologie einschliesslich Anatomie. III. Wurzel. Forts. Bot. 25:23-27.

2705. ---------- and --------. 1965. Morphologie einschliesslich Anatomie. II. Wurzel. Forts. Bot. 27:58-64.

2706. TRONCHET, A. 1927. La morphogènése de l'appareil conducteur chez les phanerogames. Bull. Soc. Roy. Bot. Belgique 59:142-159.

2707. -----------. 1928. Sur la reduction du nombre des convergents chez les phanerogames ses rapports avec la polycotylie et le développement vasculaire. Rev. Gén. Bot. 40:1-22.

2708. -----------., and MONTAUT, M.- T. 1958. La développement vasculaire dans la plantule du Viscum album et la nature morphologie du sucoir. Bull. Soc. Bot. France 105:27-31.

2709. -----------, TRONCHET, J., and JAVAUX, M. 1960. Sur le développement et la structure des Streptocarpus unifolies. Ann. Sci. Univ. Bensançon (Bot.) 15:3-11.

2710. TROUGHTON, A. 1957. The underground organs of herbage grasses. Bull. No. 44. Commonwealth Agricultural Bureaux. Farnham Royal, Bucks, England. 163 pp.

2711. TROUGHTON, J., and DONALDSON, L.A. 1972. Probing Plant Structure. McGraw-Hill Book Co., New York. 116 pp.

2712. TSCHERMAK-WOESS, E., and DOLEZAL, R. 1953. Durch Seitenwurzelbildung induzierte und spontane Mitosen in den Dauergeweben der Wurzel. Öster. Bot. Zeitschr. 100:358-402.

2713. ------------------., and HASITSCHKA, G. 1953. Über Musterbildung in der Rhizodermis und Exodermis bei einigen Angiospermen und einer Polypodiacee. Öster. Bot. Zeitschr. 100:646-651.

2714. ------------------. 1956. Karyologische Pflanzenanatomie. Protoplasma 46:798-834.

2715. TSCHIRCH, A. 1905. Über die Heterorhizie bei Dikotylen. Flora 94:68-78.

2716. TUBEUF, C. VON. 1907. Beitrag zur Biologie der Mistelkeimlinge. Naturw. Zeitschr. Forst-Landw. 5:342-349.

2717. TUBEUF, K.F. VON. 1919. Überblick über die Arten der Gattung Arceuthobium (Razoumowskia) mit besonderer Berücksichtigung ihrer Biologie und praktischen Bedeutung. Naturw. Zeitschr. Forst-Landw. 17:167-271.

2718. ----------------. 1923. Monographie der Mistel. R. Oldenbourg, München - Berlin. 832 pp.

2719. TUNMANN, O. 1907. Ueber die Bildung der Luftlücken bei den
Wurzeln der Umbelliferen. Pharm. Zentral. Deutsch. 48:885-894.

2720. ----------. 1908. Mikroscopich-pharmakognostiche Beiträge zur
Kenntnis einiger neurer Arneidrogen. Pharm. Zentral. Deutsch.
49:299-306.

2721. ----------. 1908. Über den anatomischen Bau der Rhizoma Gelsemii
(Gelsemium sempervirens Ait.). Pharm. Zentral. Deutsch. 49:
679-687.

2722. ----------. 1908. Zur Anatomie des Holzes und der Wurzel von
Morinda citrifolia L. mit besonderer Berücksichtigung der micro-
chemischen Verhältnisse. Pharm. Zentral. Deutsch. 49:1013-1017.

2723. ----------. 1914. Ueber Radix Pimpiellae inbesondere über das
Pimpinellin. Apoth.- Zeit. 29:728-730.

2724. TUPPER-CAREY, R.M., and PRIESTLEY, J.H. 1923. The composition of
the cell wall at the apical meristem of stem and root. Proc. Roy.
Soc. London (Ser. B) 95:109-131.

2725. ------------------ and --------------. 1924. The cell wall in the
radicle of Vicia faba and the shape of the meristematic cells.
New Phytol. 25:156-159.

2726. TURNER, L.M. 1934. Anatomy of the aerial roots of Vitis rotundifolia.
Bot. Gaz. 96:367-371.

2727. TUTAIUK, V.K. 1972. The Anatomy and Morphology of Plants. "Vysshaia
Shkola," Moskva. 333 pp. (in Russian)

2728. TWELE, H. 1929. Untersuchungen über regenerative Organbildung an
Wurzeln. Doctoral dissertation. Hamburg Universität. R. Noske,
Borna-Leipzig. 33 pp.

2729. UHERKOVICH, G. 1941. Beiträge zur Kenntnis der systematischen
Anatomie der Gattung Ligularia. Bot. Közlem. 38:361-363. (in
Hungarian)

2730. UNGER, W. 1925. Radix Belladonnae und Radix Sambuci Ebuli. Arch.
Pharm. 263:606-611.

2731. UPHOF, J.G. TH. 1920. Contributions toward a knowledge of the genus
Selaginella. The root. Annals Bot. 34:493-517.

2732. ---------------. 1924. The physiological anatomy of Mayaca fluviatilis.
Annals Bot. 38:389-393.

2733. ---------------. 1928. Das Florida-Arrowroot. Zeitschr. Unter. Lebens.
56:367-373.

2734. --------------. 1929. Beiträge zur Kenntnis der Burmaniacee Apteria aphylla (Nutt.) Barnhart. Öster. Bot. Zeitschr. 78:71-80.

2735. --------------., and HUMMEL, K. 1962. Plant hairs. In: Handbuch der Pflanzenanatomie. (ed.) Linsbauer, K. Band IV. Teil 5. Gebrüder Borntraeger, Berlin. 292 pp.

2736. URBAN, I. 1966. Die Frühstadien der Adventivwurzelbildung bei Calystegia sepium (L.) R. Br. Flora 156:388-394.

2737. URSPRUNG, A. 1912. Über das exzentrische Dickenwachstum an Wurzel-krummungen und über die Erklärungsversuche des exzentrischen Dickenwachstums. Beih. Bot. Centralb. 29:159-218.

2738. -------------., and BLUM, G. 1928. Über die Lage der Wasserabsorptions-zone in der Wurzel. Viert. Nat. Gesel. Zürich Jahrg. 73. No. 15. pp. 162-189.

2739. UTTAMAN, P. 1950. A study of the germination of Striga seed and on the mechanism and nature of parasitism of Striga lutea (Lour.) on rice. Proc. Indian Acad. Sci. (Sect. B) 32:133-142.

2740. VALCHKOW, V.Y. 1965. Morphological and anatomical description of the absorbing roots of some trees. Vyesti Akad. Navuk Byelaruskai SSR (Syeryya Biyalahichnykh Navuk) 3:55-61. (Byelorussian with Russian summary)

2741. VALLADE, J. 1966. Organisation et fonctionnement de l'apex radiculaire de l'Elaeis guineensis Jacq. au course de la germination. Comp. Rend. Hebd. Séances Acad. Sci. (Sér. D) 263:1961-1964.

2742. VAN BREDA DE HAAN, J. 1911. De rijstplant. I. Eene anatomische bescharijving der rijstplant. Meded. Uitg. Dept. Landb. Nederl.-Indië No. 15. pp. 1-53. (in Dutch)

2743. VAN DER BYL, P.A. 1914. The anatomy of Acacia mollissima Wild. Union So. Africa Dept. Agric. Sci. Bull. No. 3. pp. 3-32.

2744. VAN DER LEK, H.A.A. 1924. Over de wortelvorming van houtige stekken. Meded. Landb. Wagen. 28:1-230. (in Dutch with English summary)

2745. VAN DER MEULEN, R.G. 1917. Welwitschia mirabilis Hook. f. Morphol-ogie van het zaad en de vegetatieve organen. Doctoral thesis. Rijks-Universteit te Groningen. M. de Waal, Groningen. 140 pp. (in Dutch)

2746. VAN FLEET, D.S. 1942. The development and distribution of the endodermis and an associated oxidase system in monocotyledonous plants. Amer. Jour. Bot. 29:1-15.

2747. --------------. 1950. A comparison of histochemical and anatomical characteristics of the hypodermis with the endodermis in vascular plants. Amer. Jour. Bot. 37:721-725.

2748. VAN TIEGHEM, P. 1902. Germination et structure de la plantule chez les Coulacées. Jour. Bot. 16:221-226.

2749. --------------. 1903. Sur la germination des Ochnacées. Bull. Mus. Hist. Nat. 9:286.

2750. --------------. 1903. Structure et affinités des Eryhthroxylacées. Bull. Mus. Hist. Nat. 9:287-295.

2751. --------------. 1906. Sur les Agialidacées. Ann. Sci. Nat., Bot. (Sér. 9) 4:223-260.

2752. --------------. 1907. Sur les Inovulées. Ann. Sci. Nat., Bot. (Sér. 9) 6:125-260.

2753. VARGA, F. 1923 (1924). Vergleichende anatomische Untersuchung der Gattungen Succisela und Succisa mit Rucksicht auf dem verwandten Gattungen. Bot. Közlem. 21:32-47; (4)-(8).

2754. VASCONCELLOS, J. DE C. 1950. La planta de trigo. Morphologia y fisiologia. Instituto Nacional de Investigaciones Agrinomicas. Madrid. 197 pp. (in Spanish)

2755. VASILEVSKAYA, N.K. 1959. Anatomical structure of embryo and seedlings of some herbaceous plants. Vest. Lenin. Univ. (Biol. Ser.) Bull. 1:3-19. (in Russian)

2756. VASILEVSKAYA, V.K. 1957. Anatomy of bud formation on roots of some woody plants. Vest. Lenin. Univ. (Biol. Ser.) Bull. 1:3-21.

2757. ------------------. 1961. Primitive characteristics of the anatomical structure of sunflower seedlings. Bot. Zhur. 46:780-789. (in Russian with English summary)

2758. ------------------., and ISCHENKO, G.E. 1965. Anatomical structure of aboveground and underground organs of a sedge (Carex physodes). Ixv. Akad. Nauk Turkmen. SSR (Ser. Biol. Nauk) 6:38-43.

2759. VASISHT, B.R. 1927. The comparative anatomy of Ophioglossum aitchisoni d'Almeida and Ophioglossum vulgatum L. Jour. Indian Bot. Soc. 6:8-30.

2760. VASISHTA, P.C. 1971. A Textbook of Plant Anatomy. 3rd ed. S. Nagin and Co., Jullunder City, India. 382 pp.

2761. VECCHIERELLO, H. 1928. A study of the origin and development of the radicle histogens of Quercus prinus L. Cath. Univ. Amer. Contrib. No. 8. pp. 7-73.

2762. VELENOVSKÝ, J. 1905-1910. Vergleichende Morphologie der Pflanzen. Teil 1. 1905; Teil 2. 1907; Teil 3. 1910. Fr. Řivnáč, Prague. 1216 pp.

2763. VENKATRAMAN, R.B.T.S., and THOMAS, R. 1929. Studies of sugarcane roots at different stages of growth. Mem. Dept. Agric. India (Bot. Ser.) 6:145-157.

2764. VERDOORN, F. (ed.). 1938. Manual of Pteridology. Martinus Nijhoff, The Hague. 640 pp.

2765. VERZÁR-PETRI, G. 1964. Anatomy of Solanum laciniatum Ait. Ann. Univ. Sci. Budap. Rol. Ëot. Nom. (Biol.) 7:241-254.

2766. VEUILLET, J.M. 1959. Contribution à l'étude morphologique et anatomique du genre Elaeoselinum au Maroc. Trav. Inst. Sci. Cherif. (Sér. Bot.) No. 8. pp. 1-63.

2767. VIDAL, L. 1905 (1906). Anatomie de la racine et de la tige de l'Eritrichium nanum. Assoc. Franç. Avan. Sci. 34:472-475.

2768. VIDAL, M.L. 1903. Contribution à l'anatomie des Valérianacées. Ann. Univ. Grenoble 15:561-605.

2769. VIGUIER, R. 1906. Recherches anatomiques sur la classification des Araliacées. Ann. Sci. Nat., Bot. (Sér. 9) 4:1-210.

2770. VIJAYARAGHAVEN, C., and RAO, V.P. 1938. The occurrence of root-hairs on aerial roots of Sorghum. Curr. Sci. 7:20.

2771. VIKHROV, V.E. 1960. Characteristics of the microscopic structure of woody roots from some coniferous species. Dokl. Akad. Nauk Belorusskoi SSR 4:74-77.

2772. ------------., and KOSTAREVA, L.V. 1960. The anatomical structure of wood in the roots of some coniferous species. Bot. Zhur. 45: 1259-1270. (in Russian with English summary)

2773. VINTEJOUX, C. 1958. Recherches sur la racine de Lemna minor L. (Lemnacées). Ann. Sci. Nat., Bot. (Sér. 11) 19:211-261.

2774. VOGT, R. 1915. The ecology and anatomy of Polygonatum commutatum. Amer. Midl. Nat. 4:1-11.

2775. VON HAYEK, A. 1905. Monographische studien über die Gattung Saxifraga. Denk. Kaiserl. Akad. Wiss. Wien (Math.- Nat. Kl.) 77:611-709.

2776. VON OVEN, E. 1904. Beiträge zur Anatomie der Cyclanthaceae. Beih. Bot. Centralb. 16:147-198.

2777. VORONIN, N.S. 1945. To the question of the evolution of the pericycle in the roots of plants. Bot. Zhur. 30:147-153. (in Russian with English summary)

2778. ------------. 1956. De l'évolution des racines des plantes. Moskov. Obshch. Ispyt. Prirody, Otd. Biol. Biul. 61:47-58. (in Russian with French summary)

2779. ------------. 1957. Evolution of plant roots. 2. Evolution of root origin. Moskov. Obshch. Ispyt. Prirody, Otd. Biol. Biul. 62: 35-49. (in Russian with English summary)

2780. ------------. 1960. Some specific characteristics of root-formation in Juncus. Bot. Zhur. 45:1359-1360.

2781. ------------. 1969. Apical meristems of the root in gymnosperms and the principles of their graphical interpretation. Bot. Zhur. 54: 67-76.

2782. VORONTSOV, L.A. 1965. Anatomical structure of primary roots. Vest. Sel'sk. Nauk (Moskow) 7:91-92.

2783. VOROSHILOVA, G.I. 1964. Structure of the embryo and seedling of wild and cultivated soybean (Glycine) in the Far East. Vest. Lenin. Univ. (Ser. Biol.) No. 9, Part 2. pp. 45-51.

2784. VOTH, P.D. 1934. A study of the vegetative phases of Ephedra. Bot. Gaz. 96:298-313.

2785. ----------., GRIESBACH, R.A., and YEAGER, J.R. 1968. Developmental anatomy and physiology in daylily. Amer. Hort. Mag. 47:121-151.

2786. VOUK, V. 1909. Anatomie und Entwicklungsgeschichte der Lentizellen an Wurzeln von Tilia sp. Sitz. Kaiser. Akad. Wiss. Wien 118: 1073-1090.

2787. VUILLEMIN, P. 1902. Les organes souterrains du Gentiana ciliata. Bull. Soc. Bot. France 49:274-280.

2788. WAGER, V.A. 1928. I. The structure and life history of the South African Lagarosiphons. Trans. Roy. Soc. So. Africa 16:191-204.

2789. WAGNER, K., and KURZ, H. 1954. Cypress: Root and stem modifications in relation to water. Florida State Univ. Stud. (Contributions to Science, No. 2). No. 13. pp. 18-47.

2790. WAGNER, N. 1930. Über die Mitosenverteilung in der Meristem der Wurzelspitzen. Planta 10:1-37.

2791. ---------. 1937. Wachstum und Teilung der Meristemzellen in Wurzelspitzen. Planta 27:550-583.

2792. ---------. 1939. Über die Entwicklungsmechanik der Wurzelhaube und des Wurzelrippenmeristems. Planta 30:21-60.

2793. WAISEL, Y. 1972. Biology of Halophytes. In: Physiological Ecology series. (ed.) Kozlowski, T.T. Academic Press, New York. 395 pp.

2794. WAKHLOO, J.L. 1964. Autecology of Rauvolfia serpentina Benth. Jour. Indian Bot. Soc. 43:374-390.

2795. WALDRON, R.A. 1919. The peanut (Arachis hypogaea) - its history, histology, physiology, and utility. Contr. Bot. Lab. Univ. Penn. 4:301-338.

2796. WALLIS, T.E., and SAUNDERS A.M. 1924. The rhizomes of Helleborus niger and H. viridis: A comparative study. Pharm. Jour. Pharm. 113:90-94.

2797. ------------- and ------------. 1924. The rhizomes of Helleborus niger and H. viridis: A comparative study. Yearb. Pharm. Trans. Brit. Pharm. Conf. (61st Annual Meeting) pp. 664-677.

2798. -----------. 1925. Practical pharmacognosy. VIII. Pharm. Jour. Pharm. 115:594-596.

2799. -----------., and SANYAL, P.K. 1948. Indian Valerian - its structural characters. Quart. Jour. Pharm. Pharm. 21:332-343.

2800. WALTER, H. 1946. Die Grundlagen des Pflanzenlebens. Verlag. Eugen Ulmer, Stuttgart. 480 pp.

2801. WALTON, J. 1944. The roots of Equisetum limosum L. New Phytol. 43:81-86.

2802. WARBURG, O. 1900. Panadanaceae. In: Das Pflanzenreich. (ed.) Engler, A. Band 4. Teil 9. Heft 3. pp. 1-7. W. Engelmann, Leipzig.

2803. WARDEN, W.M. 1935. On the structure, development and distribution of the endodermis and its associated ducts in Senecio vulgaris. New Phytol. 34:361-385.

2804. WARDLAW, C.W. 1928. Size in relation to internal morphology. No. 3. The vascular system of roots. Trans. Roy. Soc. Edinb. 56:19-55.

2805. ------------. 1965. Organization and Evolution in Plants. Longmans, Green and Co., Ltd., London. 499 pp.

189

2806. ------------. 1968. _Morphogenesis in Plants_. Methuen and Co., Ltd., London. 451 pp.

2807. WARDROP, A.B. 1959. Cell formation in root hairs. Nature 184(Suppl. 13):996-997.

2808. WARNER, L.A. 1904. The regeneration of seedling roots after splitting. Rept. Mich. Acad. Sci. (6th Annual) p. 78.

2809. WARNING, W.C. 1930. Anatomy of the parsnip root. Doctoral dissert- ation. University of Chicago. 12 pp.

2810. ------------. 1934. Anatomy of the vegetative organs of the parsnip. Bot. Gaz. 96:44-72.

2811. WARRINGTON, P.D. 1970. The natural history and parasitism of _Geocaulon lividum_ (Santalaceae). Doctoral thesis. University of British Columbia, Vancouver. 168 pp.

2812. ----------------. 1970. The haustorium of _Geocaulon lividum_; a root parasite of the Santalaceae. Canad. Jour. Bot. 48:1669-1675.

2813. WATANABE, K. 1924. Studien über die Koralloide von _Cycas revoluta_. Bot. Mag. (Tokyo) 38:165-188.

2814. -----------. 1925. Über die kontraktion und daraus verursachte Anomalie in der Wurzel von _Cycas revoluta_. Jap. Jour. Bot. 2: 293-297.

2815. WATERSTON, J. 1912. Note on the septa in root vessels of Bromeliaceae. Trans. Proc. Bot. Soc. Edinb. 24:25-26.

2816. ------------. 1912. Morphological changes induced in roots of Bromeliaceae by attack of _Heterodera_ sp. Trans. Proc. Bot. Soc. Edinb. 24:26-35.

2817. WATREL, A.A. 1962. A study of the seedling anatomy and leaf ontogeny of _Solanum melongena_ L. var. _esculentum_ Nees. (New Hampshire hybrid). Diss. Absts. 23(1):44.

2818. WATSON, E.E. 1919. On the occurrence of root hairs on old roots of _Helianthus rigidus_. Rept. Mich. Acad. Sci. (21st Annual) p. 235.

2819. WATSON, E.V. 1936. A study of the anatomy of _Trichopus zeylanicus_ Gaertn. Notes Roy. Bot. Gard. Edinb. 19:135-156.

2820. WEAVER, H.L. 1960. Vascularization of the root-hypocotyl-cotyledon axis of _Glycine max_ (L.) Merrill. Phytomorph. 10:82-86.

2821. WEAVER, J.E. 1919. The ecological relations of roots. Publ. Carn. Inst. Wash. No. 286. 128 pp.

2822. ----------. 1920. Root development in the grassland formation. A correlation of the root systems of native vegetation and crop plants. Publ. Carn. Inst. Wash. No. 292. 151 pp.

2823. ----------., JEAN, F.C., and CRIST, J.W. 1922. Development and activities of roots of crop plants. Publ. Carn. Inst. Wash. No. 316. 117 pp.

2824. ----------. 1925. Investigations on the root habits of plants. Amer. Jour. Bot. 12:502-509.

2825. ----------. 1926. Root Development of Field Crops. McGraw-Hill Book Co., New York. 291 pp.

2826. ----------., and BRUNER, W.E. 1927. Root Development of Vegetable Crops. McGraw-Hill Book Co., New York. 351 pp.

2827. WEBER, E. 1929. Entwicklungsgeschichtliche Untersuchungen über die Gattung Allium. Bot. Archiv 25:1-44.

2828. WEBER, H. **1936.** Vergleichend-morphologische Studien über die spross-bürtige Bewurzelung. Nova Acta Leop. (N.F.) 4:229-298.

2829. --------. 1944. Über die Bewurzelungsweise von Ceratocephalus orthoceras und Valeriana tuberosa. Bot. Archiv 45:248-254.

2830. --------. 1950. Morphologische und anatomische Studien über Eichhornea crassipes (Mart.) Solms. Jahrb. Akad. Wiss. Lit. Mainz (Math.- Nat. Kl.) 6:135-161.

2831. --------. 1953. Las raices internas de Navia y Vellozia. Mutisia No. 13. pp. 1-7.

2832. --------. 1954. Wurzelstudien an tropischen Pflanzen. I. Die Bewurzel-ung von Vellozia und Navia sowie der Baumfarne und einiger Palmen. Abh. Akad. Wiss. Lit. Mainz (Math.- Nat. Kl.) No. 6. pp. 211-249.

2833. --------. 1954. Die Bewurzelungsverhältnisse der Pflanzen. Herder and Co., Freiburg. 132 pp.

2834. --------. 1958. Die Wurzelverdickungen von Calathea macrosepala Schum. und einigen anderen monokotyler Pflanzen. Beitr. Biol. Pflanz. 34: 177-193.

2835. --------. 1963. Über die wuchsform von Bulbostylis paradoxa (Spreng.) Lindm. (Cyperaceae). Abh. Akad. Wiss. Lit. Mainz (Math.- Nat. Kl.) No. 5. pp. 267-284.

2836. --------., and TROLL, W. 1968. Morphologie einschliesslich Anatomie. I. Wurzel. Forts. Bot. 30:25-43.

2837. WEBER, U. 1922. Zur Anatomie und Systematik der Gattung Isoëtes. Hedwigia 63:219-262.

2838. WEBSTER, T.R., and STEEVES, T.A. 1963. Morphology and development of the root of Selaginella densa Rydb. Phytomorph. 13:367-376.

2839. -------------- and ------------. 1964. Developmental morphology of the root of Selaginella kraussiana A. Br. and Selaginella wallacei Hieron. Canad. Jour.Bot. 42:1665-1676.

2840. ------------. 1966. Morphology and development of the root in several species of Selaginella Spring. Diss. Absts. (Sect. B) 27(5):1381B.

2841. ------------., and STEEVES, T.A. 1967. Developmental morphology of the root of Selaginella martensii Spring. Canad. Jour. Bot. 45: 395-404.

2842. WEILL, G. 1903. Recherches histologiques sur la famille des Hypericacées. Doctoral thèses. Université de Paris, École Superieure de Pharmacie. A. Joanin et Cie, Paris. 189 pp.

2843. WEINHOLD, L. 1967. Histogenetische Studien zum Grenzwurzelproblem. Beitr. Biol. Pflanz. 43:367-454.

2844. WEINREICH, R. 1925. Bau und Entwicklung der Wurzeln bei den Osmundaceen in Hinsicht auf ihre systematische Stellung. Bot. Archiv 12:5-58.

2845. WEISS, F.E. 1913. The root-apex and young root of Lyginodendron. Manchest. Lit. Philos. Soc. Mem. Proc. 57:1-10.

2846. WEITZ, R. 1923. Le lyciet (Lycium vulgare Dunal): recherches botanique, chimiques et pharmacologiques. Bull. Sci. Pharm. 28:503-508; 562-568.

2847. WELCH, M.B. 1923. The occurrence of secretory canals in certain myrtaceous plants. Proc. Linn. Soc. New So. Wales 48:660-673.

2848. WENT, F.W., and DARLEY, E. 1953. Root hair development on date palms. Rept. Date Grow. Inst. 30:3-5.

2849. WERNHAM, H.F. 1910. The morphology of Phylloglossum drummondii. Annals Bot. 24:335-347.

2850. WEST, C., and TAKEDA, H. 1915. On Isoëtes japonica A. Br. Trans. Linn. Soc. London (Ser. 2) 8:333-376.

2851. -------. 1917. A contribution to the study of the Marattiaceae. Annals Bot. 31:316-414.

2852. WESTERMAIER, M. 1900. Zur Kenntnis der Pneumatophoren. Aus dem Botanischen Institut der Universität Freiburg (Schweiz). Gebrüder Fragnière, Freiburg. 53 pp.

2853. ---------------. 1900. Zur Entwicklung und Struktur einiger Pteridophyten aus Java. Aus dem Botanischen Institut der Universität Freiburg (Schweiz). 27 pp.

2854. WETTER, C. 1951. Über die Luftwurzeln von Oleandra. Planta 39: 471-475.

2855. ---------. 1952. Untersuchungen über Anordnung und Anlegung der Wurzeln bei leptosporangiaten Farnen. Doctoral dissertation. Johannes Gutenberg-Universität zu Mainz. 57 pp.

2856. WETTSTEIN, F. 1906. Entwicklung der Beiwurzeln einiger dikotylen Sumpf- und Wasserpflanzen. Beih. Bot. Centralb. 20:1-60.

2857. WHALEY, W.G., and BOWEN, J.S. 1947. Russian dandelion (kok-saghyz), an emergency source of natural rubber. U.S.D.A. Miscl. Publ. No. 618. 212 pp.

2858. -----------., MERICLE, L.W., and HEIMSCH, C. 1952. The wall of the meristematic cell. Amer. Jour. Bot. 39:20-26.

2859. WHITAKER, E.S. 1923. Root hairs and secondary thickening in the Compositae. Bot. Gaz. 76:30-59.

2860. WHITAKER, T.W., and DAVIS, G.N. 1962. Cucurbits. World Crop Books. Interscience Publishing Co., Inc., New York. pp. 19-23.

2861. WHITE, J.H. 1907. On the polystely in roots of Orchidaceae. Univ. Toronto Stud. (Biol. Ser.) No. 6. pp. 3-20.

2862. WHITE, P.R. 1927. Studies of the physiological anatomy of the strawberry. Jour. Agric. Res. 35:481-492.

2863. WHITE, R.A. 1961. Vessels in roots of Marsilea. Science 133: 1073-1074.

2864. ----------. 1963. A comparative study of the tracheary elements of the ferns. Diss. Absts. 23(9):3096-3097.

2865. ----------. 1963. Tracheary elements of the fern. II. Morphology of tracheary elements; conclusions. Amer. Jour. Bot. 50:514-522.

2866. ----------. 1969. Vegetative reproduction in the ferns. II. Root buds in Amphoradenium. Bull. Torrey Bot. Club 96:10-19.

2867. WHITING, A.G. 1938. Development and anatomy of primary structures in the seedling of Cucurbita maxima. Bot. Gaz. 99:497-528.

2868. WHITTINGTON, W.J. 1969. Root growth. In: Proceedings of the 15th Easter School in Agricultural Science. University of Nottingham (1968). Butterworths, London. 450 pp.

2869. WIEDEMANN, E. 1927. Der Wurzelbau älterer Waldbäume. Forstarchiv 3:229-233.

2870. WIGGLESWORTH, G. 1907. The young sporophytes of Lycopodium complanatum and Lycopodium clavatum. Annals Bot. 21:211-234.

2871. WIGHT, W. 1933. Radial growth of the xylem and the starch reserves of Pinus sylvestris: a preliminary survey. New Phytol. 32:77-96.

2872. WILCOX, H. 1961. The differentiation of primary vascular tissues in roots. Rec. Adv. Bot. 1:805-808.

2873. ---------. 1962. Growth studies of the root of incense cedar, Libocedrus decurrens. I. The origin and development of primary tissues. Amer. Jour. Bot. 49:221-236.

2874. ---------. 1962. Growth studies of the root of incense cedar, Libocedrus decurrens. II. Morphological features of the root systems and growth behavior. Amer. Jour. Bot. 49:237-245.

2875. ---------. 1964. Xylem in roots of Pinus resinosa Ait. in relation to heterorhizy and growth activity. In: The Formation of Wood in Forest Trees. (ed.) Zimmermann, M.H. Academic Press, New York. pp. 459-478.

2876. ---------. 1965. Morphological features and growth behavior of the roots of red pine, Pinus resinosa Ait. Israel Jour. Bot. 14:206.

2877. ---------. 1967. Seasonal patterns of root initiation and mycorrhizal development in Pinus resinosa Ait. International Union Forestry Research Organizations. XIV IUFRO Kongress. 14th Referate, München. Part 5, Section 24. pp. 29-39.

2878. WILCOX, H.E. 1954. Primary organization of active and dormant roots of noble fir, Abies procera. Amer. Jour. Bot. 41:812-821.

2879. -----------. 1955. The regeneration of injured root systems in noble fir. Bot. Gaz. 116:221-234.

2880. -----------. 1968. Morphological studies of the roots of red pine, Pinus resinosa. I. Growth characteristics and patterns of branching. Amer. Jour. Bot. 55:247-254.

2881. -----------. 1968. Morphological studies of the roots of red pine, Pinus resinosa. II. Fungal colonization of roots and the development of mycorrhizae. Amer. Jour. Bot. 55:686-700.

2882. WILDE, A. 1902. Beiträge zur Anatomie der Linaceen. Inaugural-Dissertation. Ruprecht-Karls-Universität zu Heidelberg. J. Hörning, Heidelberg. 56 pp.

2883. WILDEMAN, E. DE. 1920. Mission forestière et agricola du Compte Jaques de Briey au Mayumbe. D. Reynaert, Bruxelles. 468 pp.

2884. WILKIE, S.J. 1916. The influence of different media on the histology of roots. Trans. Proc. Bot. Soc. Edinb. 27:76-78.

2885. WILLIAMS, B.C. 1947. The structure of the meristematic root tip and the origin of the primary tissues in the roots of vascular plants. Amer. Jour. Bot. 34:455-462.

2886. -------------. 1949. The role of the meristematic endodermal layer in the formation of intercellular canals in the root tips of composite plants. Amer. Jour. Bot. 36:806. (Abstract)

2887. -------------. 1950. The occurrence of inter-cellular canals in root tips of plants in the Polygonaceae and Labiatae. Amer. Jour. Bot. 37:668. (Abstract)

2888. -------------. 1954. Observations on intercellular canals in root tips with special reference to the Compositae. Amer. Jour. Bot. 41:104-106.

2889. WILLIAMS, C.N. 1960. Sopubia ramosa, a perennating parasite on the roots of Imperata cylindrica. Jour. West African Sci. Assn. 6: 137-141.

2890. -------------. 1963. Development of Tapinanthus bangwensis (Engler and Krause) Danser and contact with the host. Annals Bot. (N.S.) 27:641-646.

2891. WILLIAMS, M.M. 1924. Anatomy of Cheilanthes tenuifolia. Bot. Gaz. 78:378-396.

2892. -------------. 1925. The anatomy of Lindsaya linearis and Lindsaya microphylla. Proc. Linn. Soc. New So. Wales 50:391-404.

2893. -------------. 1927. The anatomy of Cheilanthes vellea. Proc. Linn. Soc. New So. Wales 52:73-84.

2894. WILLIAMS, R.G. 1921. The anatomy and morphology of Marsilea. Master's thesis. Cornell University, Ithaca, New York. 19 pp.

195

2895. WILLIAMS, S. 1930. The morphology of Trichomanes aphlebioides Christ, with special reference to the aphleboid leaves. Proc. Roy. Soc. Edinb. 50:142-152.

2896. WILLIS, J.C. 1902. Studies in the morphology and ecology of the Podostemaceae of Ceylon and India. Ann. Roy. Bot. Gard., Peradeniya 1:267-465.

2897. WILSON, B.F. 1964. Structure and growth of woody roots of Acer rubrum L. Harvard For. Papers No. 11. pp. 1-14.

2898. -----------., and HORSLEY, S.B. 1970. Ontogenetic analysis of tree roots in Acer rubrum and Betula papyrifera. Amer. Jour. Bot. 57: 161-164.

2899. -----------. 1970. The Growing Tree. The University of Massachusetts Press, Amherst. 152 pp.

2900. WILSON, C.L. 1927. Adventitious roots and shoots in an introduced weed. Bull. Torrey Bot. Club 54:35-38.

2901. WILSON, E.H. 1929. Consider the root - how it grows. House Gard. 56:83-87; 162, 164.

2902. WILSON, K. 1936. The production of root hairs in relation to the development of the piliferous layer. Annals Bot. 50:121-154.

2903. ---------., and HONEY, J.N. 1966. Root contraction in Hyacinthus orientalis. Annals Bot. (N.S.) 30:47-61.

2904. WILSON, O.T. 1913. Studies on the anatomy of alfalfa. Kansas Univ. Sci. Bull. 7:291-299.

2905. WINKLER, H. 1927. Bausteine zu einer Monographie von Ficaria. Beitr. Biol. Pflanz. 15:126-128; 376-416.

2906. WINTER, C.W. 1932. Vascular system of young plants of Medicago sativa. Bot. Gaz. 94:152-167.

2907. WINTER, J.M. 1929. Some observations on the rate of mitosis in root tip meristems of Gladiolus. Trans. Amer. Micro. Soc. 48:276-291.

2908. WINTON, A.L., and WINTON, K.B. 1932-1939. The Structure and Composition of Foods. Vol. 2. Vegetables, legumes, fruits. Wiley and Sons, New York. 904 pp.

2909. WIRTH, E.H. 1926. The pharmacognosy of Ceanothus americanus. Amer. Jour. Pharm. 98:503-514.

2910. WISSELINGH, C. VAN. 1926. Beitrag zur Kenntnis der innern Endodermis. Planta 2:27-43.

2911. WITSCH, H. VON. 1939. Zum Feinbau der Zellwand in Wurzeln. Planta 29:409-418.

2912. WITTMANN, W., BERGERSEN, F.J., and KENNEDY, G.S. 1965. The coralloid roots of Macrozamia communis L. Johnson. Austral. Jour. Biol. Sci. 18:1129-1134.

2913. WOLFE, F. 1934. Origin of adventitious roots in Cotoneaster dammeri. Bot. Gaz. 95:686-694.

2914. WOODCOCK, E.F., and DE ZEEUW, R. 1921. The anatomy of the haustorial roots of Comandra. Rept. Mich. Acad. Sci. 22:189-193.

2915. WOLPERT, J. 1910. Vergleichende Anatomie und Entwicklungsgeschichte von Alnus alnobetula und Betula. Flora 100:37-67.

2916. WOODCOCK, E.F. 1941. Seed germination and seedling anatomy of snap-dragon (Antirrhinum majus L.). Papers Mich. Acad. Sci. Arts Let. (1940) 26:99-102.

2917. --------------. 1946. Latex-tube areas of roots and leaves of the Russian dandelion. Jour. Agric. Res. 72:297-300.

2918. WOODSON, R.E., JR. 1957. The botany of Rauwolfia. In: Rauwolfia, by Woodson, R.E., Youngken, H.W., Sclitter, E., and Schneider, J.A. Little, Brown Co., Boston. 149 pp.

2919. WORSDELL, W.G. 1900. The anatomical structure of Bowenia spectabilis. Annals Bot. 14:159-160.

2920. --------------. 1900. The comparative anatomy of certain species of Encephalartos Lehm. Trans. Linn. Soc. London (Ser. 2) 4:445-460.

2921. --------------. 1901. Contributions to the comparative anatomy of the Cycadaceae. Trans. Linn. Soc. London (Ser. 2) 6:109-121.

2922. --------------. 1906. Structure and origin of the Cycadaceae. Annals Bot. 20:129-159.

2923. WRIGHT, H. 1901. Observations of Dracaena reflexa Lam. Ann. Roy. Bot. Gard., Peradeniya 1:165-172.

2924. ----------. 1904. The genus Diospyros in Ceylon : its morphology, anatomy, and taxonomy. Ann. Roy. Bot. Gard., Peradeniya 2:1-106; 133-210.

2925. WRIGHT, J. 1927. Notes on strand plants. II. Cakile maritima Scop. Trans. Proc. Bot. Soc. Edinb. 29:389-401.

2926. WRIGHT, J.O. 1951. Unusual features of the root system of the oil palm in West Africa. Nature 168(4278):748.

2927. YAMPOLSKY, C. 1922. A contribution to the study of the oil palm, Elaeis guineensis Jacq. Bull. Jard. Bot. Buitenz. 5:107-174.

2928. ------------. 1924. The pneumathodes on the roots of the oil palm (Elaeis guineensis Jacq.). Amer. Jour. Bot. 11:502-512.

2929. YARBROUGH, J.A. 1949. Arachis hypogea. The seedling, its cotyledons, hypocotyl and roots. Amer. Jour. Bot. 36:758-772.

2930. YAROSHENKO, G.D. 1945. Tragacanth type of root contraction. Bot. Zhur. 30:115-124. (in Russian with English summary)

2931. YDRAC, F.- L. 1905. Recherches anatomiques sur les Lobeliacées. Doctoral thèses. Université de Paris, École Superieure de Pharmacie. Lucien Declume, Lons-Le-Saunier. 165 pp.

2932. ------------. 1905. Sur l'appareil laticifère des Lobeliacées. Jour. Bot. 19:12-20.

2933. YEO, P.F. 1961. Germination, seedlings, and the formation of haustoria in Euphrasia. Watsonia 5:11-22.

2934. YORK, H.H. 1909. The anatomy and some biological aspects of the "American Mistletoe" Phoradendron flavescens (Pursh) Nutt. Bull. Univ. Texas, Sci. Ser. No. 13. pp. 1-31.

2935. YOSHIDA, O. 1968. Seedling anatomy of Trochodendron and Euptelea. Phytomorph. 17:183-187.

2936. YOUNG, P.T. 1933. Histogenesis and morphogenesis in the primary root of Zea mays. Doctoral dissertation. Columbia University, New York. 40 pp.

2937. YOUNGKEN, H.W. 1919. The comparative morphology, taxonomy and distribution of the Myricaceae of the Eastern United States. Contr. Bot. Lab. Univ. Penn. 4:339-400.

2938. --------------. 1950. A Textbook of Pharmacognosy. 6th ed. McGraw-Hill Book co., New York. 1063 pp.

2939. YUNG, C.- T. 1938. Developmental anatomy of the seedling of the rice plant. Bot. Gaz. 99:786-802.

2940. ZABLOCKA, W. 1936. Untersuchungen über die Mykorriza bei der Gattung Viola. Bull. Inter. Acad. Polonaise Sci. Let. (Ser. b) 1:93-101.

2941. ZACH, F. 1909. Untersuchungen über die Kurzwurzeln von Sempervivum und die daselbst auftretende endotrophe Mykorriza. Sitz. Kaiser. Akad. Wiss. Wien (Erster Halfband) 118:185-200.

2942. ZEE, S.Y., and CHAMBERS, T.C. 1968. Fine structure of the primary root phloem of Pisum. Austral. Jour. Bot. 16:37-47.

2943. ZEHENDER, S.M. 1924. Über Regeneration und Richtung der Seitenwurzeln. Flora (N.F.) 117:301-343.

2944. ZENDER, J. 1924. Le comportement de haustoriums du Cuscuta europaea dans le tissues de la plant parasitée. Comp. Rend. Séances Soc. Phys. Hist. Nat. Genève 41:132-135.

2945. ---------. 1924. Les haustoriums de la Cuscute et les réactions de l'hôte. Bull. Soc. Bot. Genève (Sér. 2) 16:189-264.

2946. ZGUROVSKAÎA, L.N. 1958. Anatomic-physiologic investigation of absorbing, growing and conducting roots of woody genera. Akad. Nauk SSSR, Sib. Otd., Trudȳ, Inst. Lesa Drev. 41:5-32. (in Russian)

2947. ZIEGENSPECK, H. 1925. Ueber Zwischenprodukte des Aufbaues von Kohlen-hydrat-Zellwänden und deren mechanische Eigenschaften. Bot. Archiv 9:297-376.

2948. ZIMMERMAN, A. 1922. Die Cucurbitaceen. Beiträge zur Anatomie und Physiologie. Heft 1. G. Fischer, Jena. 204 pp.

2949. ZIMMERMANN, M. 1964. Formation of Wood in Forest Trees. Academic Press, New York. 562 pp.

2950. ZIMMERMANN, M.H., WARDROP, A.B., and TOMLINSON, P.B. 1968. Tension wood in aerial roots of Ficus benjamina L. Wood Sci. Tech. 2:95-104.

2951. ---------------., and BROWN, C.L. 1971. Trees: Structure and Function. Springer-Verlag, New York. 336 pp.

2952. ZWEIGELT, F. 1912. Vergleichende Anatomie der Asparagoideae, Ophio-pogonoideae, Aletroideae, Luzuriagoideae, und Smilacoideae nebst Bemerkungen über die Beziehungen zwischen Ophiopogonoideae und Dracaenoideae. Denk. Kaiserl. Akad. Wiss. Wien (Math.- Nat. Kl.) 88:397-476.

2953. ZWICKY, E. 1914. Ueber Channa, ein Genussmittel der Hottentotten. (Mesembrianthemum expansum L. und tortuosum L.). Doctoral dissert-ation. Eidgenössischen Hochschule in Zürich. Zürcher und Furrer, Zürich. 60 pp.

ADDENDA

LITERATURE BEFORE 1900

2954. BAILLON, M.H. 1859. Monographie des Buxacées et des Stylocerées. Librairie de Victor Masson, Paris. 89 pp.

2955. DE MARLIÈRE, G. 1891. Structure comparée des racines renflées de certaine Ombellifères. Comp. Rend. Hebd. Séances Acad. Sci. 112: 1020-1022.

2956. WEISS, G.A. 1878. Anatomie der Pflanzen. Wilhelm Braumüller, Wien. 531 pp.

2957. WIELER, A. 1891. Über Beziehungen zwischen Wurzel und Stammholz. Thar. Forst. Jahrb. 41:143-171.

LITERATURE AFTER 1900

2958. BUTLER, V., BORNMAN, C.H., and EVERT, R.F. 1973. Welwitschia mirabilis: Morphology of the seedling. Bot. Gaz. 134:52-59.

2959. ---. 1973. Welwitschia mirabilis: Vascularization of a four-week-old seedling. Bot. Gaz. 134:59-63.

2960. ---. 1973. Welwitschia mirabilis: Vascularization of a one-year-old seedling. Bot. Gaz. 134:63-73.

2961. BYRNE, J.M. 1973. The root apex of Malva sylvestris. III. Lateral root development and the quiescent center. Amer. Jour. Bot. 60: 657-662.

2962. ----------. 1974. Root morphology. In: The Plant Root and Its Environment. University of Virginia Press, Charlottesville. Part I, pp. 3-27.

2963. CHEADLE, V.I. 1963. Vessels in the Iridaceae. Phytomorph. 13:245-248.

2964. HOLDEN, H.S., and BEXON, D. 1923. On the seedling structure of Acer pseudoplatanus. Annals Bot. 37:571-594.

2965. KIMURA, Y. 1939. Ueber die anatomische untersuchung der Wurzel von einigen Polygonatum-Arten. Jour. Jap. Bot. 15:504-512.

2966. MIA, A.J., and PATHAK, S.M. 1965. Morphological and histochemical studies of primary root meristem of Rauwolfia vomitoria. Bot. Gaz. 126:204-208.

2967. PEARSON, H.H.W. 1929. Gnetales. Cambridge University Press, London. 194 pp.

2968. PULAWSKA, Z. 1973. The parenchymo-vascular cambium and its derivative tissues in stems and roots of Bougainvillaea glabra Choisy (Nyctaginaceae). Acta Soc. Bot. Poloniae 42:41-61.

2969. RAO, A.R., and SRIVASTAVA, P. 1973. On the morphology and anatomy of Belvisia spicata (L. Fil.) Mirbel Proc. Indian Acad. Sci. (Sect. B) 77:25-29.

2970. SEAGO, J.L. 1973. Developmental anatomy in roots of Ipomoea purpurea. II. Initiation and development of secondary roots. Amer. Jour. Bot. 60:607-618.

2971. STANT, M.Y. 1973. The role of the scanning electron microscope in plant anatomy. Kew Bull. 28:105-115.

2972. SÜSS, H., and MÜLLER-STOLL, W.R. 1973. Zur Anatomie des Ast-, Stamm- und Wurzelholzes von Platanus x acerifolia (Ait.) Willd. Österr. Bot. Zeitschr. Band 121. Heft 3/4. pp. 227-249.

2973. TROLL, W. 1954. Allgemeine Botanik. Ferdinand Enke, Stuttgart. pp. 320-347.

2974. WILSON, L.A., and LOWE, S.B. 1973. The anatomy of the root system in West Indian sweet potato - Ipomoea batatas (L.) Lam. cultivars. Annals Bot. 37:633-643.

2975. ŻURAWSKA, H. 1912. Kielkowanie palm. -- Über die Keimung der Palmen. Bull. Inter. Acad. Sci. Let. Cracovie (Class Sci. Math.) Sér. B, Sci. Nat. pp. 1062-1095.

SOURCES OF PERIODICAL TITLES

꧁꧂

Insamuch as periodical literature abbreviations vary from source to source, the abbreviations utilized in this bibliography do not ascribe to any particular foreordained pattern. However, they are for the most part in accord with the American Standard for Periodical Title Abbreviations and with the International List of Periodical Title Word Abbreviations. (see the section of this bibliography entitled PERIODICAL TITLES WITH ABBREVIATIONS)

B-P-H (Botanico-Periodicum-Huntianum)

Dictionary Catalog of the National Agricultural Library

Half a Century of Soviet Serials

International Catalog of Scientific Literature (Botany)

List of Serial Publications in the British Museum (Natural History)

New Serial Titles

The Bradley Bibliography

Ullrich's International Periodicals Directory

Union List of Serials

World List of Periodicals and Serials of Interest to Forestry

World List of Scientific Periodicals

PERIODICAL TITLES - WITH ABBREVIATIONS

ed. - editor, edition   pp. - pages

Mem. - Memoirs, Mémoires Sect. - Section

N.F. - Neue Folge   Ser. - Series

N.S. - New Series   Supp. - Supplement

No. - number   Vol. - volume

Abh. Akad. Wiss. Lit. Mainz
Abhandlungen der Akademie der Wissenschaften und der Literature in Mainz

Abh. Kaiser. Leop. Carol. Deutsch. Akad. Nat.
Abhandlungen der Kaiserlichen Leopoldinisch Carolinischen deutschen Akademie der Naturforscher, Nova Acta, Halle

Abh. Königl. Acad. Wiss. Berlin
Abhandlungen der Königlichen Academie der Wissenschaften in Berlin

Abh. Königl. Preuss. Akad. Wiss.
Abhandlungen der Königlich Preussischen Akademie der Wissenschaften

Abh. Nat. Gesel. Halle
Abhandlungen der Naturforschenden Gesellschaft zu Halle

Abh. Senck. Nat. Gesel.
Abhandlungen der Senckenbergischen Naturforschenden Gesellschaft

Absts. Doct. Diss. Ohio State Univ.
Abstracts of Doctoral Dissertations, The Ohio State University

Acad. Roy. Sci. Lett. Beaux-Arts Belg.
Académie Royale des Sciences des Lettres et des Beaux-Arts de Belgique

Acad. Sci. Bohême Bull. Int'l.
L'Académie des Sciences de Bohême. Bulletin International

Acta Agron.
Acta Agronomica (Budapest)

Acta Agron. Acad. Sci. Hung.
Acta Agronomica Academiae Scientarum Hungaricae

Acta Biol. Acad. Sci. Hung.
  Acta Biologica Academiae Scientarum Hungaricae

Acta Biol. Cracov.
  Acta Biologica Cracoviensia

Acta Biol. Debric.
  Acta Biologica Debricina

Acta Bot.
  Acta Botanica, Szeged

Acta Bot. Acad. Sci. Hung.
  Acta Botanica Academiae Scientarum Hungaricae

Acta Bot. Inst. Bot. Univ. Zagreb.
  Acta Botanica Instituti Botanici Universitatis Zagrebensis

Acta Bot. Neerl.
  Acta Botanica Neerlandica

Acta Bot. Sinica
  Acta Botanica Sinica

Acta Forest. Fenn.
  Acta Forestalia Fennica, Helsinki

Acta Horti Berg.
  Acta Horti Bergiani

Acta Horti Petrop.
  Acta Horti Petropitani, St. Petersburg

Acta Soc. Bot. Poloniae
  Acta Societatis Botanicorum Poloniae

Acta Univ. Carol.
  Acta Universitatis Carolinae

Acta Univ. Lund.
  Acta Universitatis Lundensis

Actes Soc. Linn. Bord.
  Actes de la Société Linnéenne de Bordeaux

Adansonia

Adv. Bot. Res.
  Advances in Botanical Research

Adv. Front. Plant Sci.
  Advancing Frontiers of Plant Sciences

Agric. Gaz. New So. Wales
  Agricultural Gazette of New South Wales

Agric. Jour. India
  Agricultural Journal of India

Agrobiology

Agron. Jour.
  Agronomy Journal

Agron. Lusit.
  Agronomia Lusitana (Portugal)

Akad. Nauk Kazak. SSR
  Akademii Nauk Kazakskoi SSR

Akad. Nauk SSSR Sib. Otd. Trudȳ Inst. Lesa Drev.
  Akademiia Nauk SSSR (Sibirskoe Otdelenie) Trudȳ, Institut Lesa i Drevesiny

Allgem. Forst-Jagdt-Zeit.
  Allgemeine Forst-und Jagdt-Zeitung

Allionia (Turin)

Amer. Hort. Mag.
  American Horticultural Magazine

Amer. Jour. Bot.
  American Journal of Botany

Amer. Jour. Pharm.
  (The) American Journal of Pharmacy

Amer. Jour. Sci.
  (The) American Journal of Science

Amer. Midl. Nat.
  (The) American Midland Naturalist

Amer. Nat.
  (The) American Naturalist

Amer. Philos. Soc. Yearb.
  American Philosophical Society Yearbook

Anais Fac. Farm Odont. Univ. São Paulo
  Anais da Faculdad de Farmácia e Odontologia, Universidade de São Paulo

Anal. Inst. Cerc. Exp. Forest.
  Analele Institutuli de Cercetări si Experimentatie Forestiara, Bucarest

Anal. Stiint. Univ. "Al. I. Cuza," Iasi
  Analele Stiintifice Ale Universitatii "Alexandru Ion Cuza" din Iasi

Angew. Bot.
  Angewandte Botanik

Ann. Agricol.
  Annal di Agricoltura (Roma)

Ann. Biol.
  Annales de Biologie

Annali Bot.
  Annali di Botanica

Annals Bot.
  Annals of Botany

Ann. Bot. Fenn.
  Annales Botanici Fennici

Ann. Bot. Soc. Zoo.- Bot. Fenn.
  Annales Botanici Societatis Zoologicae-Botanicae, Fennicae, Vanamo, Helsinki

Ann. Fac. Agraria R. Univ. Perugia
  Annali della Facultà di Agraria della R. Università degli Studi di Perugia

Ann. Inst. Colon. Marseille
  Annales de l'Institut Colonial de Marseille

Ann. Jard. Bot. Buitenz.
  Annales du Jardin Botanique de Buitenzorg

Ann. Mag. Nat. Hist.
  Annals and Magazine of Natural History

Ann. New York Acad. Sci.
  Annals of the New York Academy of Science

Ann. Rev. Plant Phys.
  Annual Review of Plant Physiology

Ann. Roy. Bot. Gard. (Calcutta)
  Annals of the Royal Botanic Garden. Calcutta

Ann. Roy. Bot. Gard. (Peradeniya)
  Annals of the Royal Botanic Gardens, Peradeniya

Ann. Sci. Nat., Bot.
Annales des Sciences Naturelles, Botanique

Ann. Sci. Univ. Bensançon
Annales Scientifiques de l'Université de Bensançon

Ann. Univ. Grenoble
Annales de l'Université de Grenoble

Ann. Univ. Sci. Budap. Rol. Eöt. Nom.
Annales Universitatis Scientarum Budapestinensis de Rolando Eötvös
Nominatae

Anz. Akad. Wiss. Wien
Anzeiger der Akademie der Wissenschaften, Wien

Apoth.- Zeit.
Apotheker-Zeitung

Arb. Biol. Reichs. Land-Forst.
Arbeiten aus der biologischen Reichsanstalt für Land- und Forstwirtschaft

Arb. Bot. Inst. Würzburg
Arbeiten des botanischen Instituts in Würzburg

Arb. Pharm. Inst. Univ. Berlin
Arbeiten aus dem Pharmaceutisches Institut der Universität Berlin

Arch. Bot.
Archives de Botanique

Arch. Bot. Nord France
Archives Botanique du Nord de la France

Arch. Inst. Bot. Univ. Liège
Archives de l'Institut de Botanique Université de Liège

Arch. Muséum Hist. Nat.
Archives de Muséum d'Histoire Naturelle (Paris)

Arch. Pharm.
Archiv der Pharmazie

Arch. Pharm. Berich. Deutsch. Pharm. Gesel.
Archiv der Pharmazie und Berichte der deutschen Pharmazeutischen
Gesellschaft

Arch. Soc. Zool. Bot. Fenn. "Vanamo"
Archivum Societatis Zoologicae Botanicae Fennicae "Vanamo"

Arch. Suiker. Nederl.- Indië
Archief voor de Suikerindustrie in Nederlandsch-Indië

Arch. Theecult. Nederl.- Indië
Archief voor de Theeculture in Nederlandsch-Indië

Assoc. Franç. Avan. Sci.
Association Française pour l'Avancement des Sciences

Atti Accad. Gioenia Sci. Nat. Catania
Atti dell'Accademia Gioenia di Scienze Naturali di Catania

Atti Ist. Bot. "Giov. Briosi" Lab. Critt. Ital. Univ. Pavia
Atti dell'Istituto Botanico "Giovanni Briosi" e Laboratorio Crittigamico
Italiano della R. Università di Pavia

Atti Ist. Bot. Univ. Pavia
Atti dell'Istituto Botanico dell'Universita di Pavia, Milano

Atti Reale Accad. Lincei
Atti della Reale Accademia dei Lincei

Atti Reale Accad. Naz. Lincei
Atti della Reale Accademia Nazionale dei Lincei

Atti Reale Accad. Sci. Toreno
Atti della Reale Accademia delle Scienze di Toreno

Atti Soc. Elvet. Sci. Nat.
Atti della Società Elvetica delle Scienze Naturali

Atti Soc. Lig. Sci. Nat. Geogr.
Atti della Società Ligustica di Scienze Naturali e Geografiche

Atti Soc. Nat. Mat.
Atti della Società dei Naturalisti e Matematici, Modena

Aus der Heimat

Ausland. Einzeldar. Auswärt. Amt
Auslandwirtschaft in Einzeldarstellungen, Auswärtigen Amt, Berlin

Austral. For.
Australian Forestry

Austral. Jour. Biol. Sci.
Australian Journal of Biological Sciences

Austral. Jour. Bot.
Australian Journal of Botany

Beih. Bot. Centralb. (Same as Beih. Bot. Zentralb.)
   Beiheft zum botanischen Centralblatt (or Zentralblatt)

Beitr. Allgem. Bot.
   Beiträge zur Allgemeinen Botanik

Beitr. Biol. Pflanz.
   Beiträge zur Biologie der Pflanzen

Beitr. Wiss. Bot.
   Beiträge zur Wissenschaftlichen Botanik

Berich. Deutsch. Bot. Gesel.
   Berichte der deutschen botanischen **Gesellschaft**

Berich. Deutsch Pharm Gesel.
   Berichte der deutschen pharmazeutischen Gesellschaft

Berich. Königl. Lehr. Wein-Obst-Garten. Geisen. Rheim
   Berichte der Königlichen Lehranstalt für Wein, - Obst - und Gartenbau
   zu Geisenheim am Rheim

Berich. Schweiz. Bot. Gesel.
   Berichte der schweizerischen botanischen Gesellschaft

Biblio. Bot.
   Bibliotheca Botanica

Biol. Plant. **(Praha)**
   Biologia Plantarum (Praha)

Biol. Rev.
   Biological Reviews

Biol. Centralb. (Same as Biol. Zentralb.)
   Biologisches Centralblatt (or Zentralblatt)

Biotropica

Birla Inst. Tech. Sci. Mag.
   Birla Institute of Technology and Science Magazine

Blumea

Bol. Fac. Filos. Cienc. Letras Univ. São Paulo
   Boletim da Faculdade de Filosofia, Ciências e Letras, Universidade
   de São Paulo

Bol. Soc. Bot. Ital.
   Bolletino della Società Botanica Italiana

Bol. Soc. Broteriana
  Boletim da Sociedade Broteriana (Coimbra)

Bombay Univ. Jour.
  Bombay University Journal

Bot. Abhandl.
  Botanische Abhandlungen

Bot. Archiv
  Botanische Archiv

Bot. Centralb. (Same as Bot. Zentralb.)
  Botanisches Centralblatt, Jena (or Zentralblatt)

Bot. Gaz.
  Botanical Gazette

Bot. Jaarb.

  Botanisch Jaarboek, Uitgegeven door het Kruidkundig Genootschap
  Dodonaea t Ghent

Bot. Jahrb.
  Botanische Jahrbucher

Bot. Jour. Linn. Soc.
  Botanical Journal of the Linnean Society

Bot. Közlem.
  Botanikai Közlemények (Budapest)

Bot. Mag.
  Botanical Magazine

Bot. Mag. (Tokyo)
  (The) Botanical Magazine (Tokyo)

Bot. Notiser
  Botaniska Notiser

Bot. Rev.
  Botanical Review

Bot. Studien
  Botanische Studien

Bot. Studier
  Botaniska Studier

Bot. Tidskr.
  Botanisk Tidskrift

Bot. Wandt.
  Botanische Wandtafeln

Bot. Zeit.
  Botanische Zeitung

Bot. Zhur.
  Botanicheskii Zhurnal SSSR

British Mycol. Soc. Trans.
  British Mycological Society Transactions

Bulet. Agric.
  Buletinul Agriculturii. Bucuresti

Bulet. Min. Agric. Ind. Com. Dom.
  Buletinul Ministerului Agriculturii, Industriei, Commerciului şi
  Domeniilor (Bucuresti)

Bulg. Akad. Nauk. Inst. Izvet.
  Bulgarska Akademiya na Naukite, Botanischeskiya Institut, Izvetiya

Bulg. Tyut.
  Bulgarski Tyutyun

Bull. Acad. Roy. Belgique
  Bulletin de la Académie Royale de Belgique

Bull. Acad. Sci. URSS
  Bulletin de l'Academie des Sciences de l'URSS

Bull. Angew. Bot.
  Bulletin für Angewandte Botanik

Bull. Assn. Chem. Sucr. Dist. Ind. Agric. France Colon.
  Bulletin de l'Association des Chemistes de Sucrerie, de Distilleri et
  des Industries Agricoles de France et des Colonies

Bull. Bot. Soc. Bengal
  Bulletin of the Botanical Society of Bengal

Bull. Coll. Agric. Tokyo Imp. Univ.
  Bulletin of the College of Agriculture, Tokyo Imperial University

Bull. Colo. State Univ. Agric. Exp. Sta.
  Bulletin. Colorado State University Agricultural Experiment Station

Bull. Fac. Agric. Kagosh. Univ.
Bulletin of the Faculty of Agriculture, Kagoshima University

Bull. For. Comm.
Bulletin of the Forestry Commission. London

Bull. Ill. Agric. Exp. Sta.
Bulletin. Illinois Agricultural Experiment Station

Bull. Inst. Fond. Afrique Noire
Bulletin de l'Institut Fondamental d'Afrique Noire

Bull. Inst. Franç. Afrique Noire
Bulletin de l'Institut Français d'Afrique Noire

Bull. Inst. Jard. Bot. Univ. Beograd
Bulletin de l'Institut et du Jardin Botaniques de l'Université de Beograd

Bull. Inter. Acad. Polonaise Sci. Let.
Bulletin International de l'Académie Polonaise des Sciences et des Lettres

Bull. Inter. Acad. Sci. Let. Cracovie
Bulletin International de Academie des Sciences et des Lettres de Cracovie

Bull. Jard. Bot. Buitenz.
Bulletin du Jardin Botanique, Buitenzorg

Bull. Jard. Bot. Kieff
Bulletin du Jardin Botanique de Kieff

Bull. Jard. (Bot.) Princ. URSS
Bulletin du Jardin (Botanique) Principal de l'URSS (Leningrad)

Bull. Kansas Agric. Exp. Sta.
Bulletin of the Kansas Agricultural Experiment Station

Bull. Lloyd Lib. Bot. Pharm. Mat. Med.
Bulletin of the Lloyd Library of Botany, Pharmacy and Materia Medica

Bull. Mus. Hist. Nat.
Bulletin du Muséum d'Histoire Naturelle, Paris

Bull. Nat. Bot. Gard.
Bulletin. National Botanic Garden, Lucknow

Bull. New York Bot. Gard.
Bulletin of the New York Botanical Garden

Bull. New York State Coll. For.
Bulletin of the New York State College of Forestry

Bull. Princ. Bot. Gard. (Moscow)
Bulletin of the Principal Botanic Garden, Moscow

Bull. Res. Council Israel
Bulletin of the Research Council of Israel

Bull. Sci. France Belg.
Bulletin Scientifique de la France et de la Belgique

Bull. Sci. Pharm.
Bulletin des Sciences Pharmacologiques

Bull. Soc. Bot. France
Bulletin de la Société Botanique de France

Bull. Soc. Bot. Geneve
Bulletin de la Société Botanique de Genève

Bull. Soc. Hist. Nat. Afrique Nord
Bulletin de la Société d'Histoire Naturelle de l'Afrique du Nord

Bull. Soc. Hist. Nat. Autun
Bulletin de la Société d'Histoire Naturelle d'Autun

Bull. Soc. Hist. Nat. Toulouse
Bulletin de la Société d'Histoire Naturelle de Toulouse

Bull. Soc. Imp. Nat. Moscov
Bulletin de la Société Impériale des Naturalistes de Moscov

Bull. Soc. Philom. Paris
Bulletin de la Société Philomatique de Paris

Bull. Soc. Roy. Bot. Belgique
Bulletin de la Société Royale de Botanique de Belgique

Bull. Soc. Sci. Méd. Ouest
Bulletin de la Société Scientifique et Médicale de l'Ouest. Rennes

Bull. Soc. Vaud. Sci. Nat.
Bulletin de la Société Vaudoise des Sciences Naturelles

Bull. Tokyo Univ. For.
Bulletin of the Tokyo University Forests

Bull. Torrey Bot. Club
Bulletin of the Torrey Botanical Club

Bull. Univ. Texas (Sci. Ser.)
Bulletin. University of Texas. Scientific Series

Cactaceae   (Jahrbücher der Deutschen Kakteen-Gesellschaft e. V., Berlin)

Caldasia

Canad. Jour. Bot.
  Canadian Journal of Botany

Caryologia

Časop. Slez. Mus. Opa.
  Časopsis Slezského Musea v Opavě, (Ser. C), Dendrology

Castanea

Cath. Univ. Amer. Contrib.
  Catholic University of America Contributions

Centralb. Bakt. Parasit. Infek.
  Centralblatt für Bakteriologie, Parasitenkunde und Infektionskrankheiten

Ceylon Jour. Sci.
  Ceylon Journal of Science

Chem. Zeitschr.
  Chemische Zeitschrift

Chromosoma

Chronica Bot.
  Chronica Botanica

Citrus Ind.
  (The) Citrus Industry

Comp. Rend. Assoc. Franç. Avan. Sci.
  Compte Rendu de Association Française pour l'Avancement des Sciences

Comp. Rend.Hebd. Séances Acad. Agric. France
  Compte Rendu Hebdomaire des Séances de l'Académie d'Agriculture de France

Comp. Rend. Hebd. Séances Acad. Sci.
  Compte Rendu Hebdomaire Séances de l'Académie des Sciences

Comp. Rend. (Dokl.) Acad. Sci. URSS
  Comptes Rendus (Doklady) de l'Académie des Sciences de l'URSS

Comp. Rend. Séances Soc. Phys. Hist. Nat. Geneve
  Compte Rendu des Séances de la Société de Physique et d'Histoire
  Naturelle de Genève

Comp. Rend. Trav. Lab. Biol. Vég. Fac. Sci. Poitiers
  Comptes Rendus des Travaux du Laboratoire de Biologie Végétale de la
  Faculté des Sciences de Poitiers

Contr. Biol. Veg.
  Contribuàioni alla Biologia Vegetale

Contr. Bot. Lab. Univ. Penn.
  Contributions from the Botanical Laboratory of the University of
  Pennsylvania

Contr. Boyce Thomp. Inst. Plant Res.
  Contributions of the Boyce Thompson Institute for Plant Research

Contr. Univ. Laval Fonda Rech. For.
  Contribution Université Laval, Fonda de Recherches Forestieres, Quebec (City)

Coton. Cult. Coton.
  Coton et Culture Cotonnière

Curr. Sci.
  Current Science

Cytologia

Daff. Tulip Yearb.
  (The) Daffodil and Tulip Yearbook, Royal Horticultural Society, London

Dansk Bot. Arkiv
  Dansk Botanisk Arkiv

Dansk Tidskr. Farmaci
  Dansk Tidsskrift for Farmaci (Kjøbenhaven)

Das Pflanzenreich

Darwinia

Denk. Akad. Wiss.
  Denkschriften der Akademie der Wissenschaften (Wien)

Denk. Kaiserl. Akad. Wiss. Wien
  Denkschriften der Kaiserlichen Akademie der Wissenschaften in Wien

Det Kong. Danske Vid. Sel. Med.
  Det Kongelige Danske Videnskabernes Selskabs Medlemmer. Copenhagen

Det Kong. Norske Vid. Sel. Skrif.
  Det Kongelige Norske Videnskabers Selskabs Skrifter

Deutsch. Apoth.- Zeit.
  Deutsche Apotheker-Zeitung

Die Kulturepf.
  Die Kulturepflanze

Diss. Absts.
  Dissertation Abstracts

Diss. Absts. Inter.
  Dissertation Abstracts International

Dokl. Akad. Nauk Belorusskoi SSR
  Dokladỹ Akademii Nauk Belorusskoi SSR

Dokl. Akad. Nauk SSSR
  Doklady Akademiîa Nauk SSSR

Dokl. Vses. Akad. Selsk. Nauk. Imeni V.I. Lenina (Moskva)
  Doklady Vsesoiuznoi Akademii Sel'skokhoziaistvennykh Nauk Imeni V.I.
  Lenina (Moskva)

East African Agric. For. Jour.
  East African Agricultural and Forestry Journal

Ecol. Monog.
  Ecological Monographs

Ecol. Rev.
  Ecological Review (Sendai, Japan)

Ecology

Econ. Bot.
  Economic Botany

Erdész. Lapok
  Erdészeti Lapok

Evolution

Flora   (Flora oder Allgemeine Botanische Zeitung)

Florida State Univ. Stud.
  Florida State University Studies

Folia Forest. Polon.
  Folia Forestalai Polonica

For. Chron.
  Forestry Chronicle (Toronto)

For. Sci.
  Forest Science

Forstarchiv

Forstw. Centralb. (Same as Forstw. Zentralb.)
  Forstwissenschaftliches Centralblatt (or Zentralblatt)

Forts. Bot.
  Fortschritte der Botanik

Gard. Chron.
  (The) Gardener's Chronicle

Gartenbauwiss.
  Gartenbauwissenschaft

Gartenflora

Gentes Herb.
  Gentes Herbarum

Gesel. Deutsch. Naturf. Aerzte
  Gesellschaft deutscher Naturforscher und Aerzte

Giorn. Bot. Ital.
  Gironale Botanico Italiano

Glas. Hrvat. Prirod. Druš.
  Glasnik Hrvatskoga Prirodoslovnoga Društva. Zagreb

Great. Brit. For. Comm.
  Great Britain Forestry Commission

Growth

Harvard For. Papers
  Harvard Forest Papers

Hedwigia

Heredity (London)

Hilgardia

Hod. Ros. Aklim. Nas.
  Hodowla Roslin Aklimatyzacja Nasiiennictwo

Holzforschung

Horticulture

House Gard.
  House and Garden

Indian Coco. Jour.
  (The) Indian Coconut Journal

Indian Ecol.
  Indian Ecologist

Indian For.
  Indian Forester, Allahabad

Indian Jour. Agric. Sci.
  Indian Journal of Agricultural Science

Indian Jour. Pharm.
  (The) Indian Journal of Pharmacy

Indus. Sacc. Ital.
  Industria Saccarifera Italiana

Iowa Agric. Exp. Sta. Res. Bull.
  Iowa Agricultural Experiment Station. Research Bulletin

Iowa State Coll. Jour. Sci.
  Iowa State College Journal of Science

Iowa State Jour. Sci.
  Iowa State Journal of Science

Israel Jour. Bot.
  Israel Journal of Botany

Istanbul Univ. Fen Fak, Mecmuasi
  Istanbul Universitesi Fen Fakultesi Mecmuasi

Izv. Akad. Nauk SSSR
  Izvestiya Akademii Nauk SSSR

Izv. Akad. Nauk Turkmen. SSR
  Izvestiya Akademii Nauk Turkmenskoj SSR

Izv. Imp. Lêsn. Inst.
  Izvestiia Imperatorskago Lêsnogo Instituta (Petrograd)

Jahrb. Akad. Wiss. Lit. Mainz
  Jahrbuch. Akademie der Wissenschaften und der Literatur. Mainz

(Jahrb.) Egyet. Termész. Szövet. (Budapest)
  (Jahrbuch) Egyetemi Természettudományi Szövetseg (Budapest)

Jahrb. Hamburg. Wiss. Anst.
  Jahrbuch der hamburgischen wissenschaftlichen Anstalten

Jahrb. Wiss. Bot.
  Jahrbücher für wissenschaftliche Botanik

Jahresb. Ver. Angew. Bot.
  Jahresbericht Vereinigung angewandte Botanik

Jap. Jour. Bot.
  Japanese Journal of Botany

Jena. Zeitschr. Naturw.
  Jenaische Zeitschrift für Naturwissenschaft

Jour. Agric. Res.
  Journal of Agricultural Research

Jour. Amer. Soc. Agron.
  Journal of the American Society of Agronomy

Jour. Appl. Ecol.
  Journal of Applied Ecology

Jour. Arnold Arb.
  Journal of the Arnold Arboretum

Jour. Biol. Sci.
  (The) Journal of Biological Sciences (Bombay)

Jour. Bombay Nat. Hist. Soc.
  (The) Journal of the Bombay Natural History Society

Jour. Bot.
  Journal de Botanique

Jour. Cell Biol.
  (The) Journal of Cell Biology

Jour. Coll. Agric. Imp. Univ. Tokyo
  Journal of the College of Agriculture, Imperial University of Tokyo

Jour. Coll. Sci. Imp. Univ. Tokyo
  Journal of the College of Science, Imperial University of Tokyo

Jour. Dept. Agric. Kyushu Imp. Univ.
  Journal. Department of Agriculture, Kyushu Imperial University

Jour. Ecol.
  Journal of Ecology

Jour. Eli. Mitchell Sci. Soc.
  Journal of the Elisha Mitchell Scientific Society

Jour. Exp. Bot.
  Journal of Experimental Botany

Jour. Fac. Sci. Imp. Univ. Tokyo
  Journal of the Faculty of Science, Imperial University of Tokyo

Jour. Forestry
  Journal of Forestry

Jour. Genetics
  Journal of Genetics

Jour. Hort. Sci.
  Journal of Horticultural Science

Jour. Indian Bot.
  (The) Journal of Indian Botany

Jour. Indian Bot. Soc.
  Journal of the Indian Botanical Society

Jour. Inst. Wood Sci.
  Journal of the Institute of Wood Science

Jour. Jap. For. Soc.
  Journal of the Japanese Forestry Society

Jour. Jap. Soc. Hort. Sci.
  Journal of the Japanese Society of Horticultural Science

Jour. Linn. Soc.
  Journal of the Linnean Society

Jour. Minn. Acad. Sci.
  Journal of the Minnesota Academy of Science

Jour. Pharm Pharm.
  Journal of Pharmacy and Pharmacology

Jour. Pomol. Hort. Soc.
  Journal of Pomology and Horticultural Science

Jour. Proc. Roy. Soc. West. Austral.
Journal and Proceedings of the Royal Society of Western Australia

Jour. Res. Agra Univ.
Journal of Research. Agra University

Jour. Roy. Hort. Soc.
Journal of the Royal Horticultural Society

Jour. Roy. Soc. West Austral.
Journal of the Royal Society of Western Australia

Jour. Shivaji Univ.
Journal of Shivaji University

Jour. Soc. Bot. Russie
Journal de la Société Botanique de Russie

Jour. So. African Bot.
Journal of South African Botany. Cape Town

Jour. Wash. Acad. Sci.
Journal of the Washington Academy of Sciences

Jour. West African Inst. Oil Palm Res.
Journal of the West African Institute for Oil Palm Research

Jour. West African Sci. Assn.
Journal of the West African Science Association

Kansas Univ. Sci. Bull.
(The) Kansas University Science Bulletin

Kew Bull.
Kew Bulletin

Kirkia

Kosmos

Kung. Svensk. Vet. Hand.
Kungliga Svenska Vetenskapakademiens Handligar

Labdev. Jour. Sci. Tech.
Labdev. Journal of Science and Technology (Kanpur, India)

La Cellule

Land. Tijds. Neder.- Indië
Landbouwkundig Tijdsschrift voor Nederlandsch-Indië. Bogor

Landw. Jahrb.
  Landwirtschaftliche Jahrbücher. Berlin

Landw. Monats.
  Landwirtschaftliche Monatsschrift

L'Année Biol.
  L'Année Biologique

La Rev. Scient.
  La Revue Scientifique

Le Botaniste

(De) Levende Natuur (Amsterdam)

Lilloa

Linnaea

Lloydia

Maatel. Aikak.
  Maataloustieteelinen Aikakauskirja. Helsinki

Madroño

Malpighia

Manchest. Lit. Philos. Soc. Mem. Proc.
  Manchester Literary and Philosophical Society Memoirs and Proceedings

Mannbl. Natuurw.
  Mannblad voor Natuurwetenschappen. Amsterdam

Medd. Grönl.
  Meddelelser om Grönland

Meded. Landb. Wagen.
  Mededelingen van de Landbouwhoogeshool te Wageningen

Meded. Uitg. Dept. Landb. Nederl.- Indië
  Mededelingen uitgeven van het Department van Landbouw in Nederlansch-Indië

Mém. Acad. Roy. Belgique
  Mémoires de la Académie Royale de Belgique

Mém. Acad. Roy. Sci. Inst. France
  Mémoires de l'Académie Royale des Sciences de l'Institute de France

Mem. Boston Soc. Nat. Hist.
Memoirs of the Boston Society of Natural History

Mem. Coll. Sci. Kyoto Imp. Univ.
Memoirs of the College of Science, Kyoto Imperial University

Mem. Cornell Univ. Agric. Exp. Sta.
Memoir. Cornell University Agricultural Experiment Station

Mém. Cour. Mém. Sav. Etrang. Acad. Roy. Sci. Let. Beaux-Arts Belgique
Mémoires Couronnés et Mémoires des Savants Étrangers, Académie Royale
des Sciences, Lettres et Beaux-Arts de Belgique

Mem. Dept. Agric. India
Memoirs of the Department of Agriculture of India

Mem. Fac. Sci. Agric. Taihoku Imp. Univ.
Memoirs of the Faculty of Science and Agriculture, Taihoku Imperial
University (Formosa)

Mem. Nat. Acad. Sci.
Memoirs of the National Academy of Sciences, Washington, D.C.

Mem. New York Bot. Gard.
Memoirs of the New York Botanical Garden

Mem. Proc. Manchester Lit. Philos. Soc.
Memoirs and Proceedings of the Manchester Literary and Philosophical Society

Mem. Soc. Cient. "Antonio Alzate"
Memorias, Sociedad Cientifica "Antonio Alzate"

Mém. Soc. Linné. Norm.
Mémoires de la Société Linnéenne de Normandie

Mém. Soc. Nat. Sci. Math. Cherbourg
Mémoires de la Société nationale des Sciences Naturelles et Mathématiques
de Cherbourg

Mém. Soc. Roy. Soc. Liège
Mémoires de la Société Royale des Sciences de Liège

Mém. Soc. Vaud. Sci. Nat.
Mémoires de la Société Vaudoise des Sciences Naturelles, Lausanne

Mem. Torrey Bot. Club
Memoirs of the Torrey Botanical Club

Merck's Rept.
Merck's Report

Mich. Bot.
  (The) Michigan Botanist

Mikroscopie

Minn. Bot. Studies
  Minnesota Botanical Studies

Mitt. Gesel. Deutsch. Naturf. Aerzte
  Mitteilungen der Gesellschaft für deutscher Naturforscher und Aerzte

Mitt. Kaiser-Wilh.-Inst. Landw. Bromberg
  Mitteilungen des Kaiser-Wilhelm-Instituts für Landwirtschaft in Bromberg

Mitt. Landw. Inst. Königl. Univ. Breslau
  Mitteilungen der Landwirtschaftlichen Institut der Königlichen Universität
  Breslau

Mitt. Schweiz. Cent. Forst. Versuchs.
  Mitteilungen der schweizerischen Centralanstalt für die Fortsliche
  Versuchswesen

Moskov. Obshch. Ispyt. Prirody Biul.
  Moskovskoye Obsshchestvo Ispytatelei Prirody. Biulleten

Musée Bot. Leide
  Musée de Botanique de Leide

Mutisia

Nat. Hist.
  Natural History

(The) Naturalist

Nature

Naturforscher (Berlin)

Naturf. Gesel.
  Naturforschende Gesellscahft, Leipzig

Naturf. Gesel. Viertel. Beibl.
  Naturforschende Gesellschaft Vierteljahrsschrift Beiblätter. Zurich

Nat. Mag.
  Nature Magazine

Naturw. Monatsh. Biol. Chem. Geogrr. Unter.
  Naturwissenschaftliche Monatshefte für den Biologischen, Chemischen,
  Geographischen Unterricht. Leipzig

Naturw. Runds.
Naturwissenschaftliche Rundschau

Naturw. Zeitschr. Forst-Landw.
Naturwissenschaftliche Zeitschrift für Forst- und Landwirtschaft

Nauch. Dokl. Vȳss. Shkolȳ
Nauchnye Doklady Vȳsshei Shkolȳ (Moskva)

New Phytol.
New Phytologist

New Zeal. Jour. Agric. Res.
New Zealnd Journal of Agricultural Research

New Zeal. Jour. Bot.
New Zealand Journal of Botany

New Zeal. Jour. Sci. Tech.
New Zealand Journal of Science and Technology

Northw. Sci.
Northwest Science

Notes Roy. Bot. Gard. Edinb.
Notes from the Royal Botanic Garden, Edinburgh

Nouv. Arch. Mus. Hist. Nat.
Nouvelles Archives du Muséum d'Histoire Naturelle

Nova Acta Abh. Kaiser. Leop.- Carol. Deutsch.Akad. Nat.
Nova Acta Abhandlungen der Kaiserlichen Leopoldinisch-Carolinischen
deutschen Akademie der Naturforscher

Nova Acta Leop.
Nova Acta Leopoldina

Növény. Közlem.
Növénytani Közlemények (Budapest)

Nuovo Giorn. Bot. Ital.
Nuovo Giornale Botanico Italiano

Nyt Mag. Natur.
Nyt Magazin for Naturvidenskaberne

Öfver. Finska Vetens.- Soc. Förhand.
Öfversigt af Finska Vetenskaps-Societetens Förhandligar

Ohio Jour. Sci.
  (The) Ohio Journal of Science

Ohio Nat.
  Ohio Naturalist

Oleagineux

Orchideeën

Öster. Bot. Zeitschr.
  Österreichische Botanische Zeitschrift

Ottawa Nat.
  Ottawa Naturalist

Pacific Sci.
  Pacific Science

Palestine Jour. Bot.
  Palestine Journal of Botany

Papers Mich. Acad. Sci. Arts Let.
  Papers from the Michigan Academy of Science, Arts and Letters

Pharm. Arch.
  Pharmaceutical Archives

Pharm. Central. Deutsch. (Same as Pharm. Zentral. Deutsch.)
  Pharmaceutische Centralhalle für Deutschland (or Zentralhalle)

Pharm. Jour.
  (The) Pharmaceutical Journal

Pharm. Jour. Pharm.
  (The) Pharmaceutical Journal and Pharmacist

Pharm. Praxis
  Pharmazeutische Praxis

Philipp. Agric.
  Philippine Agriculturist

Philipp. Jour. Sci.
  Philippine Journal of Science

Philos. Trans. Roy. Soc. London
  Philosophical Transactions of the Royal Society of London

Physiol. Plant.
  Physiologia Plantarum

Phytomorph.
  Phytomorphology

Phyton

Phytopath.
  Phytopathology

Plant Phys.
  Plant Physiology

Plant World

Planta

Planta Med.
  Planta Medica

Preslia

Principes

Proc. Acad. Nat. Sci. Phila.
  Proceedings of the Academy of Natural Sciences of Philadelphia

Proc. Amer. Philos. Soc.
  Proceedings of the American Philosophical Society

Proc. Amer. Soc. Hort. Sci.
  Proceedings. American Society for Horticultural Science

Proc. Boston Soc. Nat. Hist.
  Proceedings of the Boston Society of Natural History

Proc. Calif. Acad. Sci.
  Proceedings of the California Academy of Sciences

Proc. Cambr. Philos. Soc.
  Proceedings of the Cambridge Philosophical Society

Proc. Crop Sci. Soc. Japan
  Proceedings of the Crop Science Society of Japan

Proc. Indian Acad. Sci.
  Proceedings of the Indian Academy of Sciences

Proc. Indiana Acad. Sci.
  Proceedings of the Indiana Academy of Science

Proc. Indian Sci. Congr. Assn.
Proceedings of the Indian Science Congress Association, Calcutta

Proc. Iowa Acad. Sci.
Proceedings of the Iowa Academy of Science

Proc. Leeds Philos. Lit. Soc.
Proceedings of the Leeds Philosophical and Literary Society

Proc. Linn. Soc. London
Proceedings of the Linnean Society of London

Proc. Linn. Soc. New So. Wales
Proceedings of the Linnean Society of New South Wales

Proc. Nat. Acad. Sci. U.S.A.
Proceedings of the National Academy of Sciences of the United States of America

Proc. Nat. Inst. Sci. India
Proceedings of the National Institute of Sciences of India

Proc. No. Dak. Acad. Sci.
Proceedings of the North Dakota Academy of Science

Proc. Penn. Acad. Sci.
Proceedings of the Pennsylvania Academy of Science

Proc. Plant Prop. Soc.
Proceedings of the Plant Propagators Society

Proc. Rajas. Acad. Sci.
Proceedings of the Rajasthan Academy of Sciences

Proc. Roy. Irish Acad.
Proceedings of the Royal Irish Academy. Dublin

Proc. Roy. Soc. Edinb.
Proceedings of the Royal Society of Edinburgh

Proc. Roy. Soc. London
Proceedings of the Royal Society of London

Proc. Roy. Soc. Queensl.
Proceedings of the Royal Society of Queensland

Proc. Roy. Soc. Victoria
Proceedings of the Royal Society of Victoria

Proc. So. Dak. Acad. Sci.
  Proceedings of the South Dakota Academy of Science

Proc. West Virg. Acad. Sci.
  Proceedings of the West Virginia Academy of Science

Proc. Trans. Roy. Soc. Canada
  Proceedings and Transactions of the Royal Society of Canada

Protoplasma

Pubbl. Centro Sperim. Agric. For.
  Pubblicazioni del Centro di Sperimentazione Agricola e Forestale

Publ. Biol. École Hautes Études Vet.
  Publications Biologiques de l'École des Hautes Études Vétérinaires

Publ. Carn. Inst. Wash.
  Publication of the Carnegie Institution of Washington

Publ. Inst. Belge Amelior. Better.
  Publication de l'Institut Belge pour l'Amelioration de la Betterave

Quart. Jour. Crude Drug Res.
  Quarterly Journal of Crude Drug Research

Quart. Jour. Micro. Sci.
  Quarterly Journal of Microscopical Science

Quart. Jour. Pharm. Pharm.
  Quarterly Journal of Pharmacy and Pharmacology

Quart. Rev. Biol.
  Quarterly Review of Biology

Rajas. Univ. Studies
  Rajasthan University Studies

Rast. Resursy
  Rastitel'nye Resursy

Rec. Quel. Trav. Anat. Vég. Exéc. Liège
  Recueil de Quelques Travaux d'Anatomie Végétale Exécutés à Liège, Bruxelles

Rec. Trav. Bot. Neerl.
  Recueil des Travaux Botaniques Néerlandais

Rep. Spec. Nov. Reg. Veg.
  Repertorium Specierum Novarum Regni Vegetabilis (Berlin)

Rept. Brit. Assn. Adv. Sci.
  Report of the British Association for the Advancement of Science

Rept. Date Grow. Inst.
  Report of the Date Grower's Institute

Rept. Mich. Acad. Sci.
  Report of the Michigan Academy of Science

Rept. Nat. Inst. Genetics
  Report. National Institute of Genetics (Misima, Japan)

Rept. Wisc. Agric. Exp. Sta.
  Report of the Wisconsin Agricultural Experiment Station

Res. Bull. Agric. Exp. Sta. Iowa
  Research Bulletin, Agricultural Experiment Station, Iowa State College
  of Agriculture and Mechanical Arts

Res. Bull. Nebr. Agric. Exp. Sta.
  Research Bulletin. Nebraska Agricultural Experiment Station

Res. Bull. Panjab Univ.
  (The) Research Bulletin. Panjab University

Res. Bull. Univ. Nebr.
  Research Bulletin. University of Nebraska

Rev. Agric.
  Revista Agricultura (Piracicaba)

Rev. Argentina Agron.
  Revista Argentina de Agronomia

Rev. Bot. Appl. Agric. Trop.
  Revue de Botanique Appliquée et d'Agriculture Tropicale

Rev. Cult. Colon.
  Revue des Cultures Coloniales

Rev. Cytol. Biol. Vég.
  Revue de Cytologie et de Biologie Végétales

Rev. Fac. Farm. Bioq. Univ. São Paulo
  Revista da Faculdad de Farmácia e Bioquímica da Universided de São Paulo

Rev. Fac. Sci. Univ. Istanbul
  Revue de la Faculté des Sciences de l'Université d'Istanbul

Rev. Gén. Bot.
  Revue Générale de Botanique

Rev. Gén. Sci. Pures Appl.
  Revue Générale des Sciences Pures et Appliques

Rev. Inst. Munic. Bot.
  Revista. Instituto Municipal de Botanica. Buenos Aires

Rev. Sci.
  Revue Scientifique (Paris)

Rev. Zool. Afric.
  Revue Zoologique Africaine

Rhodora

Riz. Rizic.
  Riz et Riziculture

Sassari R. Univ. Ist. Bot. Bull.
  Sassari R. Università degli studi Istituto Botanico, Bullettino

Schr. König. Phys.- Ökon. Gesel. Konings.
  Schriften der Königlichen physikalish-ökonomischen Gesellschaft zu
  Köningsberg

Schweiz. Apoth.- Zeit.
  Schweizerische Apotheker-Zeitung

Science

Sci. Cult.
  Science and Culture

Sci. Mem. Off. Med. Sanit. Dept. Govt. India
  Scientific Memoirs by Officers of the Medical and Sanitary Departments of
  the Government of India

Sci. Pap. App. Sec. Tiflis Bot. Gard.
  Scientific Papers of the Applied Sections of the Tiflis Botanical Garden

Sci. Pap. Cent. Res. Inst. Jap. Govt. Monop. Bur.
  Scientific Papers of the Central Research Institute, Japanese Government
  Monopoly Bureau

Search

Sida

Sinesia

Sitz. Akad. Wiss. Wien (Same as Sitz. Kaiser. Akad. Wiss. Wien)
  Sitzungsberichte der Akademie der Wissenschaften in Wien

Sitz. Heidelb. Akad. Wiss.
   Sitzungsberichte der Heidelberger Akademie der Wissenschaften

Sitz. Kaiser. Akad. Wiss. Berlin (Same as Sitz. Akad. Wiss. Wien)
   Sitzungsberichte der Kaiserlichen Akademie der Wissenschaften zu Berlin

Sitz. König. Preuss. Akad. Wiss. Berlin
   Sitzungsberichte der Königlich Preussischen Akademie der Wissenschaften
   zu Berlin

Sitz. Nat. Verein Preuss. Rhein. Westf.
   Sitzungsberichte herausgegeben vom Naturhistorischen Verein der
   Preussischen Rheinlände und Westfalens

Sitz. Öster. Akad. Wiss.
   Sitsungsberichte Österreichische Akademie der Wissenschaften

So. African Jour. Sci.
   South African Journal of Science

Southw. Nat.
   (The) Southwest Naturalist

Svensk Bot. Tidskr.
   Svensk Botanisk Tidskrift

Symb. Bot. Upsal.
   Symbolae Botanicae Upsaliensis

Symp. Soc. Exp. Biol.
   Symposium of the Society for Experimental Biology

Taiwania

Tech. Bull. Mich. (State Coll.) Agric. Exp. Sta.
   Technical Bulletin, Michigan (State College) Agricultural Experiment
   Station

Tech. Rept. Fac. For. Univ. Toronto
   Technical Report. Faculty of Forestry, University of Toronto

Tectona

Tenth Inter. Bot. Congr. Absts.
   Tenth International Botanical Congress, Abstracts

Thar. Forst. Jahrb.
   Tharandter Forstliches Jahrbuch

Torreya

Trans. Amer. Micro. Soc.
  Transactions of the American Microscopical Society

Trans. Conn. Acad. Arts Sci.
  Transactions of the Connecticut Academy of Arts and Science

Trans. Kansas Acad. Sci.
  Transactions of the Kansas Academy of Science

Trans. Linn. Soc. London
  (The) Transactions of the Linnean Society of London

Trans. Nat. Inst. Sci. India
  Transactions of the National Institute of Sciences of India

Trans. Proc. Bot. Soc. Edinb.
  Transactions and Proceedings of the Botanical Society of Edinburgh

Trans. Proc. Bot. Soc. Penn.
  Transactions and Proceedings of the Botanical Society of Pennsylvania

Trans. Proc. New Zeal. Inst.
  Transactions and Proceedings of the New Zealand Institute

Trans. Roy. Scot. Arbor. Soc.
  Transactions of the Royal Scottish Arboricultural Society

Trans. Roy. Soc. Edinb.
  Transactions of the Royal Society of Edinburgh

Trans. Roy. Soc. New Zeal.
  Transactions of the Royal Society of New Zealand

Trans. Roy. Soc. So. Africa
  Transactions of the Royal Society of South Africa

Trans. Wisc. State Hort. Soc.
  Transactions of the Wisconsin State Horticultural Society

Trav. Inst. Sci. Cherif.
  Travaux de l'Institut Scientifique Cherifen (Rabat, Morocco)

Trav. Lab. For. Toulouse
  Travaux du Laboratoire de Toulouse

Trav. Lab. Mat. Méd. École Sup. Pharm. Paris
  Travaux du Laboratoire de Matière Médicale de l'École Superiure de
  Pharmacie de Paris

Trav. Mém. Univ. Lille
  Travaux et Mémoires de l'Université de Lille

Tree-Ring Bull.
  Tree-Ring Bulletin

Trop. Woods
  Tropical Woods

Trudȳ Glav. Bot. Sada Lenin.
  Trudȳ. Glavnogo Botanicheskogo, Leningrad

Trudȳ Plodoov. Inst. Michurina
  Trudȳ. Plodoovoshchnogo Instituta Imeni I.V. Michurina

Trudȳ Prikl. Bot. Genet. Selek.
  Trudȳ po Prikladnoǐ Botanike, Genetike i Selektsii

Trudȳ Sukhum. Bot. Sada
  Trudȳ Sukhumskogo Botanicheskogo Sada (Sukhum)

Trudȳ Sverdl. Selsk. Inst.
  Trudȳ Sverdlovskogo Sel'skokhozyaǐstvennogo Instituta

Tydsk. Natuurwetensk.
  Tydskrif vir Natuurwetenskappe (Pretoria)

Uch. Zap. Omsk. Gos. Ped. Inst.
  Uchenye Zapiski Omskogo Gosudarstvennogo Pedagogicheskogo Instituta

Ukray. Bot. Zhur.
  Ukrayins'kyi Botanichny Zhurnal

Union So. Africa Dept. Agric. Sci. Bull.
  Union of South Africa Department of Agriculture, Science Bulletin (Pretoria)

U.S.D.A. Bull.
  United States Department of Agriculture, Bulletin

U.S.D.A. Bur. Plant Ind. Bull.
  United States Department of Agriculture Bureau of Plant Industry. Bulletin

U.S.D.A. Prod. Res. Rept.
  United States Department of Agriculture. Production Research Report

U.S.D.A. Tech. Bull.
  United States Department of Agriculture, Technical Bulletin

Univ. Calif. Publ. Bot.
  University of California Publications in Botany

Univ. Hawaii Res. Publ.
  University of Hawaii Research Publications

Univ. Kans. Sci. Bull.
  (The) University of Kansas Science Bulletin

Univ. New Mex. Publ. Biol.
  University of New Mexico Publications in Biology

Univ. Toronto Studies
  University of Toronto Studies

Univ. Wash. Publ. Bot.
  University of Washington Publications in Botany

Untersuch. Bot. Inst. Tübingen
  Untersuchungen aus dem botanischen Institut zu Tübingen. Leipzig

Uzbek. Biol. Zhur.
  Uzbekian Biologicheskii Zhurnal. Tashkent

Verh. Bot. Ver. Prov. Brand.
  Verhandlungen des botanischen Vereins der Provinz Brandenburg

Verh. Konink. Akad. Wetens.
  Verhandelingen der Koninklije Akademie van Wetenschappen. Amsterdam

Verh. Nat. Gesel. Basel
  Verhandlungen der Naturforschenden Gesellschaft in Basel

Verh. Zoo.- Bot. Gesel. Wien
  Verhandlungen der zoologisch-botanischen Gesellscahft in Wien

Veröffent. Geobot. Inst. Rubel Zurich
  Veröffentlichungen des Geobotanischen Instituts Rübel in Zurich

Verz. Geschr. Beijer.
  Verzamelde Geschriften van M.W. Beijerinck. Delft

Vest. Lenin Univ.
  Vestnik Leningradskogo Universiteta

Vest. Moskov. Univ.
  Vestnik Moskovskogo Universiteta (Moscow)

Vest. Sel'sk. Nauk (Moscow)
  Vestnik Sel'skokhoz Nauk (Moscow)

Viden Medd. Nat. Foren. Kjøbh.
  Videnskabelige Meddelelser fra den Naturhistoriske Forening i Kjøbenhaven

Viert. Nat. Gesel. Zürich
  Vierteljahrsschrift der Naturforschenden Gesellschaft in Zürich

Vyesti Akad. Navuk Byelaruskai SSR
  Vyesti Akademiia Navuk Byelaruskai SSR

Watsonia

Weed Res.
  Weed Research (Tokyo)

Weeds

Wiad. Bot.
  Wiadomości Botaniczne

Willdenowia

Wood Sci. Tech.
  Wood Science and Technology

Yearb. Pharm. Trans. Brit. Pharm. Conf.
  Year-book of Pharmacy with Transactions of the British Pharmaceutical
  Conference

Zeitschr. Bot.
  Zeitschrift für Botanik

Zeitschr. Ferd. Tirol Vorarl.
  Zeitschrift des Ferdinandeums für Tirol und Vorarlberg (Innsbruck)

Zeitschr. Forst- Jagd.
  Zeitschrift für Forst- und Jagdwesen

Zeitschr. Gesam. Brauw.
  Zeitschrift für das Gesammte Brauwesen

Zeitschr. Naturwiss.
  Zeitschrift für Naturwissenschaften. Leipzig

Zeitschr. Unter. Lebens.
  Zeitschrift für Untersuchung der Lebensmittel

Zeitschr. Unter. Nahr.- Genuss.
  Zeitschrift für Untersuchung der Nahrungs- und Genussmittel

Zeitschr. Ver. Deutsch. Zucker-Ind.
  Zeitschrift des Vereins der deutschen Zucker-Industrie

Zeitschr. Zucker. Böhmen
  Zeitschrift für Zuckerindustrie in Böhmen

Zeitschr. Zucker. Čechosl. Repub.
  Zeitschrift für die Zuckerindustrie der Čechoslovakischen Republik. Prague

SELECTED REFERENCES

(For pertinent literature prior to 1900)

1. Annals of Botany. 1887-1900. Vols. 1-14.

2. Botanisches Centralblatt. 1880-1901. Vols. 1-88.

3. Botanisches Zeitung (Berlin). 1843-1901. Vols. 1-59.

4. DE BARY, A. 1884. Comparative Anatomy of Phanerogams and Ferns.
   Clarendon Press, Oxford.

5. GUTTENBERG, H. VON. 1941. Der primäre Bau der Gymnospermenwurzel.
   In: Handbuch der Pflanzenfamilien. (ed.) Linsbauer, K. Band 8.
   Lief 41. Gebrüder Borntraeger, Berlin.

6. ------------------. 1968. Der primäre Bau der Angiospermenwurzel.
   In: Handbuch der Pflanzenfamilien. (eds.) Zimmerman, W., Ozenda, P.,
   and Wulff, H.D. Zweite Auflage. Spezieller Teil. Gebrüder Borntraeger, Berlin. Band 8. Teil 5.

7. HABERLANDT, G. 1914. Physiological Plant Anatomy. Macmillan and Co.,
   Ltd., London.

8. JACKSON, B.D. 1880. Guide to the Literature of Botany. Longmans,
   Green and Co., London. (Facsimilie ed. 1964 Hafner Publ. Co., New York)

9. Journal of the Linnean Society of London. (Botany) 1855-1904. Vols.
   1-35.

10. Just's Botanischer Jahresbericht. 1874-1901. Gebrüder Borntraeger,
    Leipzig.

11. LUBBOCK, J. 1892. A Contribution to Our Knowledge of Seedlings.
    Vol. 2. D. Appleton and Co., London.

12. MODESTOV, A.P. 1915. Les racines des plantes herbacées. I.N.
    Kushneret, Moscow. No. 1. pp. 1-138. (in Russian with French
    summaries)

13. PRITZEL, G.A. 1950. Thesaurus Literaturae Botanicae. New edition.
    Görlich, Milano.

14. Revue Générale de Botanique. 1889-1901. Vols. 1-13.

15. SOLEREDER, H.   1908.   <u>Systematic Anatomy of the Dicotyledons</u>.
    2 vols. Clarendon Press, Oxford.

16. TROLL, W.   1943.   Vergleichende Morphologie der Höhern Pflanzen.
    Band 1. Vegetationsorgane.   Teil 3. Wurzel und Wurzelsystems.
    Gebrüder Borntraeger, Berlin.

17. VAN TIEGHEM, P., and DOULIOT, H.   1888.   Recherches comparative sur
    l'origine des membres endogenes dans les plantes vasculaires.
    Ann. Sci. Nat., Bot. (Sér. 7) 8:1-660.

INDEX TO PLANT FAMILIES

(Selected principally from citation titles. By listing number)

239

BUTOMACEAE - 259, 2355, 2401, 2534

BUXACEAE - 2954

CACTACEAE - 47, 61, 264, 447, 618, 619, 853, 880, 1067, 1092, 1237, 1307, 1724, 1911

CALLITRICHACEAE - 2423

CALOBRYACEAE - 1191

CALYCERACEAE - 788

CAMPANULACEAE - 966, 1431

CANNACEAE - 115, 259, 1102, 2055, 2137, 2511, 2651, 2653

CANNABINACEAE - 20, 61, 420, 445, 1158, 1273, 1914, 1915, 2327

CAPPARIDACEAE - 1236

CAPRIFOLIACEAE - 61, 1225, 1377, 1384, 1757, 2730

CARICACEAE - 1386

CARYOCARACEAE - 477

CARYOPHYLLACEAE - 96, 966, 1126, 1384, 1478, 1493, 1494, 1495, 1917, 2008, 2146, 2195, 2468, 2532, 2603

CASUARINACEAE - 13, 115, 2056

CELASTRACEAE - 61, 780, 1377, 1561, 2080, 2457

CENTROLEPIDACEAE - 274

CEPHALOTACEAE - 1763, 2398

CEPHALOTAXACEAE - 589, 966, 1327

CERATOPTERIDACEAE - (see PARKERIACEAE)

CHENOPODIACEAE - 63, 64, 65, 72, 316, 348, 349, 351, 352, 409, 416, 473, 561, 783, 784, 785, 803, 830, 878, 881, 966, 1120, 1258, 1273, 1382, 1395, 1416, 1450, 1591, 1685 1756, 1820, 1850, 2075, 2076, 2100 2157, 2196, 2323, 2324, 2376, 2422, 2601

CISTACEAE - 966, 1104, 1382

COMBRETACEAE - 233, 305, 781, 1467, 1766

COMMELINACEAE - 77, 259, 580, 657, 1053, 1057, 1058, 1178, 1347, 1365, 1429, 2055, 2294, 2655, 2660, 2703

COMPOSITAE - 88, 96, 103, 115, 132, 209, 307, 317, 318, 354, 355, 385, 386, 387, 440, 490, 491, 521, 543, 593, 598, 646, 670, 840, 856, 905, 939, 954, 966, 979, 1016, 1078, 1143 1196, 1235, 1261, 1262, 1273, 1339, 1375, 1378, 1384, 1388, 1405, 1411, 1540, 1562, 1671, 1672, 1687, 1711, 1716, 1734, 1743, 1748, 1759, 1778, 1782, 1801, 1823, 1912, 1966, 1995, 2001, 2003, 2044, 2045, 2053, 2115, 2142, 2190, 2205, 2221, 2238, 2309, 2311, 2319, 2320, 2348, 2351, 2360, 2466, 2506, 2517, 2532, 2585, 2586, 2608, 2616, 2729, 2757, 2803, 2818, 2857, 2859, 2886, 2917

CONVOLVULACEAE - 7, 61, 96, 111, 136, 236, 243, 282, 346, 350, 466, 509, 679, 795, 966, 1080, 1119, 1170, 1190, 1221, 1271, 1273, 1516, 1518, 1519, 1545, 1552, 1565, 1764, 1833, 1904, 1918, 1959, 2069, 2395, 2417, 2418, 2419, 2420, 2421, 2563, 2627, 2736, 2944, 2945, 2969, 2973

CORNACEAE - 966, 1015, 1377, 1475, 150 1543, 2032

CRASSULACEAE - 861, 862, 1505, 2174, 2256, 2941

CRUCIFERAE - 7, 86, 115, 158, 199, 209 271, 329, 440, 458, 594, 598, 671,

810, 915, 966, 1065, 1066, 1113, 1114, 1155, 1176, 1273, 1341, 1710, 1762, 2094, 2098, 2099, 2103, 2141, 2147, 2205, 2260, 2394, 2403, 2471, 2505, 2589, 2631, 2672, 2798, 2925

CUCURBITACEAE - 60, 61, 115, 267, 555, 903, 918, 919, 966, 1073, 1273, 1413, 1428, 1798, 2101, 2166, 2433, 2479, 2860, 2867, 2948

CUPRESSACEAE - 45, 61, 115, 388, 701, 966, 1327, 1386, 1703, 2789, 2873, 2874

CYATHEACEAE - 2021, 2022

CYCADACEAE - 45, 78, 171, 189, 297, 360, 569, 585, 689, 738, 750, 751, 925, 927, 1256, 1701, 2077, 2128, 2234, 2298, 2356, 2516, 2617, 2733, 2813, 2814, 2912, 2919, 2920, 2921, 2922

CYCLANTHACEAE - 259, 1245, 1903, 2568, 2776

CYPERACEAE - 4, 37, 106, 107, 213, 259, 800, 870, 977, 1171, 1205, 1351, 1353, 1511, 1619, 1759, 1789, 1790, 1861, 1866, 1892, 1963, 2111, 2161, 2355, 2448, 2449, 2758

DIDIEREACEAE - 2242, 2245

DILLENIACEAE - 911, 1434

DIOSCOREACEAE - 24, 247, 269, 366, 369, 370, 599, 841, 1146, 1324, 1431, 1563, 1706, 1775, 1806, 2255, 2501

DIPSACACEAE - 61, 440, 966, 2399, 2577, 2578, 2753

DROSERACEAE - 192, 428, 912, 1221, 1290, 1651, 2496

EBENACEAE - 1377, 2634, 2924

ELAEAGNACEAE - 2432, 2525

ELATINACEAE - 179, 2181

EMPETRACEAE - 1226

EPHEDRACEAE - 61, 703, 743, 900, 2130, 2139, 2638, 2784

EQUISETACEAE - 2, 45, 119, 404, 615, 697, 700, 966, 1157, 1483, 1484, 1655, 1779, 1893, 2801

ERICACEAE - 61, 300, 1308, 1375, 1380, 1382, 1384, 1485, 1679, 1777, 2292

ERIOCAULACEAE - 202, 274, 1243, 1354, 2513, 2656

ERYTHROXYLACEAE - 2750

EUPHORBIACEAE - 61, 372, 374, 493, 936, 966, 1132, 1185, 1378, 1380, 1386, 1977, 2061, 2138, 2169, 2170, 2222, 2223, 2258, 2405, 2408, 2573, 2632

EUPTELEACEAE - 2935

FAGACEAE - 288, 341, 435, 656, 660, 760, 761, 1036, 1378, 1386, 1435, 1543, 1578, 1662, 1663, 1664, 1665, 1812, 1925, 1926, 1990, 2761

FOUQUIERIACEAE - 2257, 2404

FUMARIACEAE - 1077

GENTIANACEAE - 105, 195, 1337, 1361, 1385, 1741, 1863, 2038, 2101, 2141, 2787

GERANIACEAE - 276, 1375, 1676, 2368, 2414

GESNERIACEAE - 1076, 1079, 2709

GINKGOACEAE - 381, 966, 1758, 1884, 2128, 2437

GLEICHENIACEAE - 470, 519

GNETACEAE (see WELWITSCHIACEAE) - 13,

17, 256, 951, 1678, 1773, 1804,
1805, 2966

GRAMINEAE - 36, 86, 87, 115, 221,
259, 299, 310, 336, 347, 353, 357,
362, 363, 364, 421, 425, 432, 439,
440, 486, 499, 512, 527, 528, 545,
553, 574, 586, 614, 728, 749, 763,
801, 808, 822, 873, 920, 923, 929,
934, 947, 966, 981, 982, 983,
1002, 1012, 1090, 1106, 1111,
1117, 1156, 1159, 1160, 1161,
1162, 1186, 1188, 1194, 1202,
1273, 1283, 1285, 1286, 1287,
1302, 1303, 1304, 1305, 1309,
1359, 1373, 1378, 1421, 1436,
1444, 1448, 1458, 1476, 1502,
1504, 1506, 1509, 1524, 1526,
1527, 1530, 1531, 1532, 1533,
1534, 1535, 1536, 1549, 1574,
1592, 1620, 1628, 1629, 1658,
1673, 1694, 1754, 1799, 1808,
1828, 1832, 1851, 1855, 1891,
1898, 1939, 1948, 1954, 1961,
1965, 2007, 2020, 2034, 2039,
2040, 2046, 2058, 2063, 2068,
2086, 2112, 2131, 2199, 2208,
2210, 2237, 2252, 2267, 2316,
2317, 2318, 2343, 2346, 2347,
2355, 2357, 2454, 2455, 2456,
2514, 2555, 2570, 2587, 2590,
2628, 2710, 2742, 2754, 2763,
2770, 2782, 2889, 2936, 2939

GUTTIFERAE - 557, 1355, 1367, 1765,
2527

HAEMODORACEAE - 4, 259, 721, 1431

HALORAGACEAE - 9, 115, 2363, 2372,
2423

HAMAMELIDACEAE - 1375, 1382, 1412,
1580

HAPLOMITRIACEAE - 1191

HIPPOCASTANACEAE - 61, 135, 287

HIPPOCRATEACEAE - 1867

HIPPURIDACEAE - 2363, 2423

HYDROCHARITACEAE - 38, 116, 141, 234,
259, 334, 807, 831, 835, 1277, 1508,
1803, 1935, 2571, 2788

HYDROPHYLLACEAE - 1515

HYDROSTACHYACEAE - 2088, 2365

HYMENOPHYLLACEAE - 517, 1215, 2230,
2231, 2450, 2895

HYPERICACEAE - 61, 557, 2842

IRIDACEAE - 61, 247, 259, 306, 350,
933, 1109, 1372, 2107, 2907, 2962

ISOETACEAE - 21, 45, 380, 461, 1604,
1652, 1697, 2037, 2060, 2244, 2402,
2549, 2554, 2837, 2850

JUGLANDACEAE - 1390, 1392, 1987, 1999

JUNCACEAE - 23, 110, 259, 274, 1253,
2780

LABIATAE - 7, 55, 56, 61, 966, 1371,
1375, 1377, 1378, 1380, 1382, 1425,
1605, 2205, 2353, 2887

LARDIZABALACEAE - 2249

LAURACEAE - 501, 662, 889, 1377, 1522,
1523, 1845, 1919, 2227

LECYTHIDACEAE - 309

LEGUMINOSAE - 61, 96, 116, 134, 196,
212, 242, 249, 267, 304, 319, 320,
321, 330, 378, 382, 426, 434, 440,
533, 635, 693, 741, 749, 846, 876,
930, 966, 985, 1022, 1097, 1108,
1169, 1240, 1242, 1247, 1265, 1266,
1273, 1317, 1344, 1375, 1391, 1401,
1418, 1431, 1445, 1451, 1462, 1464,
1465, 1469, 1520, 1559, 1572, 1586,
1642, 1650, 1704, 1726, 1731, 1736,
1752, 1807, 1822, 1837, 1848, 1872,
1888, 1900, 1905, 1906, 1907, 1908,
1909, 1910, 1928, 1973, 1983, 1996,
1999, 2018, 2042, 2079, 2097, 2143,
2172, 2173, 2176, 2205, 2251, 2253,
2291, 2306, 2329, 2330, 2350, 2354,

2394, 2407, 2434, 2436, 2470,
2487, 2499, 2514, 2530, 2543,
2564, 2565, 2566, 2567, 2572,
2591, 2642, 2665, 2666, 2674,
2680, 2725, 2743, 2783, 2795,
2820, 2904, 2906, 2929, 2930,
2942

LEITNERIACEAE - 2108

LEMNACEAE - 28, 91, 2355, 2773

LENNOACEAE - 1614, 1615, 2562

LENTIBULARIACEAE - 1651

LILAEACEAE - 34

LILIACEAE - 49, 61, 86, 96, 103, 207,
208, 216, 217, 219, 259, 440, 478,
489, 542, 551, 564, 612, 659, 722,
724, 745, 746, 894, 895, 896, 899,
901, 933, 946, 966, 1023, 1087,
1154, 1178, 1207, 1250, 1259,
1273, 1316, 1340, 1346, 1363,
1371, 1375, 1378, 1385, 1386,
1429, 1470, 1497, 1498, 1499,
1541, 1544, 1631, 1693, 1709,
1795, 1860, 1865, 1868, 1876,
1877, 1882, 1949, 1956, 1960,
1964, 1975, 1985, 1996, 2028,
2057, 2067, 2122, 2123, 2124,
2125, 2217, 2278, 2299, 2344,
2379, 2403, 2497, 2498, 2535,
2636, 2637, 2641, 2656, 2774,
2785, 2798, 2827, 2903, 2952,
2964

LIMNANTHACEAE - 2328

LINACEAE - 116, 344, 554, 620, 828,
891, 966, 1273, 2882

LOASACEAE - 524

LOBELIACEAE - 340, 1063, 1371, 2931,
2932

LOGANIACEAE - 114, 1362, 1375, 1398,
1702, 1938, 2635, 2721

LORANTHACEAE - 252, 279, 463, 476,
506, 559, 623, 624, 632, 633, 777,
778, 779, 975, 1089, 1238, 1278,
1280, 1294, 1295, 1296, 1297, 1298,
1299, 1310, 1569, 1570, 1607, 1608,
1609, 1610, 1611, 1612, 1613, 1616,
1723, 1826, 1844, 1870, 1878, 1879,
2081, 2254, 2475, 2529, 2588, 2620,
2622, 2623, 2624, 2625, 2626, 2708,
2716, 2717, 2718, 2752, 2890, 2934

LYCOPODIACEAE - 10, 19, 21, 155, 259,
978, 1320, 1321, 1348, 1349, 2153,
2297, 2349, 2520, 2553, 2849, 2870

LYTHRACEAE - 497

MAGNOLIACEAE - 61, 1267, 1377, 1380,
1786

MALPIGHIACEAE - 857, 1796

MALVACEAE - 61, 459, 621, 622, 626,
854, 913, 966, 1013, 1163, 1273,
1284, 1550, 1737, 1834, 2138, 2521,
2961

MARANTACEAE - 1173, 1386, 1896, 2135,
2511, 2652, 2653, 2834

MARATTIACEAE - 22, 32, 139, 142, 492,
560, 627, 628, 629, 683, 1010, 2321,
2640, 2851

MARSILIACEAE - 2, 45, 508, 699, 1197,
1481, 1482, 2373, 2863, 2894

MARTYNIACEAE - 82

MATONIACEAE - 250, 790, 2595

MAYACACEAE - 274, 2732

MELASTOMACEAE - 674, 1073, 1366, 1446,
1447

MELIACEAE - 1093, 1189, 1468

MELIANTHACEAE - 616, 617

MENISPERMACEAE - 90, 96, 422, 874, 1073, 1389, 2183, 2474, 2476

MORACEAE - 61, 127, 339, 798, 1110, 1517, 2048, 2183, 2228, 2272, 2950

MORINGACEAE - 948

MUSACEAE - 61, 221, 259, 303, 1102, 1838, 2136, 2290, 2492, 2511, 2648, 2649, 2653

MYRICACEAE - 532, 727, 2937

MYRSINACEAE - 61, 645

MYRTACEAE - 133, 393, 739, 2847

NAJADACEAE - 33, 225, 2401

NEPENTHACEAE - 1246, 1606, 2544

NYCTAGINACEAE - 906, 1328, 1916, 2547, 2967

NYMPHAEACEAE - 61, 83, 584, 796, 843, 844, 845, 1192, 1208, 1232, 1829

NYSSACEAE - 1376, 1417, 2084

OCHNACEAE - 405, 2093, 2749

OLACACEAE - 398, 2748

OLEACEAE - 55, 56, 61, 324, 757, 758, 1381, 1603, 1988, 1989

ONAGRACEAE - 81, 415, 970, 1249, 2141, 2215, 2216, 2424

OPHIOGLOSSACEAE - 14, 29, 30, 58, 201, 259, 265, 412, 546, 547, 752, 1122, 1492, 1653, 1772, 2016, 2102, 2334, 2428, 2759

ORCHIDACEAE - 35, 53, 79, 112, 116, 117, 145, 146, 147, 168, 173, 186, 187, 204, 259, 311, 440, 484, 514, 515, 653, 654, 730, 732, 832, 839, 914, 922, 953, 972, 1081, 1082, 1083, 1084, 1085, 1086, 1147, 1281, 1282, 1350, 1356, 1375, 1386, 1415, 1591, 1601, 1626, 1627, 1640, 1641, 1648, 1842, 1869, 1940, 1950, 1951, 1952, 1953, 1955, 1968, 1984, 1997, 2009, 2010, 2029, 2109, 2179, 2185, 2189, 2225, 2232, 2271, 2312, 2313, 2333, 2364, 2372, 2378, 2451, 2458, 2459, 2511, 2528, 2552, 2645, 2861

OROBANCHACEAE - 137, 138, 284, 293, 296, 500, 802, 1382, 1501, 1513, 1514, 1596, 1634, 1636, 2206, 2493, 2597

OSMUNDACEAE - 18, 30, 829, 886, 966, 1014, 2438, 2844

OXALIDACEAE - 712, 869, 938, 944, 2268, 2337, 2619, 2621

PALMAE - 41, 59, 61, 62, 67, 125, 126, 128, 170, 173, 259, 332, 485, 520, 535, 588, 604, 605, 663, 664, 665, 675, 801, 814, 871, 931, 932, 952, 1091, 1098, 1099, 1100, 1101, 1103, 1131, 1133, 1134, 1135, 1136, 1137, 1228, 1248, 1721, 1771, 1827, 1873, 1874, 1937, 1944, 2065, 2066, 2070, 2132, 2213, 2272, 2322, 2338, 2377, 2650, 2654, 2658, 2661, 2741, 2848, 2926, 2927, 2928, 2974

PANDANACEAE - 173, 259, 535, 643, 644, 663, 664, 1003, 1131, 1138, 2272, 2382, 2557, 2802

PAPAVERACEAE - 103, 169, 313, 966, 1375, 2005, 2631

PARKERIACEAE (CERATOPTERIDACEAE) - 729, 731, 733, 734, 1039, 1426, 1643, 2047

PEDALIACEAE - 82, 1621

PETERMANNIACEAE - 2659

PHILYDRACEAE - 1234

PHRYMACEAE - 1383, 1715

PHYTOLACCACEAE - 26, 1371, 1386, 1821

PINACEAE - 3, 61, 116, 210, 308, 312, 322, 323, 440, 502, 538, 539, 590, 658, 702, 705, 747, 1048, 1255, 1314, 1327, 1423, 1441, 1456, 1457, 1576, 1646, 1686, 1690, 1700, 1733, 1811, 1847, 1853, 1886, 1920, 2151, 2160, 2197, 2209, 2300, 2335, 2403, 2646, 2871, 2875, 2876, 2877, 2878, 2879, 2880, 2881

PIPERACEAE - 8, 474, 498, 735, 1319, 1325, 1326, 1439, 1480, 1976, 2141, 2489

PLANTAGINACEAE - 1921, 2548

PLATANACEAE - 2971

PLUMBAGINACEAE - 88, 2193, 2205

PODOCARPACEAE - 690, 776, 1577, 2332, 2526, 2550

PODOSTEMACEAE - 1239, 1437, 1819, 2896

POLEMONIACEAE - 966, 1362, 1588, 1913

POLYGALACEAE - 61, 154, 375, 1371, 1410, 1969, 2085, 2441, 2539

POLYGONACEAE - 450, 954, 966, 1408, 1521, 1857, 2087, 2310, 2887

POLYPODIACEAE - 45, 433, 440, 455, 681, 695, 797, 966, 1377, 1548, 1854, 1894, 2023, 2024, 2052, 2229, 2629, 2695, 2713, 2854, 2855, 2866, 2891, 2892, 2893, 2968

PONTEDERIACEAE - 4, 237, 259, 342, 723, 737, 1670, 2830

PORTULACACEAE - 6, 799, 1357

POTAMOGETONACEAE - 33, 226, 227, 259, 631, 1038, 1174, 1177, 2182, 2401, 2598, 2599

PRIMULACEAE - 123, 879, 966, 1224, 1739, 1815, 1927, 1974, 2188, 2273

PROTEACEAE - 1528, 1649, 2141, 2211

PSILOTACEAE - 44, 1040

PYROLACEAE - 108, 124, 373, 804, 805, 1308, 1336, 1377, 1760, 2219

RAFFLESIACEAE - 228, 253, 661, 1291

RANUNCULACEAE - 7, 61, 98, 149, 159, 164, 165, 166, 167, 200, 254, 444, 467, 468, 827, 966, 1025, 1026, 1030, 1055, 1056, 1059, 1124, 1167, 1168, 1219, 1230, 1264, 1370, 1371, 1374, 1375, 1380, 1382, 1384, 1385, 1474, 1555, 1622, 1623, 1624, 1718, 1719, 1745, 1858, 1864, 1890, 1945, 2002, 2138, 2239, 2429, 2536, 2542, 2796, 2797, 2829, 2905

RAPATACEAE - 649

RESEDACEAE - 1942

RHAMNACEAE - 321, 1364, 1384, 2909

RHIZOPHORACEAE - 548, 549, 550, 606, 921, 1129, 1130, 2152, 2184

ROSACEAE - 7, 25. 61, 73, 74, 86, 394, 410, 417, 597, 651, 973, 992, 1144, 1375, 1377, 1378, 1740, 1792, 1793, 1794, 1825, 1839, 1871, 1971, 1972, 1992, 1993, 2006, 2207, 2266, 2276, 2461, 2531, 2532, 2631, 2862, 2913

ROXBURGHIACEAE (see STEMONACEAE)

RUBIACEAE - 96, 261, 376, 966, 1368, 1386, 1681, 1682, 1774, 1830, 1924, 2126, 2722

RUSCACEAE - 723

RUTACEAE - 55, 56, 61, 96, 193, 406,

496, 817, 1127, 1272, 1637, 2375, 2478

SALICACEAE - 116, 571, 652, 655, 884, 1449, 2325, 2430, 2602

SALVADORACEAE - 2473

SALVINIACEAE - 31, 205, 696, 708, 887, 1659, 1660, 1735, 2220

SANTALACEAE - 144, 174, 395, 396, 397, 399, 400, 441, 1024, 1027, 1028, 1029, 1064, 1311, 1312, 1313, 1400, 1632, 1633, 1946, 2119, 2154, 2233, 2235, 2389, 2472, 2811, 2812, 2914

SAPINDACEAE - 61, 287

SAPOTACEAE - 371, 684, 685, 913, 2502

SARRACENIACEAE - 2398

SAURURACEAE - 1231, 1360, 1406

SAXIFRAGACEAE - 567, 676, 1165, 1166, 1382, 1740, 2572, 2775

SCHEUZERIACEAE - 694, 1322, 2062, 2512

SCHIZAEACEAE - 407, 408, 518, 566, 759, 1981, 2593

SCROPHULARIACEAE - 54, 61, 92, 93, 94, 95, 140, 144, 209, 289, 500, 503, 523, 634, 641, 686, 687, 725, 754, 966, 1096, 1105, 1274, 1279, 1289, 1293, 1300, 1380, 1384, 1385, 1393, 1396, 1397, 1399, 1404, 1431, 1618, 1635, 1751, 1783, 1816, 1824, 1863, 1979, 2030, 2031, 2117, 2118, 2120, 2121, 2141, 2302, 2518, 2523, 2540, 2541, 2675, 2676, 2739, 2889, 2916, 2933

SELAGINELLACEAE - 45, 197, 224, 260, 577, 578, 579, 1146, 1251, 2731, 2838, 2839, 2840, 2841

SIMAROUBACEAE - 1, 61, 2205

SOLANACEAE - 61, 86, 134, 257, 345, 365, 775, 794, 809, 852, 904, 943, 966, 991, 1031, 1187, 1227, 1273, 1378, 1384, 1553, 1677, 1692, 1980, 2101, 2339, 2392, 2443, 2465, 2559, 2606, 2614, 2615, 2730, 2765, 2817, 2846

SONNERATIACEAE - 1889, 2682, 2683

STEMONACEAE (ROXBURGHIACEAE) - 367, 1358, 1720, 2657

STERCULIACEAE - 568, 688, 1189, 1922, 1962

TACCACEAE - 259

TAMARICACEAE - 1141

TAXACEAE - 255, 736, 1327, 1335, 2194

TAXODIACEAE - 494, 966, 1455

TERNSTROEMIACEAE (see THEACEAE) - 424

THEACEAE (see TERNSTROEMIACEAE) - 883

THURNIACEAE - 833

THYMELIACEAE - 1394, 1490, 2488

TILIACEAE - 61, 1345, 1962, 2425, 2426, 2427, 2786

TRAPACEAE - 2218

TROCHODENDRONACEAE - 2935

TROPAEOLACEAE - 61, 162, 966, 2495

TYPHACEAE - 259, 1759, 1897, 2111, 2512

UMBELLIFERAE - 42, 68, 69, 70, 96, 101, 198, 302, 442, 537, 575, 576, 753, 902, 924, 958, 966, 989, 990, 1000, 1127, 1220, 1263, 1273, 1352, 1377, 1385, 1431, 1582,

1656, 1680, 1683, 1698, 1747,
2091, 2301, 2307, 2445, 2462,
2605, 2613, 2719, 2723, 2766,
2798, 2809, 2955, 2970

URTICACEAE - 966, 1193, 1407, 1442,
2477

VALERIANACEAE - 96, 359, 954, 966,
1072, 1306, 1414, 1862, 2243, 2580,
2581, 2582, 2583, 2584, 2768, 2799,
2829

VELLOZIACEAE - 104, 368, 2831, 2832

VERBENACEAE - 233, 562, 682, 1835,
2043, 2681

VIOLACEAE - 1507, 2940

VITACEAE - 122, 315, 384, 451, 475,
1073, 1380, 2183, 2352, 2726

WELWITSCHIACEAE (see GNETACEAE) - 15,
16, 547, 2576, 2745, 2958, 2959,
2960, 2965

XANTHORRHOEACEAE - 1006, 1007

XYRIDACEAE - 274, 647, 2134

ZINGIBERACEAE - 66, 259, 551, 2111,
2133, 2511, 2647, 2653, 2660

ZYGOPHYLLACEAE - 913, 2059, 2751

INDEX TO PLANT GENERA BY FAMILY
(By listing number)

ᴄ᷈ᷤ⟨◉ᴴ◉⟩ᴄ᷈ᷤ

ACANTHACEAE
    Acanthus - 1846, 2600
    Dianthera - 1369, 1486
    Ruellia - 1362, 1369
    Spigelia - 1362

ACERACEAE
    Acer - 1543, 1547, 1755, 1856,
       2162, 2271, 2897, 2898, 2963

ACTINIDIACEAE
    Actinidia - 1525

ADOXACEAE
    Adoxa - 2561

AGAVACEAE
    Cordyline - 2663
    Dracaena - 1708, 1791, 2467,
    Nolina - 1260
    Phormium - 358, 2211
    Yucca - 343

AIZOACEAE
    Glinus - 5
    Mesembryanthemum - 84, 1538, 1539,
       1934, 2366, 2953
    Mollugo - 1379, 2078

ALISMACEAE
    Alisma - 1207
    Sagittaria - 1172, 1669, 2435,
       2515, 2533

AMARANTHACEAE
    Achyranthes - 1489
    Alternanthera - 1488
    Cyathula - 1489
    Pupalia - 1489

AMARYLLIDACEAE
    Alstroemeria - 391, 787
    Cooperia - 756
    Crinum - 530, 892, 950
    Narcissus - 673, 726, 745, 945,
       959, 1712
    Sternbergia - 328

ANACARDIACEAE
    Cotinus - 1164
    Lannea - 1852
    Podoon - 178
    Pistacia - 1986
    Rhus - 175, 603, 1378, 1849
    Schinus - 890

ANNONACEAE
    Xylopia - 1466

APOCYNACEAE
    Apocynum - 1378
    Carissa - 464
    Hederanthera - 2072
    Nerium - 2101
    Peltastes - 1433
    Pleiocarpa - 2071
    Plumeria - 383
    Rauwolfia - 680, 860, 1459, 1684,
       1902, 2794, 2918, 2965
    Strophanthus - 1125

AQUIFOLIACEAE
    Ilex - 1380

ARACEAE
    Acorus - 1380, 1551, 1780
    Anthurium - 877
    Arisaema - 1380, 2116, 2263
    Arum - 215, 248
    Calla - 942
    Lysichiton - 2308
    Monstera - 487, 488, 2485
    Philodendron - 2180
    Pistia - 141
    Sauromatum - 2415
    Symplocarpus - 2308
    Syngonium - 451, 1088

ARALIACEAE
    Aralia - 913, 1387
    Hedera - 1142
    Heptapleurum - 2618
    Kolopanax - 331

Panax - 1387

ARAUCARIACEAE
Araucaria - 230, 295, 485, 2453

ARISTOLOCHIACEAE
Aristolochia - 377, 1371
Asarum - 1380, 1675, 2397

ASCLEPIADACEAE
Cryptostegia - 356
Dischidia - 80, 246
Hemidesmus - 2192
Menabea - 1275, 2090

BALANOPHORACEAE
Balanophora - 960, 1004, 1292, 2560
Dactylanthus - 1936
Mystropetalon - 1252
Thonningia - 1787

BALSAMINACEAE
Impatiens - 11, 97, 452, 460, 842, 1342, 1343, 1382, 1831, 1901

BEGONIACEAE
Begonia - 2494

BERBERIDACEAE
Caulophyllum - 610, 1371
Diphylleia - 1623
Jeffersonia - 1371
Podophyllum - 109, 1219, 1371, 1623
Ranzania - 1625

BETULACEAE
Alnus - 1813, 2525, 2915
Betula - 1419, 1420, 1647, 2250, 2336, 2915

BIGNONIACEAE
Anemopaegma - 1432
Bignonia - 100
Chilopsis - 2406
Incarvillea - 2280
Jacaranda - 2054

BOMBACACEAE
Ceiba - 1943

BORAGINACEAE
Eritrichium - 2767
Mertensia - 2491

BROMELIACEAE
Ananas - 786, 1386, 1595
Tillandsia - 472, 525

BURMANNIACEAE
Apteria - 2734
Burmannia - 987, 988
Thismia - 448, 449, 2106

CACTACEAE
Grusonia - 1237
Mamillaria - 853
Melocactus - 47, 264
Neomamillaria - 1307
Opuntia - 1067, 1092, 1724
Rhipsalis - 1911

CALOBRYACEAE
Takakia - 1191

CANNACEAE
Canna - 2651

CANNABINACEAE
Cannabis - 20, 445, 1273
Humulus - 420, 1158, 1914, 1915, 2327

CAPPARIDACEAE
Gynandropsis - 1236

CAPRIFOLIACEAE
Linnea - 1225
Sambucus - 1377, 1757, 2730
Viburnum - 1384

CARICACEAE
Carica - 1386

CARYOCARACEAE
Caryocar - 477

CARYOPHYLLACEAE
Arenaria - 1495, 2195
Dianthus - 2532, 2603
Gypsophila - 2146
Paronychia - 1478
Polycarpon - 1478

Pteranthus - 1478
Saponaria - 1126, 1384, 2008, 2468
Scleranthus - 1478
Silene - 1917
Spergula - 1478, 1493
Thylacospermum - 1494, 1495

CASUARINACEAE
Casuarina - 2056

CELASTRACEAE
Euonymus - 29, 1561, 2457

CEPHALOTACEAE
Cephalotus - 2398

CEPHALOTAXACEAE
Cephalotaxus - 589

CHENOPODIACEAE
Anabasis - 1258
Arthrocnemum - 416
Atriplex - 830
Beta - 316, 348, 349, 351, 352,
    409, 561, 783, 784, 785, 878,
    1120, 1273, 1416, 1450, 1756,
    1820, 2075, 2076, 2100, 2157,
    2323, 2324, 2376, 2422, 2601
Camphorosma - 1685
Chenopodium - 1382, 1395, 2601
Corispermum - 1685
Cycloloma - 2196
Salicornia - 803, 881
Salsola - 63, 1850
Suaeda - 63

CISTACEAE
Helianthemum - 1382

COMBRETACEAE
Laguncularia - 233, 1467
Terminalia - 305, 781, 1766

COMMELINACEAE
Cochliostema - 2702
Commelina - 1347
Rhoeo - 1429
Tradescantia - 77, 1053, 1058,
    1178, 2294
Triceratella - 2655

COMPOSITAE
Ambrosia - 1388
Arctium -2045
Artemisia - 1912
Atractylis - 132
Balsamorrhiza - 1016
Berardia - 1801
Calendula - 386, 939
Carlina - 132
Chrysanthemum - 1711, 2532
Cichorium - 1562, 2585, 2586
Cirsium - 1235, 1748
Cnicus - 2309
Cosmos - 385
Cousinia - 2045
Cynara - 2115
Dahlia - 1262
Enhydra - 1782
Eupatorium - 1375
Fitchia - 646
Gaertneria - 1734
Grindelia - 1143, 1378
Guizotia - 1823
Helianthus - 317, 318, 521, 543,
    2360, 2466, 2616, 2757, 2818
Inula - 954
Krigia - 103, 1405
Lactuca - 1273
Ligularia - 2729
Lygodesmia - 2517
Megalodonta - 1339
Microseris - 2506
Parthenium - 354, 1716, 1966, 2311
Saussurea - 2190
Scorzonera - 2221
Senecio - 88, 2351, 2803
Solidago - 1384, 2053
Soliva - 670
Tagetes - 387
Taraxacum - 355, 490, 491, 593,
    1078, 1196, 1540, 1995, 2001,
    2319, 2320, 2348, 2857, 2917
Thrincia - 979, 1778, 2003
Townsendia - 598
Tragopogon - 1261
Tridax - 2044
Veronica - 1687
Wedelia - 2238
Xanthium - 840, 856

CONVOLVULACEAE

Calystegia - 2736
Convolvulus - 111, 236, 282, 509, 1519, 1545
Cuscuta - 136, 1080, 1119, 1221, 1565, 1764, 1918, 2069, 2395, 2627, 2944, 2945
Evolvulus - 1518
Ipomoea - 243, 346, 350, 679, 795, 1170, 1190, 1271, 1273, 1516, 1833, 1904, 1959, 2418, 2420, 2421, 2969, 2973
Merremia - 2563

CORNACEAE
Cornus - 1377, 1500, 1543

CRASSULACEAE
Bryophyllum - 1505
Kalanchoe - 861, 862, 2174
Sempervivum - 2941

CRUCIFERAE
Armoracia - 1710
Brassica - 158, 199, 458, 671, 810, 2098, 2099, 2505
Cakile - 2925
Cheiranthus - 1341
Descurainia - 915
Draba - 2147
Erysimum - 2094
Lepidium - 1065, 2471
Lithodraba - 329
Neobeckia - 1762
Raphanus - 199, 1066, 1113, 1114, 1155, 1176, 1273, 2103, 2589, 2672
Sinapsis - 199, 458, 1066, 2260

CUPRESSACEAE
Cupressus - 2789
Juniperus - 1386, 1703
Libocedrus - 2873, 2874
Thuja - 388, 701, 1386

CYATHEACEAE
Cyathea - 2021, 2022

CYCADACEAE
Bowenia - 189, 2919
Ceratozamia - 925
Cycas - 78, 171, 360, 585, 689, 1701, 2234, 2813, 2814

Dioon - 927, 2516, 2617
Encephalartos - 2920
Macrozamia - 569, 925, 1256, 2298, 2912
Microcycas - 750
Stangeria - 738, 2077
Zamia - 689, 2733

CYCLANTHACEAE
Carludovica - 1903

CYPERACEAE
Bulbostylis - 1171
Carex - 106, 1351, 1511, 2758
Cladium - 800
Cyperus - 1789, 1790, 1866
Dulichium - 107
Eleocharis - 1861
Fimbristylis - 2449
Kobresia - 2448
Schoenodendron - 977
Scirpus - 1963
Uncinia - 1619   .

DILLENIACEAE
Acrotrema - 911
Davilla - 1434

DIOSCOREACEAE
Dioscorea - 369, 370, 841, 1324, 1384, 1706, 1806, 2501
Epipetrum - 2255
Tamus - 1775
Trichopus - 366

DIPSACACEAE
Dipsacus - 2399
Knautia - 2577, 2578
Succisa - 2753
Succisela - 2753

DROSERACEAE
Byblis - 1651
Dionaea - 2496
Drosera - 428, 912, 1221, 1290
Drosophyllum - 192

EBENACEAE
Diospyros - 1377, 2924

ELAEAGNACEAE
Elaeagnus - 2525

ELATINACEAE
  Elatine - 179, 2181

EMPETRACEAE
  Empetrum - 1226

EPHEDRACEAE
  Ephedra - 703, 743, 900, 2130,
    2638, 2784

EQUISETACEAE
  Equisetum - 2, 119, 404, 615, 697,
    700, 1157, 1483, 1484, 1655,
    1779, 1893, 2801

ERICACEAE
  Arbutus - 2292
  Arctostaphylos - 1380
  Calluna - 1485
  Epigaea - 1384
  Gaultheria - 1375
  Kalmia - 1382
  Rhododendron - 300
  Vaccinium - 1777

ERIOCAULACEAE
  Eriocaulon - 1243, 1354, 2513

EUPHORBIACEAE
  Bridelia - 2061
  Euphorbia - 372, 374, 936, 1378,
    1977, 2222, 2223, 2258, 2573
  Hevea - 493
  Hura - 1132
  Jatropha - 1386, 2169, 2170
  Ricinus - 2405, 2408, 2632
  Stillingia - 1380

EUPTELEACEAE
  Euptelea - 2935

FAGACEAE
  Castanea - 1386, 1664
  Fagus - 760, 761, 1543
  Nothofagus - 341
  Quercus - 288, 435, 656, 1378,
    1435, 1578, 1662, 1663, 1665,
    1812, 1925, 1926, 1990, 2761

FOUQUIERIACEAE
  Fouquieria - 2257, 2404

FUMARIACEAE
  Corydalis - 1077

GENTIANACEAE
  Bartonia - 1361
  Gentiana - 1863, 2038, 2787
  Menyanthes - 2101
  Obolaria - 105
  Sabbatia - 1385

GERANIACEAE
  Geranium - 1375
  Pelargonium - 2414
  Sarcocaulon - 2368

GESNERIACEAE
  Streptocarpus - 1079, 2709

GINKGOACEAE
  Ginkgo - 381, 1758, 1884, 2128,
    2437

GLEICHENIACEAE
  Stromatopteris - 470

GNETACEAE (see WELWITSCHIACEAE)
  Gnetum - 17, 256, 951, 1678, 1773,
    1804, 1805, 2966

GRAMINEAE
  Agropyrum - 1378
  Agrostis - 1156, 2112
  Alopecurus - 1592
  Ammophila - 2210
  Andropogon - 310, 527, 528, 545
  Arundo - 2034
  Avena - 364, 512, 553, 1186, 1436
  Bambusa - 221, 1832, 2058, 2454,
    2455, 2590
  Bouteloua - 425
  Brachypodium - 1965
  Bromus - 2063
  Cynosurus - 2628
  Dactylis - 1574
  Digitaria - 2039
  Euchlaena - 924
  Holcus - 728
  Hordeum - 586, 587, 1186, 1287,
    1444, 1506, 1509, 1524, 1855
  Imperata - 2889
  Molinia - 1458

252

Munroa - 1359
Nardus - 822
Oryza - 499, 873, 1188, 1283,
    1502, 1526, 1530, 1531, 1532,
    1533, 1534, 1535, 1536, 1658,
    1939, 2020, 2347, 2456, 2742,
    2939
Panicum - 439, 2316
Phalaris - 486, 1111
Phleum - 362, 808, 1159, 1160,
    1302, 1303, 1948, 2007
Poa - 1202, 2007
Saccharum - 347, 353, 1002, 2763
Setaria - 2007
Sorghum - 299, 357, 923, 2131,
    2267, 2770
Spartina - 2570
Spinifex - 1090, 1961
Sporobolus - 808
Stipa - 2063
Themeda - 1161
Triticum - 364, 553, 614, 929,
    1106, 1273, 1304, 1828, 2046,
    2086, 2317, 2754
Zea - 364, 432, 920, 934, 981,
    982, 983, 1117, 1273, 1285,
    1286, 1504, 1527, 1549, 1620,
    1628, 1629, 1673, 1754, 1808,
    1851, 1898, 2040, 2208, 2346,
    2936
Zizania - 2555

GUTTIFERAE
    Garcinia - 1367, 2527
    Symphonia - 1765
    Triadenum - 1355

HALORAGACEAE
    Gunnera - 9, 2372

HAMAMELIDACEAE
    Hamamelis - 1382, 1412
    Liquidambar - 1375, 1580

HAPLOMITRIACEAE
    Haplomitrium - 1191

HIPPOCASTANACEAE
    Aesculus - 135

HYDROCHARITACEAE

Elodea - 807, 1508
Enhalus - 831, 2571
Hydrocharis - 38, 835
Lagarosiphons - 2788
Stratioites - 234, 334, 1277, 1803
Vallisneria - 141

HYDROPHYLLACEAE
    Hydrolea - 1515

HYDROSTACHYACEAE
    Hydrostachys - 2088, 2365

HYMENOPHYLLACEAE
    Hymenophyllum - 2450
    Loxsoma - 1215
    Mecodium - 2231
    Meringium - 2230
    Trichomanes - 2895

IRIDACEAE
    Gladiolus - 350, 1109, 2107, 2907
    Sisyrinchium - 1372

ISOETACEAE
    Isoetes - 21, 380, 461, 1604, 1652,
        1697, 2037, 2060, 2244, 2402,
        2549, 2554, 2837, 2850
    Stylites - 1604

JUGLANDACEAE
    Carya - 1392
    Juglans - 1390, 1392, 1987, 1999

JUNCACEAE
    Juncus - 110, 2780
    Prionium - 23

LABIATAE
    Coleus - 2353
    Collinsonia - 1377
    Cunila - 1371
    Glechoma - 1378
    Hedeoma - 1375
    Lycopus - 1382
    Mentha - 1425, 1605
    Monarda - 1380, 2226

LARDIZABALACEAE
    Holboellia - 2249

LAURACEAE
  Cassytha - 501, 662, 889, 1845, 2227
  Laurus - 1522, 1523
  Sassafras - 1377
  Umbellularia - 1522, 1523

LECYTHIDACEAE
  Barringtonia - 309

LEGUMINOSAE
  Acacia - 321, 2743
  Aeschynomene - 1465, 2329
  Afzelia - 1464
  Anthocleista - 1462, 1469
  Apios - 1401
  Arachis - 196, 212, 2251, 2795 2929
  Astragalus - 2291
  Baptisia - 1375
  Bauhinia - 1240
  Caragana - 2572
  Cassia - 378, 1265, 1266
  Cercidium - 2407
  Cicer - 1108, 1391
  Clitorea - 1520
  Coronilla - 2350
  Cyamopsis - 1983, 2143
  Derris - 249, 2042, 2330
  Eperua - 1445
  Gleditsia - 1837
  Glottidium - 1108
  Glycine (see Soja) - 330, 1910, 2565, 2566, 2783, 2820
  Glycyrrhiza - 1736
  Hedysarum - 2436
  Herminiera - 134
  Hesperastragalus - 1451
  Hymenaea - 876
  Krameria - 635, 2543
  Lotus - 1908
  Lupinus - 426, 741, 1344, 1999, 2176
  Medicago - 1097, 1273, 1586, 1642, 1928, 1973, 2470, 2904, 2906
  Melilotus - 319, 320, 533, 985, 1807, 1848
  Mimosa - 1822, 1999
  Neptunia - 1888
  Parkinsonia - 2407
  Phaseolus - 693, 930, 1242, 1247, 1726, 1996, 2567

Pisum - 1169, 1273, 2079, 2172, 2173, 2253, 2487, 2665, 2666, 2942
Prosopis - 1900, 2680
Pueraria - 2306
Robinia - 1022
Sesbania - 242, 846, 1888, 2530
Soja (see Glycine) - 434
Spartium - 2354
Tephrosia - 2018
Vicia - 304, 321, 382, 1572, 1704, 1731, 2499, 2674, 2725
Viminaria - 1650

LEITNERIACEAE
  Leitneria - 2108

LEMNACEAE
  Lemna - 28, 91, 2773

LENNOACEAE
  Lennoa - 2562
  Pholisma - 1614, 1615

LENTIBULARIACEAE
  Polypompholyx - 1651

LILAEACEAE
  Lilaea - 34

LILIACEAE
  Aletris - 1346, 1378, 2952
  Allium - 216, 612, 946, 1273, 1340, 1470, 1544, 1795, 1876, 1877, 1975, 2636, 2827
  Aloe - 1259, 1709
  Anthericum - 895
  Aphyllanthus - 2656
  Asparagus - 478, 1429, 1497, 1498, 1499, 1949, 1964, 2379
  Asphodelus - 542, 1868, 1882, 1985, 2057
  Brodiaea - 2497, 2498
  Chamaelirium - 1385
  Chlorophytum - 895, 1178, 1960
  Colchicum - 217, 1023
  Convallaria - 238, 1001, 1378
  Endymion - 745
  Erythronium - 489, 1371, 2028, 2299
  Gagea - 2067
  Galtonia - 1207
  Hemerocallis - 542, 2785

Hyacinthus - 1541, 2903
Lilium - 219
Medeola - 1375
Oakesia - 103
Paris - 551
Petrosavia - 2535
Polygonatum - 2774, 2964
Ruscus - 1693
Scilla - 745
Smilax - 2641
Thysanotus - 564
Trillium - 2344, 2637
Tulipa - 2299
Urginea - 659
Uvularia - 103, 208, 2217
Veratrum - 49, 1386

LIMNANTHACEAE
Floerkea - 2328

LINACEAE
Linum - 344, 554, 620, 828, 891,
1273, 2882

LOASACEAE
Mentzelia - 524

LOBELIACEAE
Lobelia - 340, 1063, 1371, 2931

LOGANIACEAE
Gelsemium - 1375, 1938, 2635,
2721
Polypremum - 1398
Spigelia - 1362, 1938
Strychnos - 114, 1702

LORANTHACEAE
Antidaphne - 1613
Arceuthobium - 463, 506, 777, 778,
779, 1278, 1294, 1295, 1297,
1298, 1609, 2081, 2529, 2588,
2620, 2717
Atkinsonia - 1879
Gaiadendron - 1610, 1612
Korthalsella - 2625
Loranthus - 1826, 1878
Notothixos - 1844
Nuytsia - 1310
Phoradendron - 623, 624, 632, 633,
2624, 2934

Phthirusa - 1613
Phrygilanthus - 2254
Psittacanthus - 1616
Struthanthus - 1280
Tapinanthus - 2890
Viscum - 975, 1723, 1870, 2622,
2623, 2625, 2626, 2708

LYCOPODIACEAE
Lycopodium - 21, 155, 978, 1320,
1321, 1348, 1349, 2153, 2297,
2349, 2520, 2553, 2870
Phylloglossum - 10, 19, 2849

LYTHRACEAE
Lythrum - 497

MAGNOLIACEAE
Liriodendron - 30, 1786
Magnolia - 1267, 1380, 1786

MALPIGHIACEAE
Byrsonima - 857
Heteropteris - 1796

MALVACEAE
Abutilon - 1737
Althaea - 459
Gossypium - 854, 1013, 1163, 1273
1284, 2521
Hibiscus - 913, 1550
Malva - 621, 622, 626, 1834, 2961

MARANTACEAE
Calathea - 2834
Maranta - 1386, 1896, 2135, 2652
Schumannianthus - 1173

MARATTIACEAE
Angiopteris - 139, 492, 627, 1010
Danaea - 560
Kaulfussia - 627
Marattia - 22, 32, 683

MARSILIACEAE
Marsilea - 2, 699, 1197, 2373,
2863, 2894
Pilularia - 508, 1481
Regnellidium - 1482

Perularia - 1627
Phalaenopsis - 1282
Platanthera - 1627
Pogonia - 653, 1350
Satyrium - 1953
Spiranthes - 1984
Taeniophyllum - 922, 1968
Trimos - 2458
Vanda - 53
Vanilla - 311, 1386, 1997

OROBANCHACEAE
Aeginetia - 1501, 1634, 1636
Aphyllon - 2493
Christisonia - 296
Conopholis - 293
Epiphegus - 802, 1382
Orobanche - 284, 1513, 1514,
1596, 2597
Phelipaea - 2206

OSMUNDACEAE
Osmunda - 18, 30, 829
Todea - 18, 886, 2438

OXALIDACEAE
Oxalis - 869, 938, 944, 2337,
2619, 2621

PALMAE
Acanthoriza - 62, 675
Borassus - 952
Calamus - 1137
Chamaedorea - 665
Chamaerops - 1133, 1135
Cocos - 871, 1873, 1874, 2070,
2338, 2974
Copernicia - 1228
Elaeis - 332, 588, 1248, 1937,
2213, 2322, 2741, 2926, 2927,
2928
Iriartea - 535
Juania - 2661
Phoenix - 59, 485, 604, 605, 1091,
1134, 2848
Rhapis - 2658
Salacca - 1944
Washingtonia - 1136

PANDANACEAE
Pandanus - 535, 1003, 1138, 2382,
2802

PAPAVERACEAE
Diclytra - 103
Papaver - 169, 313
Sanguinaria - 1375, 2005

PARKERIACEAE (CERATOPTERIDACEAE)
Ceratopteris - 729, 731, 733, 734,
1039, 1426, 1643, 2047

PEDALIACEAE
Sesamum - 1621

PHRYMACEAE
Phryma - 1383, 1715

PHYTOLACCACEAE
Ercilla - 26
Petivera - 1386
Phytolacca - 1371, 1821

PINACEAE
Abies - 3, 1314, 1456, 1811, 2878,
2879
Cedrus - 590, 1733
Larix - 1423
Picea - 2160
Pinus - 210, 308, 312, 702, 747,
1048, 1255, 1441, 1576, 1646,
1686, 1690, 1700, 1853, 1920,
2151, 2197, 2209, 2300, 2335,
2646, 2871, 2875, 2876, 2877,
2880, 2881
Pseudotsuga - 322, 323, 502, 1847

PIPERACEAE
Peperomia - 8, 1319, 1480, 1976,
2489
Piper - 474, 498, 735

PLANTAGINACEAE
Plantago - 1921, 2548

PLATANACEAE
Platanus - 2971

PLUMBAGINACEAE
Plumbago - 2193
Statice - 882

PODOCARPACEAE
Acmopyle - 2332
Podocarpus - 690, 776, 1577, 2550

PODOSTEMACEAE
   Hydrobium - 1437
   Podostemon - 1239

POLEMONIACEAE
   Phlox - 1362, 1588, 1913

POLYGALACEAE
   Epirrhizanthes - 2085
   Polygala - 154, 375, 1371, 1410,
     1969, 2441, 2539

POLYGONACEAE
   Fagopyrum - 1521
   Polygonum - 1408
   Rheum - 954, 1857, 2310

POLYPODIACEAE
   Achrostichum - 2629
   Adiantum - 1377
   Amphoradenium - 2866
   Asplenium - 2695
   Belvisia - 2968
   Camptosaurus - 1854
   Cheilanthes - 2891, 2893
   Diplazium - 2023
   Elaphoglossum - 433
   Lindsaya - 2892
   Niphobolus - 2052
   Oleandra - 2024, 2854
   Polypodium - 1894
   Pteris - 681
   Tectaria - 1548, 2229

PONTEDERIACEAE
   Eichhornea - 342, 1670, 2830
   Pontederia - 737

PORTULACACEAE
   Claytonia - 1357
   Portulaca - 799

POTAMOGETONACEAE
   Cymodocea - 227
   Posidonia - 227
   Potamogeton - 226, 631, 1038,
     1174
   Ruppia - 1177
   Zannichellia - 33
   Zostera - 227, 2598, 2599

PRIMULACEAE
   Cyclamen - 1224
   Glaux - 1927
   Hottonia - 2188
   Primula - 1974

PROTEACEAE
   Grevillea - 1528
   Hakea - 1649

PSILOTACEAE
   Psilotum - 1040
   Tmesipteris - 44

PYROLACEAE
   Allotropa - 804
   Chimaphila - 1377
   Monotropa - 124, 1336, 2219
   Monotropis - 805
   Pterospora - 373
   Pyrola - 108

RAFFLESIACEAE
   Brugmansia - 661
   Rafflesia - 228, 253

RANUNCULACEAE
   Aconitum - 467, 468, 1167, 1371,
     1624, 1864, 2536
   Adonis - 1025
   Anemone - 98, 1858
   Anemonella - 1370
   Caltha - 1945
   Ceratocephalus - 2829
   Cimifuga - 1375
   Clematis - 254
   Coptis - 1380
   Delphinium - 149, 1555
   Eranthis - 1219
   Glaucidium - 1623
   Helleborus - 2796, 2797
   Hepatica - 1385
   Hydrastis - 200, 1059, 1384, 1623
   Isopyrum - 159, 1374
   Pulsatilla - 2239
   Ranunculus - 444, 827, 1030, 1055,
     1056, 1124, 1168, 1230, 1384,
     1622, 1745, 2002, 2905
   Thalictrum - 164, 1026
   Xanthorrhiza - 1382

RHAMNACEAE
  Ceanothus - 1364, 2909
  Rhamnus - 1384
  Zizyphus - 321

RHIZOPHORACEAE
  Bruguiera - 1889
  Carallia - 606
  Rhizophora - 548, 549, 550, 921,
    1129, 1130, 2152, 2184

ROSACEAE
  Amygdalus - 410
  Chaenomeles - 1144
  Cotoneaster - 2913
  Cydonia - 1144
  Dryas - 1871
  Fragaria - 73, 1792, 1793, 1794,
    1992, 1993, 2862
  Gillenia - 1375
  Malus - 417, 1839, 1971, 1972,
    2006, 2276, 2461
  Prunus - 1377, 1825, 2006
  Pyrus - 992
  Rosa - 651, 2532
  Rubus - 973, 1378, 2266
  Sibiraea - 394

RUBIACEAE
  Cephalanthus - 1368
  Coffea - 1386, 1924
  Diodia - 1368
  Galium - 1368
  Houstonia - 1368
  Manettia - 1774
  Mitchellia - 1368
  Mitragyna - 1830
  Morinda - 2722
  Myrmecodia - 261
  Oldenlandia - 1368
  Putoria - 2126

RUTACEAE
  Citrus - 406, 817, 1272, 2375
  Limonia - 2478

SALICACEAE
  Populus - 571, 884, 1449, 2325,
    2430
  Salix - 652, 655, 2602

SALVADORACEAE
  Salvadora - 2473

SALVINIACEAE
  Azolla - 31, 696, 708, 887, 1659,
    1660, 2220
  Salvinia - 205, 1735

SANTALACEAE
  Buckleya - 1632, 1633
  Cansjera - 400
  Comandra - 1400, 1946, 2119, 2914
  Exocarpus - 441, 1027, 1028, 1029
  Geocaulon - 2811, 2812
  Mida - 2472
  Osyris - 1024, 1064, 2154
  Santalum - 395, 396, 397, 1312,
    2233, 2235
  Thesium - 174, 2389

SAPOTACEAE
  Illipis - 913
  Synsepalum - 371

SARRACENIACEAE
  Sarracenia - 2398

SAURURACEAE
  Anemiopsis - 1360
  Saururus - 1231, 1406

SAXIFRAGACEAE
  Carpodetus - 567
  Donatia - 676
  Heuchera - 1382
  Ribes - 1165, 1166, 2572
  Saxifraga - 2775

SCHEUZERIACEAE
  Triglochin - 694, 1322, 2062

SCHIZAEACEAE
  Lygodium - 759, 1981
  Schizaea - 407, 408, 518, 566, 2593

SCROPHULARIACEAE
  Alectrolophus - 95
  Antirrhinum - 2916
  Bartschia - 1289
  Buttonia - 523

Castilleja - 1274, 1618, 1783
Chamaegigas - 1279
Chelone - 1396
Chionophila - 1393
Cordylanthus - 754
Craterostigma - 289
Dasistoma - 2118
Digitalis - 2675, 2676
Dopatrium - 1979
Euphrasia - 94, 95, 2933
Gratiola - 1397
Ilsanthes - 1399
Lathraea - 92, 93, 140, 725, 1300
Leptandra - 1384, 1404
Limosella - 54
Linaria - 686, 687, 1399
Melampyrum - 641, 1105, 1293,
    2117
Odontites - 94, 95
Orthantha - 94
Orthocarpus - 634
Pedicularis - 503, 1824, 2120,
    2121, 2523
Picrorhiza - 1863
Scrophularia - 1399
Scutellaria - 1380
Siphonostegia - 1635
Sophronanthe - 1397
Sopubia - 2889
Striga - 1751, 2030, 2031, 2302,
    2540, 2541, 2739
Tozzia - 1289
Verbascum - 1385
Veronica - 1096, 1304

SELAGINELLACEAE
    Selaginella - 197, 224, 260, 577,
        578, 579, 1146, 1251, 2731,
        2838, 2839, 2840, 2841

SIMAROUBACEAE
    Ailanthus - 1

SOLANACEAE
    Atropa - 2730
    Brunfelsia - 1227
    Capsicum - 775
    Datura - 1031, 1384, 2339
    Hyoscyamus - 852, 2465
    Lycopersicon - 809, 943, 1273,
        1553, 2559

Lycium - 2846
Nicotiana - 257, 365, 991, 1187,
    1980
Scopolia - 2392
Solanum - 134, 345, 1273, 1378,
    1677, 2101, 2606, 2614, 2765,
    2817
Thielavia - 794
Withania - 904, 2443

SONNERATIACEAE
    Sonneratia - 1889, 2683

STEMONACEAE
    Croomia - 1358, 2657
    Stemona - 1720

STERCULIACEAE
    Abroma - 1922
    Heritiera - 1189
    Melochia - 1962
    Theobroma - 568

TAMARICACEAE
    Tamarix - 1141

TAXACEAE
    Taxus - 1335, 2194
    Torreya - 736

TAXODIACEAE
    Metasequoia - 494
    Sequoia - 1455

THEACEAE
    Thea - 883

THYMELIACEAE
    Daphne - 2488
    Dirca - 1394
    Stellera - 1490

TILIACEAE
    Corchorus - 1962
    Tilia - 1345, 2425, 2426, 2427,
        2786

TRAPACEAE
    Trapa - 2218

TROCHODENDRONACEAE
   Trochodendron - 2935

TROPAEOLACEAE
   Tropaeolum - 162, 2495

TYPHACEAE
   Typha - 1897

UMBELLIFERAE
   Achillea - 1385
   Aegopodium - 1582
   Angelica - 1000
   Apium - 1273
   Azorella - 1683, 2605
   Bupleurum - 1656
   Cicuta - 1377, 2462
   Conopodium - 69
   Daucus - 989, 990, 1263, 2613
   Elaeoselinum - 2766
   Erigenia - 1352
   Eryngium - 198, 537
   Ferula - 1680
   Foeniculum - 575, 576
   Heracleum - 2970
   Ligusticum - 1747
   Myrrhidendron - 2301
   Pastinaca - 2809
   Perideridia - 753
   Pimpinella - 2723
   Prangos - 302
   Ptilimnium - 958

URTICACEAE
   Boehmeria - 1407
   Streblus - 1442, 2477

VALERIANACEAE
   Nardostachys - 1862
   Stangea - 2243
   Valeriana - 359, 954, 1072, 1306,
      1414, 2580, 2581, 2582, 2583,
      2584, 2768, 2829

VELLOZIACEAE
   Barbacenopsis - 368
   Vellozia - 2831, 2832

VERBENACEAE
   Avicennia - 233, 562, 682, 1835,
      2043, 2681

VIOLACEAE
   Crocion - 1507
   Viola - 2940

VITACEAE
   Ampelopsis - 1380
   Cissus - 122, 451, 475, 2352

WELWITSCHIACEAE (see GNETACEAE)
   Welwitschia - 15, 16, 2576, 2745,
      2958, 2959, 2960, 2965

XYRIDACEAE
   Abolboda - 647
   Achlyphila - 647
   Orectanthe - 647

ZINGIBERACEAE
   Costus - 551

ZYGOPHYLLACEAE
   Tribulus - 913

᭢᯶᷾᷾᯶

Astragalus - 2291
Atkinsonia - 1879
Atractylis - 132
Atropa - 2730
Avena - 364, 512, 553, 1186, 1436
Avicennia - 233, 562, 682, 1835,
    2043, 2681
Azolla - 31, 696, 708, 887, 1659,
    1660, 2220
Azorella - 1683, 2605
Balanophora - 960, 1004, 1292, 2560
Balsamorrhiza - 1016
Bambusa - 221, 1832, 2058, 2454,
    2455, 2590
Baptisia - 1375
Barbacenopsis - 368
Barringtonia - 309
Bartonia - 1361
Bartschia - 1289
Bauhinia - 1240
Begonia - 2494
Belvisia - 2968
Berardia - 1801
Beta - 316, 348, 349, 351, 352, 409,
    561, 783, 784, 785, 878, 1120,
    1273, 1416, 1450, 1756, 1820,
    2075, 2076, 2100, 2157, 2323,
    2324, 2376, 2422, 2601
Betula - 1419, 1420, 1647, 2250,
    2336, 2915
Bignonia - 100
Boerhaavia - 906
Boehmeria - 1407
Borassus - 952
Botrychium - 29, 30, 412
Bougainvillea - 2547, 2967
Bouteloua - 425
Bowenia - 189, 2919
Brachypodium - 1965
Brassica - 158, 199, 458, 671, 810,
    2098, 2099, 2505
Bridelia - 2061
Brodiaea - 2497, 2498
Bromus - 2063
Brugmansia - 661
Bruguiera - 1889
Brunfelsia - 1227
Bryophyllum - 1505
Buckleya - 1632, 1633
Bulbostylis - 1171
Bupleurum - 1656

Burmannia - 987, 988
Buttonia - 523
Byblis - 1651
Byrsonima - 857
Cakile - 2925
Calamus - 1137
Calathea - 2834
Calendula - 386, 939
Calla - 942
Calluna - 1485
Calopogon - 654
Caltha - 1945
Calystegia - 2736
Camphorosma - 1685
Camptosaurus - 1854
Canna - 2651
Cannabis - 20, 445, 1273
Cansjera - 400
Capsicum - 775
Caragana - 2572
Carallia - 606
Carapa - 1189
Carex - 106, 1351, 1511, 2758
Carica - 1386
Carissa - 464
Carlina - 132
Carludovica - 1903
Carpodetus - 567
Carya - 1392
Caryocar - 477
Cassia - 378, 1265, 1266
Cassytha - 501, 662, 889, 1845, 2227
Castanea - 1386, 1664
Castilleja - 1274, 1618, 1783
Casuarina - 2056
Cattleya - 2458
Caulophyllum - 610, 1371
Ceanothus - 1364, 2909
Cecropia - 127
Cedrus - 590, 1733
Ceiba - 1943
Cephalanthera - 1084
Cephalanthus - 1368
Cephalotaxus - 589
Cephalotus - 2398
Ceratocephalus - 2829
Ceratopteris - 729, 731, 733, 734,
    1039, 1426, 1643, 2047
Ceratozamia - 925
Cercidium - 2407
Chaenomeles - 1144

Chamaedorea - 665
Chamaegigas - 1279
Chamaelirium - 1385
Chamaenerion - 970
Chamaerops - 1133, 1135
Cheilanthes - 2891, 2893
Cheiranthus - 1341
Chelone - 1396
Chenopodium - 1382, 1395, 2601
Chilopsis - 2406
Chimaphila - 1377
Chionophila - 1393
Chlorophytum - 895, 1178, 1960
Christisonia - 296
Chrysanthemum - 1711, 2532
Cichorium - 1562, 2585, 2586
Cicuta - 1377, 2462
Cimifuga - 1375
Cirsium - 1235, 1748
Cissampelos - 1389
Cissus - 122, 451, 475, 2352
Citrus - 406, 817, 1272, 2375
Cladium - 800
Claytonia - 1357
Clematis - 254
Clitorea - 1520
Cocculus - 90, 422, 2476
Cochliostema - 2702
Cocos - 871, 1873, 1874, 2070, 2338,
    2974
Coffea - 1386, 1924
Colchicum - 217, 1023
Coleus - 2353
Collinsonia - 1377
Comandra - 1400, 1946, 2119, 2914
Commelina - 1347
Conopholis - 293
Conopodium - 69
Convallaria - 238, 1001, 1378
Convolvulus - 111, 236, 282, 509,
    1519, 1545
Cooperia - 756
Copernicia - 1228
Coptis - 1380
Corchorus - 1962
Cordylanthus - 754
Corispermum - 1685
Cornus - 1377, 1500, 1543
Coronilla - 2350
Corydalis - 1077
Cosmos - 385

Costus - 551
Cotinus - 1164
Cotoneaster - 2913
Cousinia - 2045
Craterostigma - 289
Crinum - 530, 892, 950
Crocion - 1507
Croomia - 1358, 2657
Cryptostegia - 356
Cunila - 1371
Cupressus - 2789
Cuscuta - 136, 1080, 1119, 1221,
    1565, 1764, 1918, 2069, 2395,
    2627, 2944, 2945
Cyathea - 2021, 2022
Cyathula - 1489
Cycas - 78, 171, 360, 585, 689, 1701,
    2234, 2813, 2814
Cyclamen - 1224
Cycloloma - 2196
Cydonia - 1144
Cymbidium - 1415
Cymodocea - 227
Cyamopsis - 1983, 2143
Cynara - 2115
Cynosurus - 2628
Cyperus - 1789, 1790, 1866
Cypripedium - 1084, 1375, 2312, 2313
Dactylanthus - 1936
Dactylis - 1574
Dahlia - 1262
Danaea - 560
Daphne - 2488
Dasistoma - 2118
Datura - 1031, 1384, 2339
Daucus - 989, 990, 1263, 2613
Davilla - 1434
Delphinium - 149, 1555
Dendrobium - 730, 1869, 1940
Derris - 249, 2042, 2330
Descurainia - 915
Dianthera - 1369, 1486
Dianthus - 2532, 2603
Diclytra - 103
Digitalis - 2675, 2676
Digitaria - 2039
Diodia - 1368
Dionaea - 2496
Dioon - 927, 2516, 2617
Dioscorea - 369, 370, 841, 1324, 1384,
    1706, 1806, 2501

Diospyros - 1377, 2924
Diphylleia - 1623
Diplazium - 2023
Dipodium - 1842
Dipsacus - 2399
Dirca - 1394
Dischidia - 80, 246
Donatia - 676
Dopatrium - 1979
Draba - 2147
Dracaena - 1708, 1791, 2467, 2923, 2952
Drosera - 428, 912, 1221, 1290
Drosophyllum - 192
Dryas - 1871
Dulichium - 107
Eichhornea - 342, 1670, 2830
Elaeagnus - 2525
Elaeis - 332, 588, 1248, 1937, 2213, 2322, 2741, 2926, 2927, 2928
Elaeoselinum - 2766
Elaphoglossum - 433
Elatine - 179, 2181
Eleocharis - 1861
Elodea - 807, 1508
Empetrum - 1226
Encephalartos - 2920
Endymion - 745
Enhalus - 831, 2571
Enhydra - 1782
Entandrophragma - 1468
Eperua - 1445
Ephedra - 703, 743, 900, 2130, 2638, 2784
Epigaea - 1384
Epilobium - 415
Epinetrum - 874
Epipetrum - 2255
Epiphegus - 802, 1382
Epirrhizanthes - 2085
Equisetum - 2, 119, 404, 615, 697, 700, 1157, 1483, 1484, 1655, 1779, 1893, 2801
Eranthis - 1219
Ercilla - 26
Erigenia - 1352
Eryngium - 198, 537
Erysimum - 2094
Erythronium - 489, 1371, 2028, 2299
Eucalyptus - 739
Euchlaena - 924

Eulophia - 1648
Euonymus - 29, 1561, 2457
Eupatorium - 1375
Euphorbia - 372, 374, 936, 1378, 1977, 2222, 2223, 2258, 2573
Euphrasia - 94, 95, 2933
Euptelea - 2935
Evolvulus - 1518
Exocarpus - 441, 1027, 1028, 1029
Fagopyrum - 1521
Fagus - 760, 761, 1543
Ferula - 1680
Ficus - 339, 798, 1110, 1517, 2228, 2950
Fimbristylis - 2449
Floerkia - 2328
Foeniculum - 575, 576
Fouquieria - 2257, 2404
Fragaria - 73, 1792, 1793, 1794, 1992, 1992, 1993, 2862
Fraxinus - 1603
Gaertneria - 1734
Gagea - 2067
Gaiadendron - 1610, 1612
Galium - 1368
Galtonia - 1207
Garcinia - 1367, 2527
Gaultheria - 1375
Gelsemium - 1375, 1938, 2635, 2721
Gentiana - 1863, 2038, 2787
Geocaulon - 2811, 2812
Geranium - 1375
Gillenia - 1375
Ginkgo - 381, 1758, 1884, 2128, 2437
Gladiolus - 350, 1109, 2107, 2907
Glaucidium - 1623
Glaux - 1927
Glechoma - 1378
Gleditsia - 1837
Glinus - 5
Glottidium - 1108
Glycine (see Soja) - 330, 1910, 2565, 2566, 2783, 2820
Glycyrrhiza - 1736
Gnetum - 17, 256, 951, 1678, 1773, 1804, 1805, 2966
Goodyera - 1085
Gossypium - 854, 1013, 1163, 1273, 1284, 2521
Gratiola - 1397
Grevillea - 1528

Grindelia - 1143, 1378
Grusonia - 1237
Guizotia - 1823
Gunnera - 9, 2372
Gynandropsis - 1236
Gypsophila - 2146
Haemaria - 2371
Hakea - 1649
Hamamelis - 1382, 1412
Haplomitrium - 1191
Hartwegia - 145
Hedeoma - 1375
Hedera - 1142
Hederanthera - 2072
Hedysarum - 2436
Helianthemum - 1382
Helianthus - 317, 318, 521, 543,
    2360, 2466, 2616, 2757, 2818
Helleborine - 1084
Helleborus - 2796, 2797
Helminthostachys - 58, 1653
Hemerocallis - 542, 2785
Hemidesmus - 2192
Hepatica - 1385
Heptapleurum - 2618
Heracleum - 2970
Heritiera - 1189
Herminiera - 134
Hesperastragalus - 1451
Heteropteris - 1796
Heuchera - 1382
Hevea - 493
Hibiscus - 913, 1550
Holboellia - 2249
Holcus - 728
Hordeum - 586, 587, 1186, 1287,
    1444, 1506, 1509, 1524, 1855
Hottonia - 2188
Houstonia - 1368
Humulus - 420, 1158, 1914, 1915,
    2327
Hura - 1132
Hyacinthus - 1541, 2903
Hydrastis - 200, 1059, 1384, 1623
Hydrobium - 1437
Hydrocharis - 38, 835
Hydrolea - 1515
Hydrostachys - 2088, 2365
Hymenaea - 876
Hymenophyllum - 2450
Hyoscyamus - 852, 2465

Ilex - 1380
Illipis - 913
Ilsanthes - 1399
Impatiens - 11, 97, 452, 460, 842,
    1342, 1343, 1382, 1831, 1901
Imperata - 2889
Incarvillea - 2280
Inula - 954
Ipomoea - 243, 346, 350, 679, 795,
    1170, 1190, 1271, 1273, 1516, 1833,
    1904, 1959, 2418, 2420, 2421, 2969,
    2973
Iriartea - 535
Isoetes - 21, 380, 461, 1604, 1652,
    1697, 2037, 2060, 2244, 2402, 2549,
    2554, 2837, 2850
Isopyrum - 159, 1374
Jacaranda - 2054
Jatropha - 1386, 2169, 2170
Jeffersonia - 1371
Juania - 2661
Juglans - 1390, 1392, 1987, 1999
Juncus - 110, 2780
Juniperus - 1386, 1703
Jussieua - 1249, 2424
Kalanchoe - 861, 862, 2174
Kalmia - 1382
Kaulfussia - 627
Knautia - 2577, 2578
Kobresia - 2448
Kolopanax - 331
Krameria - 635, 2543
Krigia - 103, 1405
Lactuca - 1273
Lagarosiphons - 2788
Laguncularia - 233, 1467
Lannea - 1852
Larix - 1423
Lathraea - 92, 93, 140, 725, 1300
Laurus - 1522, 1523
Leitneria - 2108
Lemna - 28, 91, 2773
Lennoa - 2562
Lepidium - 1065, 2471
Leptandra - 1384, 1404
Libocedrus - 2873, 2874
Ligularia - 2729
Ligusticum - 1747
Lilaea - 34
Lilium - 219
Limodorum - 1084

Limonia - 2478
Limosella - 54
Linaria - 686, 687, 1399
Lindsaya - 2892
Linnea - 1225
Linum - 344, 554, 620, 828, 891,
    1273, 2882
Liquidambar - 1375, 1580
Liriodendron - 30, 1786
Listera - 1085
Lithodraba - 329
Lobelia - 340, 1063, 1371, 2931
Lophira - 2093
Loranthus - 1826, 1878
Lotus - 1908
Loxsoma - 1215
Lupinus - 426, 741, 1344, 1999, 2176
Lycium - 2846
Lycopersicon - 809, 943, 1273, 1553,
    2559
Lycopodium - 21, 155, 978, 1320,
    1321, 1348, 1349, 2153, 2297,
    2349, 2520, 2553, 2870
Lycopus - 1382
Lygodesmia - 2517
Lygodium - 759, 1981
Lysichiton - 2308
Lythrum - 497
Macrozamia - 569, 925, 1256, 2298,
    2912
Magnolia - 1267, 1380, 1786
Malus - 417, 1839, 1971, 1972,
    2006, 2276, 2461
Malva - 621, 622, 626, 1834, 2961
Mamillaria - 853
Manettia - 1774
Maranta - 1386, 1896, 2135, 2652
Marattia - 22, 32, 683
Marsilea - 2, 699, 1197, 2373,
    2863, 2894
Matonia - 250, 790, 2595
Mayaca - 2732
Mecodium - 2231
Medeola - 1375
Medicago - 1097, 1273, 1586, 1642,
    1928, 1973, 2470, 2904, 2906
Megalodonta - 1339
Melampyrum - 641, 1105, 1293, 2117
Melilotus - 319, 320, 533, 985,
    1807, 1848
Melochia - 1962

Menabea - 1275, 2090
Mentha - 1425, 1605
Mentzelia - 524
Menyanthes - 2101
Meringium - 2230
Merremia - 2563
Mertensia - 2491
Mesembryanthemum - 84, 1538, 1539,
    1934, 2366, 2953
Metasequoia - 494
Microseris - 2506
Microcycas - 750
Mida - 2472
Mimosa - 1822, 1999
Mirabilis - 1916
Mitchellia - 1368
Mitragyna - 1830
Molinia - 1458
Mollugo - 1379, 2078
Monarda - 1380, 2226
Monotropa - 124, 1336, 2219
Monotropis - 805
Monstera - 487, 488, 2485
Morinda - 2722
Morus - 2048
Munroa - 1359
Musa - 303, 1838, 2290, 2492
Myrica - 532
Myrmecodia - 261
Myrrhidendron - 2301
Mystropetalon - 1252
Najas - 33, 225
Narcissus - 673, 726, 745, 945, 959,
    1712
Nardostachys - 1862
Nardus - 822
Nelumbo - 1232
Neobeckia - 1762
Neomamillaria - 1307
Neottia - 204, 1085
Nepenthes - 1246, 1606
Neptunia - 1888
Nerium - 2101
Nicotiana - 257, 365, 991, 1187, 1980
Niphobolus - 2052
Nolina - 1260
Nothofagus - 341
Notothixos - 1844
Nuphar - 1208
Nuytsia - 1310
Nymphaea - 796, 843, 1829

Nyssa - 1376, 1417, 2084
Oakesia - 103
Obolaria - 105
Odontites - 94, 95
Olax - 398
Oldenlandia - 1368
Olea - 324, 757, 758, 1381, 1988, 1989
Oleandra - 2024, 2854
Ophioglossum - 546, 547, 752, 1122, 1492, 1772, 2102, 2334, 2428, 2759
Opuntia - 1067, 1092, 1724
Orchis - 1081, 1082, 2451
Orectanthe - 647
Orobanche - 284, 1513, 1514, 1596, 2597
Ortantha - 94
Orthocarpus - 634
Oryza - 499, 873, 1188, 1283, 1503, 1526, 1530, 1531, 1532, 1533, 1534, 1535, 1536, 1658, 1939, 2020, 2347, 2456, 2742, 2939
Osmunda - 18, 30, 829
Osyris - 1024, 1064, 2154
Oxalis - 869, 938, 944, 2337, 2619, 2621
Panax - 1387
Pandanus - 535, 1003, 1138, 2382, 2802
Panicum - 439, 2316
Paris - 551
Papaver - 169, 313
Parkinsonia - 2407
Paronychia - 1478
Parthenium - 354, 1716, 1966, 2311
Pastinaca - 2809
Pecteilis - 1626
Pedicularis - 503, 1824, 2120, 2121, 2523
Pelargonium - 2414
Peltastes - 1433
Peperomia - 8, 1319, 1480, 1976, 2489
Perideridia - 753
Perularia - 1627
Petivera - 1386
Petrosavia - 2535
Phalaenopsis - 1282
Phalaris - 486, 1111

Phaseolus - 693, 930, 1242, 1247, 1726, 1996, 2567
Phelipaea - 2206
Phenakospermum - 2649
Philodendron - 2180
Phleum - 362, 808, 1159, 1160, 1302, 1303, 1948, 2007
Phlox - 1362, 1588, 1913
Phoenix - 59, 485, 604, 605, 1134, 2848
Pholisma - 1614, 1615
Phoradendron - 623, 624, 632, 633, 2624, 2934
Phormium - 358, 2211
Phrygilanthus - 2254
Phryma - 1383, 1715
Phthirusa - 1613
Phylloglossum - 10, 19, 2849
Phytolacca - 1371, 1821
Picea - 2160
Picrorhiza - 1863
Pilularia - 508, 1481
Pimpinella - 2723
Pinus - 210, 308, 312, 702, 747, 1048, 1255, 1441, 1576, 1646, 1686, 1690, 1700, 1853, 1920, 2151, 2197, 2209, 2300, 2335, 2646, 2871, 2875, 2876, 2877, 2880, 2881
Piper - 474, 498, 735
Pistachia - 1986
Pistia - 141
Pisum - 1169, 1273, 2079, 2172, 2173, 2253, 2487, 2665, 2666, 2942
Plantago - 1921, 2548
Platanthera - 1627
Platanus - 2971
Pleiocarpa - 2071
Plumbago - 2193
Plumeria - 383
Poa - 1202, 2007
Podocarpus - 690, 776, 1577, 2550
Podoon - 178
Podophyllum - 109, 1219, 1371, 1623
Podostemon - 1239
Pogonia - 653, 1350
Polycarpon - 1478
Polygala - 154, 375, 1371, 1410, 1969, 2441, 2539
Polygonatum - 2774, 2964
Polygonum - 1408
Polypodium - 1894

Polypompholyx - 1651
Polypremum - 1398
Pontederia - 737
Populus - 571, 884, 1449, 2325, 2430
Posidonia - 227
Potamogeton - 226, 631, 1038, 1174
Prangos - 302
Primula - 1974
Prionium - 23
Prosopis - 1900, 2680
Prunus - 1377, 1825, 2006
Pseudotsuga - 322, 323, 502, 1847
Psidium - 133
Psilotum - 1040
Psittacanthus - 1616
Pteranthus - 1478
Pteris - 681
Pterospora - 373
Ptilimnium - 958
Pueraria - 2306
Pulsatilla - 2239
Pupalia - 1489
Putoria - 2126
Pyrola - 108
Pyrus - 992
Quercus - 288, 435, 656, 1378, 1435,
    1578, 1662, 1663, 1665, 1812,
    1925, 1926, 1990, 2761
Rafflesia - 228, 253
Ranunculus - 444, 827, 1030, 1055,
    1056, 1124, 1168, 1230, 1384,
    1622, 1745, 2002, 2905
Ranzania - 1625
Raphanus - 199, 1066, 1113, 1114,
    1155, 1176, 1273, 2103, 2589,
    2672
Rauwolfia - 680, 860, 1459, 1684,
    1902, 2794, 2918, 2965
Regnellidium - 1482
Rhamnus - 1384
Rhapis - 2658
Rheum - 954, 1857, 2310
Rhexia - 1366
Rhipsalis - 1911
Rhizophora - 548, 549, 550, 921, 1129,
    1130, 2152, 2184
Rhododendron - 300
Rhoeo - 1429
Rhus - 175, 603, 1378, 1849
Ribes - 1165, 1166, 2572
Ricinus - 2405, 2408, 2632

Robinia - 1022
Rosa - 651, 2532
Ruellia - 1362, 1369
Ruppia - 1177
Ruscus - 1693
Sabbatia - 1385
Saccharum - 347, 353, 1002, 2763
Sagittaria - 1172, 1669, 2435, 2515
    2533
Salacca - 1944
Salicornia - 803, 881
Salix - 652, 655, 2602
Salsola - 63, 1850
Salvadora - 2473
Salvinia - 205, 1735
Sambucus - 1377, 1757, 2730
Sanguinaria - 1375, 2005
Santalum - 395, 396, 397, 1312, 2233,
    2235
Saponaria - 1126, 1384, 2008, 2468
Sarcocaulon - 2368
Sarracenia - 2398
Sassafras - 1377
Satyrium - 1953
Sauromatum - 2415
Saururus - 1231, 1406
Saussurea - 2190
Saxifraga - 2775
Schinus - 890
Schizaea - 407, 408, 518, 566, 2593
Schoenodendron - 977
Schumannianthus - 1173
Scilla - 745
Scirpus - 1963
Scleranthus - 1478
Scopolia - 2392
Scorzonera - 2221
Scrophularia - 1399
Scutellaria - 1380
Selaginella - 197, 224, 260, 577, 579,
    1146, 1251, 2731, 2838, 2839, 2840,
    2841
Sempervivum - 2941
Senecio - 88, 2351, 2803
Sequoia - 1455
Sesamum - 1621
Sesbania - 242, 846, 1888, 2530
Setaria - 2007
Sibiraea - 394
Silene - 1917
Sinapsis - 199, 458, 1066, 2260

Siphonostegia - 1635
Sisyrinchium - 1372
Smilax - 2641
Soja (see Glycine) - 434
Solanum - 134, 345, 1273, 1378,
    1677, 2101, 2606, 2614, 2765,
    2817
Solidago - 1384, 2053
Soliva - 670
Sonneratia - 1889, 2683
Sophronanthe - 1397
Sopubia - 2889
Sorghum - 299, 357, 923, 2131, 2267,
    2770
Spartina - 2570
Spartium - 2354
Spergula - 1478, 1493
Spigelia - 1362, 1938
Spinifex - 1090, 1961
Spiranthes - 1984
Sporobolus - 808
Stangea - 2243
Stangeria - 738, 2077
Statice - 882
Stellera - 1490
Stemona - 1455
Sternbergia - 328
Stillingia - 1380
Stipa - 2063
Stratioites - 234, 334, 1277, 1803
Streptocarpus - 1079, 2709
Streblus - 1442, 2477
Striga - 1751, 2030, 2031, 2302,
    2540, 2541, 2739
Stromatopteris - 470
Strophanthus - 1125
Struttanthus - 1280
Strychnos - 114, 1702
Stylites - 1604
Suaeda - 63
Succisa - 2753
Succisela - 2753
Symphonia - 1765
Symplocarpus - 2308
Syngonium - 451, 1088
Synsepalum - 371
Taeniophyllum - 922, 1968
Tagetes - 387
Takakia - 1191
Tamarix - 1141
Tamus - 1775
Tapinanthus - 2890

Taraxacum - 355, 490, 491, 593, 1078,
    1196, 1540, 1995, 2001, 2319, 2320,
    2348, 2857, 2917
Taxus - 1335, 2194
Tectaria - 1548, 2229
Tephrosia - 2018
Terminalia - 305, 781, 1766
Thalictrum - 164, 1026
Thea - 883
Themeda - 1161
Theobroma - 568
Thesium - 174, 2389
Thielavia - 794
Thismia - 448, 449, 2106
Thonningia - 1787
Thrincia - 979, 1778, 2003
Thuja - 388, 701, 1386
Thylacospermum - 1494, 1495
Thysanotus - 564
Tibouchina - 674
Tilia - 1345, 2425, 2426, 2427, 2786
Tiliacora - 2474
Tillandsia - 448, 449, 2106
Tinospora - 90, 422
Tmesipteris - 44
Todea - 18, 886, 2438
Torreya - 736
Townsendia - 598
Tozzia - 1289
Tradescantia - 77, 1053, 1058, 1178,
    2294
Tragopogon - 1261
Trapa - 2218
Triadenum - 1355
Tribulus - 913
Triceratella - 2655
Trichomanes - 2895
Trichopus - 366
Tridax - 2044
Triglochin - 694, 1322, 2062
Trillium - 2344, 2637
Trimos - 2458
Triticum - 364, 553, 614, 929, 1106,
    1273, 1304, 1828, 2046, 2086, 2317,
    2754
Trochodendron - 2935
Tropaeolum - 162, 2495
Tulipa - 2299
Typha - 1897
Umbellularia - 1522, 1523
Uncinia - 1619
Urginea - 659

Uvularia - 103, 208, 2217
Vaccinium - 1777
Valeriana - 359, 954, 1072, 1306,
    1414, 2580, 2581, 2582, 2583,
    2584, 2768, 2829
Vallisneria - 141
Vanda - 53
Vanilla - 311, 1386, 1997
Vellozia - 2831, 2832
Veratrum - 49, 1386
Verbascum - 1385
Veronica - 1096, 1304
Viburnum - 1384
Vicia - 304, 321, 382, 1572, 1704,
    1731, 2499, 2674, 2725
Viminaria - 1650
Viola - 2940
Viscum - 975, 1723, 1870, 2622,
    2623, 2625, 2626, 2708
Vitis - 315, 384, 2726
Washingtonia - 1136
Wedelia - 2238
Welwitschia - 15, 16, 2576, 2745,
    2958, 2959, 2960, 2965
Withania - 2443
Xanthium - 840, 856
Xanthorrhiza - 1382
Xylopia - 1466
Yucca - 343
Zamia - 689, 2733
Zannichellia - 33
Zea - 364, 432, 920, 934, 981, 982,
    983, 1117, 1273, 1285, 1286,
    1504, 1527, 1549, 1620, 1628,
    1629, 1673, 1754, 1808, 1851,
    1898, 2040, 2208, 2346, 2936
Zizania - 2555
Zizyphus - 321
Zostera - 227, 2598, 2599